NEW WORLD PRIMATES

Warren G. Kinzey (1935–1994)
Copyright © 1989 Yousuf Karsh

NEW WORLD PRIMATES

Ecology, Evolution, and Behavior

Warren G. Kinzey

Editor

ALDINE DE GRUYTER

New York

About the Editor

The late **Warren G. Kinzey** (1935–1994) was head of the anthropology department at City College of the City University of New York; he also taught at the CUNY Graduate Center and the CUNY Medical School. His research and field studies on the primates of the Peruvian Amazon focused on the relationship between diet and distinctive tooth wear patterns. Among his earlier writings were an edited volume, *Evolution of Human Behavior: Primate Models,* and numerous contributions to professional journals in anthropology and primatology.

ALDINE DE GRUYTER
A division of Walter de Gruyter, Inc.
200 Saw Mill River Road
Hawthorne, New York 10532

This publication is printed on acid free paper ⊗

Library of Congress Cataloging-in-Publication Data
New world primates : ecology, evolution, and behavior / Warren G.
    Kinzey, editor.
        p.   cm. — (Foundations of human behavior)
    Most of the papers were presented in a symposium on Nov. 19, 1988
at the annual meeting of the American Anthropological Association in
Phoenix, Arizona.
    Includes bibliographical references (p.   ) and index.
    ISBN 0-202-01185-2 (alk. paper). — ISBN 0-202-01186-0 (pbk. :
alk. paper)
    1. Cebidae—Ecology—Congresses.   2. Cebidae—Evolution—
Congresses.   3. Cebidae—Behavior—Congresses.   I. Kinzey, Warren
G., 1935–1994.   II. American Anthropological Association. Meeting
(87th : 1988 : Phoenix, Ariz.)   III. Series.
QL737.P925N48   1997
599.8098—dc20
                                                            96-38566
                                                            CIP

Manufactured in the United States of America
10 9 8 7 6 5 4 3 2 1

Contents

MAPS
Distribution Maps of South and Central America by Genus

Contributors

Connie M. Anderson

Department of Anthropology
Hartwick College

Este Armstrong

Chesapeake Information Systems
Gaithersburg, Md. and
    USUHS, Bethesda, Md.

John Fleagle

Department of Anatomical
    Sciences
State University of New York
Stony Brook

Gerald H. Jacobs

Department of Psychology
University of California
Santa Barbara

Richard Kay

Department of Biological
    Anthropology and Anatomy
Duke University Medical Center

William McGrew

Department of Sociology and
    Anthropology
Miami University
Ohio

Charles T. Snowdon

Department of Psychology
University of Wisconsin
Madison

Mary Anne Shea

USUHS
Bethesda, Md.

Leslie E. Sponsel

Department of Anthropology
University of Hawaii
Honolulu

Karen Strier

Department of Anthropology
University of Wisconsin
Madison

Patricia C. Wright

Executive Director
Institute for the Conservation
  of Tropical Environments
State University of New York
Stony Brook

Acknowledgments

Warren Kinzey, the editor of this volume, unfortunately could not live to see it published. He succumbed in the early autumn of 1994 to amyotrophic lateral sclerosis, a devastating illness that gradually prevented him from putting pen to paper but did not deter him from getting on with the job. His mind remained clear to the very end, and his graduate students took upon themselves the task of transcribing his dictation. That this book has appeared at all is a tribute to the dedication of Professor Kinzey and the loyalty of his students and colleagues. The loss of Kinzey at the age of 58 was a severe one both for physical anthropology and for primatology. *New World Primates* is his legacy to colleagues and younger students who know him only by his reputation and the work of the two generations he trained.

Until now the demanding task of completing this complex book, still unfinished at the time of Kinzey's death, has been thankless. It is time to acknowledge the good work of many people whose collaboration in this instance was absolutely essential, and who never lost faith in the undertaking.

All but two of the contributors to this volume, including Kinzey himself, initially presented the papers that make up Part I in a symposium held in Arizona some eight years ago. Their chapters—and the two by Jacobs and Wright added later—have been through several subsequent revisions and redactions, incorporating wherever possible citations to more recent field research. Pursuing the contributors in the field, across continents, has been a challenge for all concerned.

Part II, the synopsis of the 16 genera of new world monkeys to whom nearly half the book has been devoted, was Warren Kinzey's own work, but in collaboration with the students in a graduate seminar offered at the City University of New York in the autumn of 1991, on Kinzey's return from two years on loan to the National Science Foundation. The seminar participants, who searched through the vast literature and drafted a preliminary version of the introduction to every genus except for one, are not mentioned in Professor Kinzey's final text. They should therefore receive due credit here: Jennifer Ashley (for *Chiropotes*); Sylvia Atsalis (for *Brachyteles*); Douglas Broadfield (for *Leontopithecus*); Linda Brown (for *Callithrix*); Ana Bueno (for *Callimico*); Elena Cunningham (for

Pithecia); Milton Herrera (for *Aotus*); Berriam Hobby (for *Cacajao*); Kelley McFarland (for *Ateles*); Eti Mann (for *Callicebus*); Sam Marquez (for *Cebuella*); Reiko Matsuda (for *Alouatta*); Laura Robinson (for *Lagothrix*); Ximena Valderrama (for *Cebus*); and Lodewijk Werre (for *Saimiri*). (The introduction to *Saguinus* was wholly Kinzey's from the outset.)

With the exceptions of maps 8 and 12, all of the maps in Part II, carefully delineating the habitats of these genera in Central America and the Southern Hemisphere, are the further work of Elena Cunningham. Maps 8, showing the habitats of the two marmoset genera, and 12, showing the habitats of the two tamarins, are taken from R. W. Sussman and W. G. Kinzey, "The ecological role of the Callitrichidae: A review," *American Journal of Physical Anthropology* 64 (4), 419–449, copyright © 1984 Wiley-Liss, Inc., and used with the permission of Professor Sussman. The classification schema that constitutes chapter 10 is derived from the cited works by Dr. Alfred L. Rosenberger and colleagues and by Professor Russell A. Mittermeier and colleagues, respectively. The first pass at controlling the daunting master reference list, with its various cross-references, detours, and updatings, was made by Linda Brown.

Three of Warren Kinzey's colleagues deserve especial mention here. Professor Robert W. Sussman of Washington University, St. Louis, looked over the introductory chapter at an earlier stage. Dr. Alfred L. Rosenberger of the Smithsonian Institution and Professor John F. Oates of the City University Graduate Center both offered help and encouragement in dark moments, when the project itself seemed to founder.

Our final indebtedness goes beyond all that we have mentioned to date. Professor Marilyn A. Norconk of Kent State University set aside many weeks of her own research time, and devoted immense amounts of energy, to reviewing everything in the volume that would have been Warren Kinzey's own responsibility, had he lived. Errors and lapses, unavoidable as they are, should not, however, be attributed to Professor Norconk's vigilant eye. And Julianne Kelly generously made all of her late husband's papers and files on this project available to us at the most difficult and stressful time, helping without hesitation to bring it to long-delayed fruition. The modesty attending on scholarly acknowledgment is inadequate to our purpose in thanking them both for their time, patience, and human warmth.

                                                              The Publisher

Introduction

Most of the papers in the first part of this book were presented in a symposium, "Anthropological Perspectives on New World Primates," on November 19, 1988 at the annual meeting of the American Anthropological Association in Phoenix, Arizona. The papers by Jacobs and by Wright were added subsequently. The theme of this symposium was the diversity of neotropical (New World) primates and their relevance for anthropology. General evolutionary patterns as seen in the neotropical ecosystem are emphasized in Part I of the book. "A very new world" of primates has striking parallels to the Old World, even though in the past it is their differences that have been emphasized. Many of these parallels are noted throughout the book.

Anthropologists (and other social scientists) have traditionally ignored the New World primates, largely because they are not in the mainstream of human evolution. Yet, they provide a unique source of information, exhibiting parallel adaptations to apes and Old World monkeys, tool-use, a wider variety of mating systems, and parallels to human language. The knowledge of this diversity and relevance to anthropological questions should be made more readily available to all anthropologists and behavioral scientists.

Cebus monkeys use tools and provide an interesting parallel to chimpanzees; several species of New World primate, including night monkeys and titi monkeys, have monogamous mating systems; the *only* primates known to have a polyandrous mating system are found in South and Central America; the pygmy marmoset has been shown to have several parallels to human language; and there is a much wider diversity of types of arboreal locomotor behavior in New World monkeys than in Old World monkeys. Perhaps most important, from the point of view of behavioral scientists, as the commentator at our symposium, cultural anthropologist Ray Hames stated at our symposium, "primatologists are way ahead of other anthropologists in describing, behaviorally and statistically, patterns of social interaction." Human behavioral scientists have a lot to learn from primatologists.

Experimental behavioral scientists often use New World primates in their research. Currently, there is no readily available summary source of the *natural* behavior of these animals, especially regarding *differences* among them that may be pertinent to other types of research. For example, neuroscientists, working on the visual system, need to know the kinds of visual phenomena that the animals confront under natural conditions: whether the animals eat leaves or fruits, the colors of fruits, and types of arboreal pathways, etc. Thus, this book provides both a series of up-to-date chapters on state-of-the art research, and summarizes what is known, without jargon, of the field and naturalistic behavior of each genus.

A brief synopsis of each chapter follows. Fleagle and Kay summarize the morphological and biogeographic evidence supporting an African origin for the New World primates. This is particularly significant for anthropologists because the earliest anthropoids from the Old World (the Fayum) are remarkably like platyrrhines in many aspects. This is, admittedly, a one-sided view, but it is necessary to counter the erroneous view, long-held by most anthropologists, that the platyrrhines came from North America.

Armstrong and Shea review both recent and classical data on the brains of New World monkeys and catarrhines. By permitting our order to have flexible, intelligent behaviors, the brain has been central to primate evolution. At the same time the brain is remarkably conservative, showing small differences between Old and New World monkeys. The authors' comparison focuses on neural differences between these two large primate groups to help us understand how brains evolve when selection is *not* viewed as being related to tools or language. Thus, the results should be of interest to anthropologists, primatologists, psychologists, and neurobiologists.

Snowdon looks for linguistic parallels with human language in the spontaneous communication of primates. One of the most striking features of the Neotropical primates, in contrast with those of the Old World, is that they are *all* arboreal. Snowdon points out that as a result of ceboid primates therefore having greater reliance on vocal as opposed to visual cues, they are more relevant to understanding the evolution of language than the chimpanzee and gorilla studies using signs and symbols. Within the Neotropical primates there are many phenomena that are strikingly parallel to human linguistic phenomena, and there are several aspects of human language that are not seen. The comparison of similarities and of differences indicates where certain aspects of language first appeared. Snowdon makes a strong case for the study of vocalization in New World primates because "several phenomena that were once thought to be exclusive to human speech and language have been documented in New World primates." "For each of these phenomena," he argues, "New World primates present some of the most compelling examples of parallels to human speech and language."

McGrew discusses the controversy over the mating system of marmosets and tamarins and demonstrates that there are laboratory data as well as field data that support the idea that these animals are *not* monogamous. (Thus far, researchers have tended to believe that while some of the field data support this concept, none of the lab data does so.) The importance of this contribution lies in showing how laboratory studies can contribute significantly to our understanding of behavior. He focuses on sex differences in rearing to set up a framework for understanding sexual division of labor in human family life—not to draw an analogy from one to the other, but to demonstrate processes.

Strier compares the effect of subtle cues of social relations (e.g., maintenance and avoidance of proximity) with aggression and overt competition. One of the important differences between Old World and New World monkeys is that in the latter, dominance relationships are much less common, reflecting less frequent agonistic behavior. Strier demonstrates how fitness is enhanced by affiliative interactions in muriquis, the largest of the Neotropical primates. This is an excellent example of how an affiliative coalition among *related* males serves to reduce tension in competitive situations. It is in marked contrast to the affiliative coalitions formed in *unrelated* baboons, among whom behavioral interactions are generally tense and competitive. Even though her sample size is small, Strier provides evidence for the importance of male–male coalitions in reducing the frequency of agonistic encounters. She suggests that similar coalitions are also important in reducing aggression among humans.

Anderson's chapter is written specifically for anthropologists whose main understanding of primate behavior has come from baboons and chimpanzees. It is of particular relevance to cultural anthropologists who teach courses on history and theory in anthropology, as it will help them to understand what kinds of primate studies have been carried out, how to interpret the data, and especially what to make of the theories that have resulted. It is written for an audience familiar mainly with the conclusions of primate studies, rather than their methods.

Wright, one of few field workers to study primates in both the New and the Old World tropics, compares foraging adaptations in Old and New World species and answers questions such as why New World monkeys are not more like prosimians.

Sponsel emphasizes the relationship between primate ecology and cultural ecology in exploring the indigenous human niche in Amazonian ecosystems. He introduces the concept of ethnoprimatology—the interface between human and primate ecology, using Amazonia as an example. By avoiding jargon, his chapter is accessible and relevant to anthropologists who are not primatologists, and to primatologists as well.

In Part II of the book, I provide a synopsis of each of the New World primate genera, describing basic known behaviors of each, including what field work has been done, where, by whom, and what particularly significant aspects of the behavior of each would be of interest behaviorally to anthropologists and other nonprimatologists. There is a map of the distribution of each genus, a taxonomy of living New World primates to the level of species, and a bibliography. This section is extensively cross-referenced to the first nine articles in the book.

The first attempt to describe the similarities and differences in behavior of all known New World primates was by Moynihan (1976b). Previous to this, the only available description of the behavior of these primates was by the mammalogist, Ivan Sanderson (1957). Although most of Sanderson's data were from anecdotal reports, and his book is now completely out of date, it was published just in time to become the text for the first graduate seminar in primate behavior taught in the United States, by S. L. Washburn at the University of Chicago in 1958 (see Haraway 1989: footnote 34, p. 218). Moynihan's book is very readable (as Sanderson's had been). Despite the paucity of field data, it was very insightful for its time. For example, Moynihan pointed out similarities in ecology between squirrel monkeys and the African talapoin monkeys. Since most of our knowledge of platyrrhine primates in their natural habitat has come since the 1970s, Moynihan's book is now incomplete.

In 1977 Hershkovitz published the first volume of an extensive treatise on platyrrhine primates. Thus far only volume I (Hershkovitz, 1977), on marmosets and tamarins, has been published. Subsequent accounts of many genera have been published separately, however, and are cited in the appropriate chapters. In 1981 Coimbra-Filho and Mittermeier (1981) inaugurated a two-volume series on the ecology and behavior of neotropical primates. The descriptions of the genera *Aotus, Cacajao, Callicebus, Callimico, Cebus, Chiropotes, Pithecia,* and *Saimiri,* which were included in the first volume, are now quite out of date. Descriptions of the remaining genera covered in Volume II (Mittermeier et al. 1988a) are much more current. Nevertheless, there are many new data on all genera, including descriptions of four new species, that provide more up-to-date accounts for all 16 genera in this volume.

Readers who are interested in less detailed accounts of all living primates, not just those in the New World, will find several up-to-date works. For example, "The Encyclopedia of Mammals" (MacDonald, 1984) has a synopsis of each primate group. Kavanaugh (1983), and more recently Preston-Mafham (1992), provides descriptions and colored photographs of all the living primates. For a description of each primate group, together with detailed topical expositions on many categories in behavior and ecology, the reader should consult Smuts et al. (1987).

Descriptions of all the primate fossils may be found in Szalay and Delson (1979), which is currently being revised. An excellent account of the conservation status of every living primate species may be found in Wolfheim (1983).

Lest I overlook important historical accounts, the following should be mentioned. Napier and Napier (1967) were the first to describe the behavior and ecology of each primate genus in a uniform format. Even though this book is out of date, it provides an excellent introduction to the early literature. Another work frequently used beginning in the 1950s as a supplement in many introductory anthropology courses was Clark's very readable "History of the Primates," first published in 1949, most recently available in the fifth edition (1966), but out of print in 1993. W. C. Osman Hill began a series (Hill, 1953b) on the anatomy, behavior, and fossil record of the primates that he did not complete before his death. The volumes on the New World primates (Hill, 1957, 1960, 1962) were completed. Although the anatomical descriptions are very useful, the maps of primate distributions are not in accord with current taxonomy.

There are several guides to field work. Bernardes (1988) provides a bibliography of all primate field work conducted in Brazil prior to that date. Wolfe (1987) published a bibliography of field studies of all nonhuman primates. Baldwin et al. (1977) provide a bibliography of all New World primate field studies prior to that date.

Previously (Kinzey 1986a, 1996) I have discussed the history of field studies of the New World primates. In Kinzey (1986a) there is a table of major field sites, citing the references for each. At that time there were many species of platyrrhine primate as yet not studied. As of this writing *Ateles fusciceps* and *Callithrix geoffroyi* are the only species of neotropical primate that have not yet been studied in the wild.

Perhaps the most difficult aspect of any reference book on the primates is the problem of taxonomy. Very few taxonomists agree. My own starting point is Rosenberger et al. (1990). To this I have added the four species that have been described since that paper. In addition, several new subspecies are referred to in the text. In most of the generic descriptions I refer to other commonly held systematic accounts.

Those who are unfamiliar with the New World primates should realize that there are very few characters that clearly distinguish New World monkeys from Old World monkeys. The two most striking features are the broad and flat (platyrrhine) nose, and the presence of three bicuspid (premolar) teeth on each side of each jaw; all the Old World primates, including humans, have only two. For those interested in a general survey of the New World monkeys, I recommend reading the introductory sections to each of the 16 chapters in Part II of the book.

I

PERSPECTIVES ON NEW WORLD PRIMATES

1

Platyrrhines, Catarrhines, and the Fossil Record

JOHN G. FLEAGLE and RICHARD F. KAY

Introduction

With 16 genera, over 70 species, and up to 16 sympatric species, living platyrrhines comprise a taxonomically and adaptively diverse radiation. Considering the extensive living radiation in the Neotropics today and the relatively good fossil record for other South American mammals, the fossil record of New World monkeys is relatively poor. Until recently, a large shoe box could contain the primate fossils from all of South America and the Caribbean from the last 30 million years. However, in the past two decades the platyrrhine fossil record has been expanding rapidly and provides a broad overview and many tantalizing hints about the evolutionary history of the group. In this review, we will examine the platyrrhine fossil record, discuss the clues it provides about the geographic, phylogenetic, and adaptive history of the group, and outline some of the major unresolved issues in platyrrhine evolution. For convenience, fossil platyrrhines are divided into four groups on the basis of age and geography: (1) the earliest platyrrhine fossils from a single late Oligocene locality in Bolivia; (2) several difficult to interpret genera from the early and middle Miocene of southern Argentina and Chile; (3) an array of relatively modern genera from the late Miocene of Colombia; and (4) a number of unusual species from Pleistocene or Recent deposits in the Carribean and Brazil (Figures 1.1 and 1.2).

The Earliest Platyrrhines

The earliest platyrrhine fossils are *Branisella* and *Szalatavus*, all specimens of which come from a single stratigraphic level of late Oligocene (Deseadan Land Mammal Age) rocks near the village of Salla, Bolivia.

Figure 1.1.   Map showing the location and age of sites yielding fossil platyrrhines.

Today this locality is more than 13,000 feet above sea level, but at the
time of deposition was probably at around 3,000 feet. *Branisella* is a small
monkey, the size of an owl monkey. The low, rounded cusps of the
molars suggest a frugivorous diet and the small P2 and the shape of the
mandible suggest a short-faced monkey. No clear indications of a relationship of *Branisella* to any particular group of later platyrrhines have
been demonstrated.

The possible presence of a second monkey of similar size named
Szalatavus has been proposed recently by Rosenberger and colleagues
(Rosenberger et al., 1991). Recovery of many new specimens of monkeys
from Salla in the past 3 years by Japanese, American, and Bolivian scientists suggest *Branisella* and *Szalatavus* may be the same species (Kay and

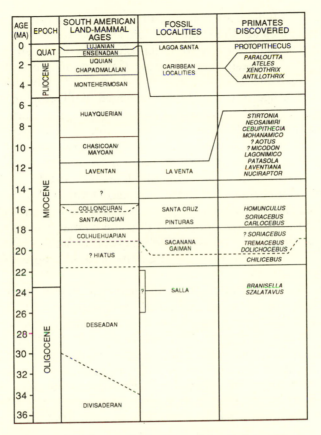

AGE (MA)	EPOCH	SOUTH AMERICAN LAND-MAMMAL AGES	FOSSIL LOCALITIES	PRIMATES DISCOVERED
0	QUAT	LUJANIAN / ENSENADAN	LAGOA SANTA	PROTOPITHECUS
2	PLIOCENE	UQUIAN / CHAPADMALALAN	CARIBBEAN LOCALITIES	*PARALOUTTA* / *ATELES* / *XENOTHRIX* / *ANTILLOTHRIX*
4		MONTEHERMOSAN		
6	MIOCENE	HUAYQUERIAN		*STIRTONIA* / *NEOSAIMIRI*
8				*CEBUPITHECIA* / *MOHANAMICO*
10		CHASICOAN/ MAYOAN		*? AOTUS* / *? MICODON* / *LAGONIMICO* / *PATASOLA*
12		LAVENTAN	LA VENTA	*LAVENTIANA* / *NUCIRAPTOR*
14		?		
16		COLLONCURAN	SANTA CRUZ	*HOMUNCULUS*
		SANTACRUCIAN	PINTURAS	*SORIACEBUS* / *CARLOCEBUS*
18		COLHUEHUAPIAN		*? SORIACEBUS*
20			SACANANA GAIMAN	*TREMACEBUS* / *DOLICHOCEBUS*
		? HIATUS		*CHILICEBUS*
22				
24	OLIGOCENE		SALLA	*BRANISELLA* / *SZALATAVUS*
26		DESEADAN		
28				
30				
32				
34				
36		DIVISADERAN		

Figure 1.2. The geochronology of fossil platyrrhines.

Williams, 1995). Moreover, the recovery of a 32-million-year-old Chilean rodent (another immigrant element of the South American biota frequently linked with primates as having entered South America from Africa in the middle Cenozoic) suggests platyrrhines may have a longer undocumented history in South America than presently documented (Wyss et al., 1993).

The Patagonian Platyrrhines

In the early and middle Miocene of southern Argentina, informally known as Patagonia, there were over a half dozen genera and species of platyrrhines—part of a rich fauna dominated by rodents, endemic ungulates, sloths, and marsupials. These monkeys are now known from hundreds of fossils, mostly isolated dental and postcranial remains.

However, the phyletic relationships of these fossil monkeys to particular living subfamilies are the subject of considerable debate.

Dolichocebus gaimanensis is from deposits of the Colhuehuapian Land mammal Age (earliest Miocene) near Gaiman, in Chubut Province, southern Argentina. *Dolichocebus* is a medium-sized platyrrhine (2–3 kg) known from a nearly complete but damaged skull, numerous isolated teeth, and a talus. It has dimorphic canines, three premolars, and upper molars that resemble the same teeth in *Saimiri*, *Callicebus*, or *Aotus*. The molar morphology of *Dolichocebus* suggests a frugivorous diet (Fleagle et al., 1996). The skull of *Dolichocebus* has a narrow, posteriorly widening snout, complete postorbital closure, moderate-size orbits, a very narrow interorbital dimension, and relatively large tooth roots. Rosenberger (1979, 1982) has argued that *Dolichocebus* had an interorbital foramen linking the right and left orbits—an unusual cranial feature found only in *Saimiri* among living primates. The talus of *Dolichocebus* is most similar to that of *Cebus* or *Saimiri*, suggesting either a rapid arboreal quadruped or a leaper.

On the basis of the interorbital foramen and several other aspects of the cranial morphology of *Dolichocebus*, Rosenberger (1979) argued that this genus is the sister group of the living squirrel monkey. However, Hershkovitz argued that the Oligocene monkey is too distinctive in other cranial features, such as the palate shape, and molar root morphology to bear any close relationship to living platyrrhines. Analysis of the isolated teeth found in association with *Dolichocebus* yield a similarly dichotomous picture of the relationships of this genus; it is either closely related to the squirrel monkey, or the sister group of all living platyrrhines (Fleagle and Kay, 1989).

Tremacebus harringtoni is from the Colhuehuapian (early Miocene) locality of Sacanana, also in Chubut, Argentina. It was a smaller (1–2 kg) monkey than *Dolichocebus*. The type specimen and only fossil clearly attributable to this species is a nearly complete but broken skull with a relatively short, broad snout and larger orbits than any diurnal platyrrhines. *Tremacebus* shows greatest cranial similarities to the extant platyrrhines *Callicebus* and *Aotus*. On the basis of the moderately enlarged orbits, Rosenberger (1984) suggested that it is an ancestor of the living owl monkey, and other authorities have noted similarities to *Callicebus*. A difficulty is the lack of any dental evidence—the skull of *Tremacebus* has only tooth roots and a few broken crowns—or postcranial bones.

Chilicebus carrascoensis is a newly described fossil monkey from early Miocene (~20 mya) deposits in the Andes of Chile (Flynn et al., 1995). It was a medium-sized monkey (1200 g) with square upper molars suggesting a frugivorous–folivorous diet.

From slightly younger deposits of the Santacrucian Land Mammal

Age (early–middle Miocene) in Santa Cruz Province, Argentina there are several additional genera and species of fossil platyrrhines. These fossil monkeys come from two main geographic and geologic areas—the slightly older Pinturas Formation (17.5–16.5 mya) in the West and the younger Santa Cruz Formation (16.5–16.0 mya) in the East.

The Pinturas Formation preserved in the western part of southern Argentina in the foothills of the Andes has yielded an abundant fauna of fossil birds, reptiles, and mammals, including two genera and four primate species (Fleagle, 1990; Fleagle et al., 1996). Evidence from the sediments, fossil pollen, fossil birds, and abundant nests of fossil insects indicate that the Pinturas primates lived in a forested habitat in what must have been a time of climatic fluctuations, with periods of relative wetness separated by periods of desiccation (Bown and Larriestra, 1990).

The best known, and most unusual of the Pinturas primates is *Soriacebus*, with two species—the saki-sized *S. ameghinorum* and the tamarin-sized *S. adrianae*. *Soriacebus* has large procumbent styliform lower incisors that form a continuous arcade with the large canine, a tall P2, tiny posterior premolars, narrow lower molars, and a deep jaw. The upper teeth have large dagger-like canines, broad premolars, and small triangular molars. The facial skeleton is very deep. It was probably frugivorous and may have used its large front teeth for seed predation, as is the case for the Pitheciinae, sakis and uakaries of the Amazon and Orinoco Basins. The few postcranial elements of *Soriacebus* suggest quadrupedal running and leaping habits, with some clinging. The affinities of *Soriacebus* have been debated: some regard it as a basal pitheciine on the basis of the large incisors and the deep mandible (Kinzey, 1992; Rosenberger, 1992). Others suggest these similarities to be the result of adaptive convergence for seed predation (Kay, 1990).

The other fossil platyrrhine from Pinturas is *Carlocebus*, also with two species—the saki-sized *C. intermedius* and the larger *C. carmenensis*. The dentition of *Carlocebus* is more generalized than that of *Soriacebus*, with small vertical incisors, a small canine, and relative larger premolars and molars. As in *Soriacebus*, the mandible is relatively deep. The upper dentition is characterized by very broad premolars and molars (Fleagle, 1990). Dentally, *Carlocebus* appears to have been frugivorous with some folivory. Its dentition is most comparable to that of *Callicebus*, the titi monkey. Postcranial remains of *Carlocebus* suggest arboreal quadrupedal habits (Meldrum, 1993).

Homunculus patagonicus, from the early–middle Miocene (16.5 million years ago) Santa Cruz Formation on the Atlantic coast of southern Argentina, was one of the earliest fossil platyrrhines discovered (Ameghino, 1891), and for many years all fossil platyrrhines were placed in this genus. It was a medium-size monkey, with the largest individuals probably weighing nearly 3 kg. The dental formula is 2.1.3.3. The lower incisors are narrow and spatulate; the canines are probably sexually dimorphic, and the molars are characterized by relatively small cusps connected by long shearing crests; they have a small, square trigonid and a broader talonid with a prominent cristid obliqua. The mandible is relatively shallow compared with that of the Pinturas primates. *Homunculus* was probably frugivorous and folivorous. The cranium of *Homunculus* has procumbent upper incisors, a relatively short face and a long, gracile brain case with no sagittal crest (Tauber, 1991).

The limb elements resemble those of a callitrichid (Ciochon and Corruccini, 1975), and suggest that *Homunculus* was partly saltatory in its locomotion; however the hindlimb was not particularly long and it seems more likely that *Homunculus* was largely quadrupedal (Meldrum, 1993). In some details of its limbs, such as the great size of the lesser trochanter on the femur, *Homunculus* resembles the early anthropoids from Egypt in what are probably primitive features.

As the name indicates, Ameghino (1891) originally thought *Homunculus* was in the ancestry of humans; it is not. Most later studies have noted either the unique features of the genus or dental similarities to *Aotus*, *Callicebus*, or *Alouatta* (Bluntschli, 1931). In describing a new cranium of *Homunculus*, Tauber (1991) noted many similarities to pitheciines. The similarities in the dentitions of *Homunculus* and *Carlocebus* are striking, and seem to indicate that the two are closely related, whatever their relationship to later primate radiations.

All the Patagonian platyrrhines show unusual combinations of features linking them with what are, today, distinct clades of platyrrhines. In phylogenetic analyses each has been placed with a distinct extant clade (e.g., *Dolichocebus* with *Saimiri*; *Tremacebus* with *Aotus*, *Soriacebus* with pitheciines) or viewed as outgroups to all modern subfamilies. There are several possible explanations for this phenomenon, which are not mutually exclusive. In an "ecological vicar model" the Patagonian primates are viewed as a geographically isolated radiation of early platyrrhines that is collateral to the evolution of all extant platyrrhines. This hypothesis is compatible with both their morphological distinctiveness and geographic origin near the tip of South America. Patagonia has a long history as a separate biogeographic area within South America containing a distinct flora and invertebrate fauna. An alternative hypothesis asserts that the living subfamilies of platyrrhines diverged early in platyrrhine evolution to fill many of the ecological niches that they occupy today, and that Patagonian Miocene forms are primitive representatives of those clades. In this latter view, the unusual combination of features reflects the fact that many of these fossil taxa are very near the initial split of modern clades, before most acquired the larger suite of features that characterizes their living members. This hypothesis is consistent with their age. There is no doubt that part of our inability to clearly resolve the phylogenetic position of these monkeys is that we lack a clear understanding of the polarity of the features characterizing extant platyrrhines—which features are primitive retentions and which are derived specializations in different groups. However, neither of these hypotheses can be adequately tested without similar-aged primates from elsewhere in the continent.

A More Modern Community

The oldest fossil platyrrhines from more tropical parts of South America come from La Venta in Colombia in deposits approximately 11.5 to 13.5 million years old. Compared with the Patagonian fossil platyrrhines, many of the fossil monkeys from La Venta are strikingly similar to modern platyrrhines and clearly belong in living subfamilies. Comparison of the La Venta fauna with modern South American faunas indicates a wet tropical environment for this region in the late Miocene (Kay and Madden, 1996).

Several taxa from La Venta belong to the Pitheciinae, the sakis and uakaries that today inhabit wet tropical forests. Living pitheciines eat fruits and are capable of opening very tough exocarps to extract nutritious seeds that are unavailable as food for other monkeys. *Cebupithecia sarmientoi* was similar in size (2–3 kg) and many aspects of anatomy to living pitheciines, with its stout canines, procumbent incisors, and flat cheek teeth with little cusp relief. Like living pitheciines, *Cebupithecia* probably ate mainly fruit and used its large anterior dentition for opening seeds. The *Cebupithecia* skeleton shows more similarities to the leaping *Pithecia* than to the more quadrupedal sakis such as *Chiropotes*, but retains many skeletal features found in other platyrrhine subfamilies while lacking some shared derived features of living pitheciines (Meldrum and Lemelin, 1991). It is probably very near the origin of the modern pitheciine radiation (Kay, 1990).

A recently described pitheciine, *Nuciraptor*, was similar in body size to *Cebupithecia*. It has procumbent and styliform lower incisors as in the living taxa, but its canines were less specialized for prying open tough fruits and its cheek teeth suggest more soft fruit in its diet (Meldrum and Kay, 1996).

Mohanamico hershkovitzi is a small (1 kg) fossil monkey that is known from a single mandible (Luchterhand et al., 1986). It has been placed near the base of the evolutionary radiation of the pitheciines, certainly before *Cebupithecia* or *Nuciraptor*, on the basis of its large lateral incisor and the structure of the canine and anterior premolar. It was probably frugivorous.

Setoguchi and Rosenberger (1987) described a new species of owl monkey, *Aotus dindensis*, from La Venta. A small facial fragment suggests that the Miocene species could have had large orbits similar to those of the nocturnal owl monkey (but see Kay, 1990). There has been some debate about the similarities between *Aotus dindensis* and *Mohanamico*, illustrating the conservative nature of mandibular dentition in small-sized frugivorous platyrrhines.

Stirtonia, the largest (6 kg) of the La Venta primates, is known from two species, a larger, older species, *Stirtonia victoriae*, and a younger smaller one, *S. tatacoensis*. The latter has many dental similarities in its upper and lower dentition to the living howling monkey (*Alouatta*), including long lower molars with relatively small trigonids and very large upper molars with well-developed shearing crests and styles. It was a folivore. Isolated molars that resemble *Stirtonia* (and *Alouatta*) also have been recovered from late Miocene deposits at Rio Acre in western Brazil (Kay and Frailey, 1993).

Neosaimiri fieldsi is very similar in size and morphology to the living squirrel monkey. It differs most clearly in incisor proportions and in having less developed molar shearing, suggesting it was possibly more frugivorous and less insectivorous than *Saimiri*. An isolated humerus from the same deposits is indistinguishable from the same bone in *Saimiri*. Recently named *Laventiana annectens* (Rosenberger et al., 1991) is very similar to *Neosaimiri*, differing only in the variable development of a postentoconid notch, an unusual character not seen in any living platyrrhines. Many others now place *L. annectens* in the same genus or even species with *Neosaimiri fieldsi* (Kay and Meldrum, 1996; Takai, 1994).

One of the most obvious gaps in the platyrrhine fossil record for many years has been the absence of any evidence for the ancestry of callitrichines—marmosets, tamarins, and *Callimico*. In recent decades and years several putative fossil callitrichines have been described, each with different morphologies and different implications for the origin of the group. One thing most paleontologists agree about is a link between callitrichines and squirrel monkeys (e.g., Kay, 1994; Kay and Meldrum, 1996; Rosenberger, 1980; Rosenberger et al., 1991; Takai, 1994). In this regard, two newly described La Venta species are particularly important.

Patasola magdalena is a new species from La Venta that is slightly smaller than the living squirrel monkey. It shares most dental features with *Callimico*, *Saimiri*, and *Neosaimiri*, but the deciduous premolars bear a closer resemblance to some callitrichines such as *Saguinus* or *Leon-*

topithecus. On the basis of the latter synapomorphies, *Patasola* is identified by its describers as a callitrichine, more closely related to marmosets and tamarins than is *Callimico* (Kay and Meldrum, 1996).

Lagonimico conclucatus is another new species from La Venta. It was roughly the size of an owl monkey and has been described by Kay (1994) as a giant tamarin. Based on a different set of features, mainly upper molar shape, he placed it with marmosets and tamarins. While the phyletic position of *Lagonimico* and *Patasola* is similar with respect to living callitrichines, both are placed between marmosets and tamarins on the one hand and *Callimico* on the other, their morphology is very different. *Patasola* much more closely resembles squirrel monkeys and *Callimico*, whereas *Lagonimico* far more closely resembles tamarins. The absence of upper molar hypocones in a monkey the size of *Lagonimico* suggests that acquisition of the distinctive marmoset and tamarin molar morphology did not necessarily evolve in conjunction with small size.

Micodon kiotensis is a poorly known species from La Venta that is based on three small, isolated teeth (Setoguchi and Rosenberger, 1985). It has been described as a fossil marmoset, primarily on the basis of size. The validity of the species has been challenged, in particular, whether the collection of three teeth even belong to the same taxon, let alone whether it has affinities with callitrichines.

The presence of several putative fossil callitrichines at La Venta is exciting, but clearly demands further analysis, since each offers a somewhat different picture of the origin and early evolution of the group. It is almost certain that the origin of this group involved a rather bushy phylogeny as evidenced by the parallel and convergent features found in marmosets, tamarins, and *Callimico*. The fossils reinforce this view. Analyses of these fossils also appear to support the view that callitrichines are the sister taxon of the squirrel monkey, *Saimiri*, as suggested by Rosenberger (1981, 1992). However, we are still far from understanding many details of the origin and radiation of these very successful platyrrhines.

Despite the debated taxa (*Mohanamico*, *Micodon*) and various contradictions in the proposed phylogenetic relationships among some of the platyrrhines, the fauna from LaVenta indicates that the major groups of extant platyrrhines were clearly differentiated and present in central Colombia by 13.5 million years ago. Among all this diversity, the absence of any putative relatives of either *Cebus* or *Callicebus*, two of the most widespread modern genera, and the absence of any spider or woolly monkeys are notable. The modernness of the La Venta fauna may reflect its relatively late age, the geographic location of La Venta closer than other fossil localities to the Amazon Basin, where living New

World monkeys are most abundant, or, most probably, both of these relationships. In the absence of other faunas of comparable age from elsewhere, these issues cannot be answered.

Aside from a few isolated teeth from the latest Miocene in the upper reaches of the Amazon (Kay and Frailey, 1993), and some remains from latest Pleistocene or Recent cave faunas in Brazil described below, the La-Venta fauna is the youngest fossil deposit yielding platyrrhines from all of Central and South America. Thus, while the pre-Pleistocene fossil record of South America provides documentation of the major diversification of some subfamilies and major clades, we have very little knowledge of platyrrhine species diversity or biogeography in the past, since fossil platyrrhines of different ages tend to be from different geographic areas.

Pleistocene Platyrrhines

The lack of any clear understanding about geographic and temporal patterns of platyrrhine diversity is underscored by the fact that the youngest fossil platyrrhines, those from Pleistocene and Recent caves of the Caribbean and Brazil, are among the most unusual. One of the earliest fossil primates ever recovered and recognized to be unlike living forms was found in a Brazilian cave in the 1830s by the Danish naturalist Peter Wilhelm Lund. In 1836, alongside remains of *Homo*, *Callithrix*, *Cebus*, and *Alouatta*, Lund found a proximal femur and distal humerus of an ateline-like primate of Latest Pleistocene or Recent age (8,000–12,000 years old) that was nearly 50% larger than the same bones in the largest living platyrrhines and probably had a body weight nearly two and a half times as large (Figure 1.3). The distinctiveness of this fossil, *Protopithecus brasilensis*, which more than doubles the known body size of New World monkeys, has only recently been appreciated (Hartwig, 1995). Recent discoveries of additional fossil monkeys from Pleistocene caves in Brazil (Cartelle, 1992) promise to reveal an even greater unsuspected diversity of New World monkeys in the very recent past.

Xenothrix mcgregori is a latest Pleistocene or Recent primate from the island of Jamaica, where there are no extant nonhuman primates. It is a medium-sized platyrrmine (2 kg) known from postcranial material and a mandible with a dental formula of 2.1.3.2 resembling that of marmosets and tamarins. However, the molars are very different in having large bulbous cusps, and a second molar larger than the first. *Xenothrix* was probably a frugivorous species, or may have fed on insect larvae, like the aye-aye of Madagascar. Postcranial remains attributed to *Xenothrix* evidence an unusual type of slow quadrupedal locomotion that has no counterpart among living platyrrhines (MacPhee and Fleagle, 1991).

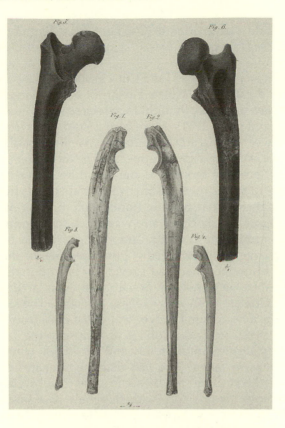

Figure 1.3. Primate remains collected in 1836 by Lund from the Pleistocene/Recent cave of Lagoa Santa in Brazil, with a modern titi monkey for comparison. From top to bottom, *Protopithecus brasiliensis* (fossil), *Alouatta* sp. (fossil), *Callicebus personatus*. Courtesy of Walter Hartwig.

In recent years, it has become clear that *Xenothrix* is just one of several platyrrhines that lived in the Caribbean prior to the first appearance of humans several thousand years ago. Other fragmentary platyrrhine fossils have been found in two other Jamaican caves. Both specimens are proximal femora and are quite different from *Xenothrix*, suggesting at least three primates on that island (Ford, 1990).

Numerous, largely undescribed, dental specimens and a tibia from Pleistocene and Recent cave deposits in Haiti and the Dominican Republic have been recently assigned to the species *Antillothrix bernensis* (Mac-Phee et al., 1995). The dental remains, which may be as much as 100,000 years old, indicate a large primate (2–3 kg) with a dentition reminiscent of living *Callicebus* or possible *Cebus*, suggesting a diet of hard fruit or seeds.

New Pleistocene platyrrhines from Cuba include a well-preserved skull, a mandible, and numerous isolated teeth of a very large platyrrhine, *Paralouatta varonai*, originally thought to be related to the howling monkey (Rivero and Arredondo, 1991). More recent studies have shown many differences between *Paralouatta* and *Alouatta*. Another fossil primate from Cuba *"Montanea anthropomorphus"* appears to be the remains of a modern spider monkey, possibly brought to the island by native peoples.

The recovery of many, quite distinctive primate fossils from the Carribean, often from sites that predate human colonization of the islands, demonstrates quite clearly that there was an endemic primate fauna in the Caribbean until quite recently. Interestingly, island biogeographic studies of the larger islands "predict" the presence of more small- to medium-sized frugivores than had previously been described. This unveiling of an extensive Caribbean fauna raises even larger issues about the origin and ultimate extinction of these primates. The simplest explanation of the Caribbean fauna is over-water dispersal from nearby parts of Central and South America; Cuba is very close to the Yucatan Peninsula, where primates are found today and Venezuela where there is an even more extensive fauna. The fact that the primates on Cuba and Hispaniola have been suggested by some to be closely related to *Ateles*, *Alouatta*, and *Cebus* accords well with this view of a simple dispersal, perhaps relatively recently. However, detailed study of the Antillean primates has demonstrated some very distinctive monkeys—especially *Xenothrix*, but also *Paralouatta*, and *Antillothrix bernensis*. It seems more likely that the Carribean primate may have been separated from other platyrrhines for many millions of years. In the absence of a better knowledge of platyrrhines from elsewhere, it is virtually impossible to calibrate the origin of the Caribbean platyrrhines, but a recent report of a Miocene talus from Cuba that resembles the *Dolichocebus* talus from the early Miocene of Argentina accords with a long period of endemism for the Caribbean monkeys (MacPhee and Iturralde-Vinent, 1995).

Summary of Fossil Platyrrhines

The current fossil record of platyrrhine evolution provides a number of insights into the history of New World monkeys as well as the limitations of the current record. Overall, the current record documents three phenomena: (1) an array of distinctive fossil monkeys from the late Oligocene and early Miocene of the southern part of the continent (Bolivia, Chile, and Argentina) that are unquestionable platyrrhines, but cannot be clearly placed in modern subfamilies; (2) a great diversity of fossil platyrrhines from the middle-to-late Miocene (13.5–11.5 mya) of Colombia that are clearly attributable to extant subfamilies, or even genera; and (3) an increasing number of Pleistocene–Recent fossils from the Caribbean and Brazil that are very distictive from either modern New World monkeys or the middle–late Miocene fossils. Thus, while the fossil record provides evidence that many aspects of the extant Platyrrhine radiation were present by the later part of the Miocene, it also provides evidence of a much greater diversity of New World monkeys from both earlier and later times in disparate geographic regions, dispelling any simple view that the evolution of New World monkeys has been largely static for the past 20 million years (e.g., Delson and Rosenberger, 1984). Among the outstanding questions begging to be answered are the relationships of the Patagonian and Carribean primates to the extant radiation, and especially the Pleistocene history of Platyrrhines. Are the early Miocene monkeys from Patagonia "missing links" at the base of the extant (and later Miocene) radiation, or a "dead end" group of early, possibly geographically isolated monkeys, collataral to the modern radiation? Are the extinct monkeys from the Carribean the remnants of an ancient endemic radiation on those islands or a collection of waif dispersals from the "Neotropical mainland"? How much more diverse were Pleistocene Platyrrhines than those alive today? In many ways the most striking feature of the fossil record of New World monkeys is the fact that it is quite clearly three geographically and temporally separated glimpses into a largely undocumented radiation of primates. To more fully appreciate platyrrhines in a broader evolutionary perspective, and to address the question of the origin of platyrrhines, it is instructive to compare the radiation of Platyrrhines with that of early anthropoids from the Old World.

Early Anthropoid Evolution in the Old World

Since the first decade of this century, most of our knowledge of early anthropoid evolution has come from the Eocene/Oligocene deposits in the Fayum of Egypt (Simons and Rasmussen, 1991). Until most recently, the fossil primates from the Fayum could be rather clearly allocated into one of two families of early anthropoids. Propliopiithecids (*Aegyptopithecus* and *Propliopithecus*) are moderate-sized (4–6 kg) fruit and leaf-eating early catarrhines that have the dental formula (2.1.2.3/2.1.2.3) of later Old World monkeys and apes, but retain primitive platyrrhine features in their ear region and most aspects of their postcranial anatomy (Fleagle,

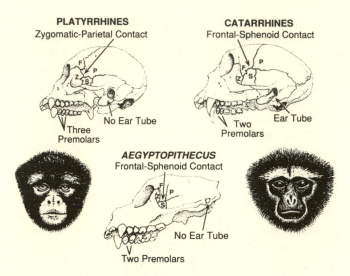

Figure 1.4.   Comparison of the cranial features that characterize extant platyrrhines, extant catarrhines, and the early
catarrhine *Aegyptopithecus*, showing the intermediate anatomical mossiac in *Aegyptopithecus*.

1988; Figure 1.4). Most authorities recognize them as stem catarrhines that
postdate the platyrrhine/catarrhine split, but precede the divergence of
hominoids and cercopithecoids (Fleagle and Kay, 1983; Figure 1.5).

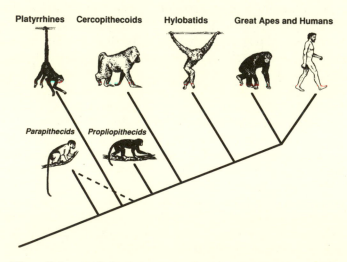

Figure 1.5.   A cladogram showing the phyletic relationships
of parapithecids and propliopithecids relative to extant
anthropoid groups.

The parapithecids, the other family, are a much more primitive group. They are platyrrhine-like in many aspects of their anatomy including their dental formula and many aspects of cranial and dental anatomy, but also show a number of primitive "prosimian" features of the femur and dentition not found in any later anthropoids. Their exact phylogenetic position is less clear, but they probably precede the platyrrhine–catarrhine divergence or, alternately, are the most primitive catarrhines (Fleagle and Kay, 1987; Figure 1.5). Although distinct from all later Old World catarrhines, and in many ways intermediate between living prosimians, platyrrhines, and catarrhines, propliopithecids and parapithecids are nevertheless clearly anthropoid in most aspects of their anatomy (Figure 1.5). They blur the geographic and anatomical distinctions of platyrrhines and catarrhines, but do do not stretch the "anatomical space" of anthropoids much beyond its present range.

In the past 5 years our understanding of early anthropoid evolution has changed in two ways: through the discovery of new, more primitive anthropoids from Egypt and from a diversity of early anthropoids from other localities, and, most recently, other continents. Propliopithecids and parapithecids are primarily known from the upper levels of the Jebel Qatrani Formation in the Fayum in deposits that are most probably early Oligocene in age. However, intense work at Quarry L-41 in the lowest, probably late Eocene, deposits of the same formation have yielded a totally different fauna of early anthropoids that greatly expands the diversity of the suborder and has muddled the issue of anthropoid origins (Simons, 1992; Simons and Rasmussen, 1995).

These new primates from the Eocene levels of the Fayum seem to fall into two major groups (Simons et al., 1994; Simons and Rasmussen, 1995). Best known are the oligopithecines, formerly represented only by the poorly known *Oligopithecus*. *Catopithecus*, a slightly smaller genus (1200 g) that resembles *Oligopithecus* in dental anatomy, is known from several complete skulls and various postcranial remains (e.g., Simons, 1990, 1995). These demonstrate that while *Catopithecus* is clearly anthropoid in having a fused frontal bone, postorbital closure, a platyrrhine-like ear region, and a catarrhine-like dental formula (2.1.2.3.), its dentition shows greater morphological similarities to adapid prosimians than to later anthropoids in molar and premolar anatomy (Rasmussen and Simons, 1992; Simons et al., 1994).

The oligopithecines are generally regarded as the lineal ancestors of the propliopithecids, but if so, they indicate that many dental features of platyrrhines, catarrhines, and probably parapithecids were acquired independently (e.g., Kay and Williams, 1994).

The other new Fayum primates show even more unusual features for early anthropoids. On the basis of their possession of three premolars and some striking postcranial similarities to later parapithecids, they have been placed loosely in the suprfamily parapithecoidea (Simons et al., 1994). *Serapia eocaena* is a tiny (200 g), generalized parapithecid that also shows some premolar similarities to oligopithecines and later anthropoids. *Arsinoea* is a tiny species with crenulated molars, perhaps related to seed-eating. *Proteopithecus* is yet another tiny anthropoid with extremely broad molars and a jaw that lacked a fused symphysis. Its phylogenetic affinities are unclear, but it shows many dental similarities to the early platyrrhine *Branisella* (Kay and Williams, 1995). Today there are no marmoset-sized anthropoids in the Old World, and it is generally throught that the tiny marmosets of the neotropics are secondarily "dwarfed" from much larger ancestors. However, the new Fayum remains suggest that tiny size may have been characteristic of the earliest anthropoids.

In addition to the new material from the Eocene levels of the Fayum, there are numerous other new early anthropioids from other parts of North Africa that present a similar picture (Figure 1.6). These include *Biretia*, a single lower molar from Bir el Ater in Algeria, as well as *Tabelia* and *Algeripithecus* from Glib Zegdou in Algeria (Godinot, 1994). In addition, there are abundant remains of anthropids similar to those in the Fayum from several localities in Oman. All of these new taxa show dental similarities to later Fayum anthropoids, but are very much smaller. While only preliminary reports of these taxa are available at present, they suggest a very diverse adaptive radiation of anthropoids in the Eocene of Africa that was quite different in both body size and ecological adaptations from the anthropoiods of the early Oligocene or today.

Most recently, Beard and colleagues (1994, 1996) described a new primate fauna, including adapid, omomyid, and tarsiid prosimians, as well as purported early anthropoids from the Eocene site of Shanghaung in China (Figure 1.5). The proposed anthropoid, *Eosimias*, is another tiny primate with primitive, tarsier-like molars and relatively large, possibly sexually dimorphic canines. They also noted dental similarities to platyrrhines.

EARLY ANTHROPOID FOSSIL LOCALITIES

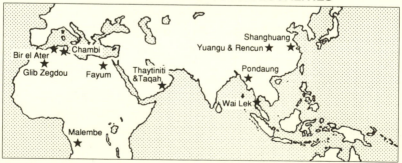

Figure 1.6.   A map showing localities yielding fossil anthropoids or putative anthropoids from the Eocene and early Oligocene of Africa and Asia.

Overall, the most striking characteristics of the early anthropoids from the Old World are the many differences from either extant prosimians or extant catarrhines, and their persistent similarity to platyrrhines, in most nondental aspects of their anatomy (e.g., Fleagle and Kay, 1994). The anatomical similarities between Egyptian Oligocene anthropoids and platyrrhines were first noticed many years ago (e.g., Simons, 1967b; Fleagle et al., 1975) and this gave rise to the view that propliopithecids and parapithecids preceded the divergence of the modern catarrhines groups—Old World monkeys and apes (Fleagle and Kay, 1983; Fleagle, 1986). However, from a platyrrhine perspective, the morphological similarities between early African anthropoids and extant platyrrhines suggest that many aspects of platyrrhine cranial and postcranial anatomy are primitive retentions from an early anthropoid condition.

Platyrrhine Origins

Few topics in primate evolution are subject to more diverse hypotheses and unsatisfactory alternatives than the problem of platyrrhine origins (e.g., Ciochon and Chiarelli, 1980; Hoffstetter, 1980). For most of the Cenozoic Era, South America was an island continent, with its own fauna distinct from that on other continental areas (e.g., Simpson, 1980). Thus, edentates and marsupials were diverse and abundant, the ungulates were astrapotheres, pyrotheres, notoungulates, and litopterns rather than artiodactyls and perissodactyls, and the predators were marsupials and giant birds rather than creodonts and carnivores. Primates first appear in the fossil record of South America in the late Oligocene, with no evidence of the group prior to that time, despite an abundant fossil record of other mammals from the southern part of the continent (e.g., Hoffstetter, 1980; Simpson, 1980; Hartwig, 1994). The problem of platyrrhine origins is both a phylogenetic question and a biogeographical one. What did the last common ancestor of platyrrhines and catarrhines, or the first platyrrhine after the split look like and how did it get to South America? These issues are far from being resolved to everyone's satisfaction. However, the record of early anthropoid evolution in Africa and our present knowledge of primate evolution on other continents can be used to evaluate the alternatives.

The Phylogenetic Origin of Platyrrhines

In the early part of this century, it was generally believed that platyrrhines evolved from some group of North American prosimians; probably the notharctine adapids and catarrhines were evolved (probably independently) from some Old World prosimians (e.g., Gregory, 1922). Thus, the last common ancestor of platyrrhines would not be recognized structurally as an anthropoid primate. This view still has some adherents, but seems most unlikely for several reasons. Almost all molecular studies indicate that platyrrhines and catarrhines shared a long period of common ancestry separate from any extant group of prosimians (Sarich and Cronin, 1980; Miyamoto and Goodman, 1990). Furthermore, all fossil and extant platyrrhines, all fossil and living catarrhines, and the Oligocene parapithecids of Africa show evidence of extensive postorbital closure, a striking derived feature uniting anthropoids (Cartmill, 1980; Ross, 1994). This feature was surely in the common ancestor of the group. Despite dental and some postcranial similarities shown by various Paleogene fossil prosimians to early anthropoids (Rasmussen, 1990, 1994), none shows any evidence of postorbital closure. Thus, if platyrrhines and catarrhines evolved independently from separate North American and European adapid ancestors, they must have evolved postorbital closure independently—a most unlikely event. It seems much more likely that the last common ancestor of platyrrhines and catarrhines was itself an anthropoid, with postorbital closure (Cartmill, 1980).

Moreover, the discovery that the earliest anthropoids in the Old World (and to a lesser degree the earliest catarrhines) are essentially platyrrhine-like in many aspects of cranial and postcranial morphology suggests that their common features are not parallelisms, but characteristics of the earliest anthropoids. This is also consistent with the platyrrhine fossil record, which suggests that platyrrhines have changed far less during the last 25 million years than have catarrhines (e.g., Delson and Rosenberger, 1984). On the basis of our understanding of both platyrrhine and catarrhine evolution, it seems most likely that the earliest platyrrhine and the last common ancestor of platyrrhines and catarrhines was an animal very much like a small platyrrhine or parapithecid in both cranial and postcranial anatomy (e.g., Kay, 1980; Cartmill, 1980; Fleagle and Kay, 1987; Ford, 1990).

On the basis of our understanding of paleogeography, it is clear that the immigration of primates to South America must have involved some type of long distance rafting or island hopping across substantial water barriers, regardless of the ultimate source area. Such dispersal is almost certainly rare, but must have taken place if we are to account for the presence of land vertebrates on other islands, such as the Carribbean (e.g., MacPhee and Woods, 1982). North America and South America were separated throughout the Cenozoic until Central America slid into place about 5 million years ago to form the Panama Land Bridge and initiate the Great Faunal Interchange (Stehli and Webb, 1985). Africa and South America have been separated by the South Atlantic Ocean since the middle Cretaceous, well before the first appearance of primates (e.g., Tarling, 1980; Stehli and Webb, 1985). Even though the Atlantic Ocean has been increasing in width for the past 160 million years, the distances in the Eocene between the west coast of Africa and the eastern coast of South America were not appreciably shorter than they are today (e.g., MacFadden, 1990). There was no obvious bridge to South America during the early Cenozoic; however it is probably significant that the lowest drop in sea level during the past 500 million years appears to have taken place during the early Oligocene, thus appreciably narrowing water barriers (Haq et al., 1987). However, given the isolation of South America and the rarity of immigration from other continents prior to 5 million years ago, our best available evidence for reconstructing the source of platyrrhines is our current understanding of the biogeography of the immigrants themselves.

For many years, paleontologists argued that Primates and most other groups of mammals in South America were immigrants that rafted from North America. Such a scenario is not unreasonable, for many of the early ungulates (e.g., condylarths) that are found in the early Cenozoic of South America have North American relatives. However, the two groups of mammals that first appear in South America in the Oligocene, higher primates and caviamorph rodents, are not known to have ever existed in North America prior to very recent times (Hoffsetter, 1980; Savage and Russell, 1983; Wyss, 1993). Thus, hypothesizing North America as the proximate source area for these groups requires a second unsubstantiated hypothesis that they were, in fact, on that continent, but have not yet been discovered in the fossil record, and that they immigrated to South America at a time when none of the mammalian groups known to have been in North America made the journey. The same is true for postulating any other continent, including Antarctica, as the source for platyrrhine origins. However, primitive anthropoids strikingly similar to platyrrhines, and caviamorph rodents are abundant in the fossil record of Africa 5 to 10 million years prior to their first appearance in South America (Hoffstetter, 1980; Fleagle et al., 1986; Simons and Rasmussen, 1995). Until early anthropoids are discovered on any other continent, Africa must be considered the most likely ultimate and proximate source of origin for platyrrhines (Hoffstetter, 1980; Simons and Kay, 1986; Fleagle and Kay, 1987; Fleagle, 1988; Wyss et al., 1993; Flynn et al., 1995).

Implications for the Study of Platyrrhine Biology

A broad consideration of the fossil record of platyrrhine and early anthropoid evolution raises a number of interesting questions about the modern platyrrhine radiation. Early African anthropoids are, in much of their anatomy, African platyrrhines; or, alternately, many of the characteristics of platyrrhine morphology and ecology seem to be features that characterized all anthropoids 30 million years ago. Attempts to understand the functional anatomy and reconstruct the habits of these early anthropoids must be based on a knowledge of the extant species with the most similar anatomy, i.e., extant platyrrhines. However, aside from the practical problems of making functional sense of a fossil bone from the Oligocene of Egypt, and the biogeographic questions of how early anthropoids found their way between these two continents, there are some much broader issues raised by the similarities between early Old World anthropoids and living and fossil platyrrhines.

Why have platyrrhines retained the small size, frugivorous, arboreal habits of the Oligocene anthropoids while later catarrhines evolved larger size, more folivory, and terrestriality in several lineages? Are the differences due to environmental differences between the available habitats in South America and the Old World? Certainly the flora and range of habitats are more diverse and the gross geographic area is greater in Africa and Eurasia where catarrhine evolution has taken place than in South America where platyrrhines have evolved. Then too, there are differences in their mammalian competitors and predators. When platyrrhines first appeared in South America, the continent was filled with abundant insectivorous and frugivorous arboreal marsupials and insectivorous and folivorous edentates as well as large terrestrial predaceous birds. In addition, they seem to have arrived either contemporaneously with or slightly after the caviamorph rodents, which rapidly expanded into a very extensive array of arboreal and terrestrial species of all sizes (e.g., Simpson, 1980). In addition, there were no prosimians in the New World. Are the differences due to some intrinsic constraint in the nature of the early anthropoids themselves?

In the Old World, new discoveries from the Eocene of the Fayum have revealed an extraordinary diversity of Eocene early anthropoids, both more generalized and more specialized than the better known parapithecids and propliopithecids known from the Oligocene. According to current interpretations, the parapithecids appear to have gone extinct in the Old World while the propliopithecids gave rise to later catarrhines. On the other hand, the platyrrhines appear to be derived from a parapithecid-like ancestor. Is this just evolutionary serendipity? These are major issues in primatology that can be addressed only through comparative study of platyrrhines and the other major primate radiations using both neontological and paleontological perspectives.

Acknowledgments

We thank the late Warren Kinzey for soliciting this paper and offering many insights and suggestions, and for his relentless badgering over many years that led to its completion. Luci Betti drew most of the illustrations. Much of the work described in this paper was originally done in conjunction with E. L. Simons. Walter Hartwig, Kaye Reed, Kelly McNeese, and Roberto Fajardo provided valuable comments and suggestions on the contents of the manuscript and bibliographic assistance. Joan Kelly showed her usual unbounded patience. This work was funded in part by grants from the National Science Foundation (BNS 8606796, BNS 9012154, and BRS8614533), The LSB Leakey Foundation, and the National Geographic Society.

2

Brains of New World and Old World Monkeys

ESTE ARMSTRONG and MARY ANNE SHEA

Introduction

New World monkeys have had a separate evolutionary history, probably since the lower Oligocene, a time that is about four times longer than the hominid stock has been separate from other primates (see Fleagle and Kay, this volume). Given the centrality of the brain to primate evolution (Clark, 1959), it is reasonable to expect that the long separate phylogenies produced differences in the central nervous system, particularly if brains evolve relatively quickly. If, on the other hand, the brain is under strong selective pressures to maintain a particular organization and size, few differences between Old and New World monkeys are likely to be found. To examine this question, as well as document differences between New and Old World monkeys, two aspects of neuroanatomy are reviewed here: size and organizational differences that are independent of size.

The taxonomic relationships among New World monkeys have undergone recent reanalysis (Rosenberger, 1984; Schneider et al., 1993, Harad et al., 1995). In this chapter, the superfamily of extant New World monkeys, Ceboidea, is considered to have two families: the family Cebidae, which is composed of three subfamilies, Cebinae (*Cebus* and *Saimiri*), Aotinae (*Aotus*), and Callitrichinae (marmosets and tamarins), and the family Atelidae, which is composed of Atelinae and Pitheciinae. Species of *Cebus*, *Saimiri*, and *Aotus* remain controversial in their taxonomic affinities.

Brain Size

Primates are characterized by having large brain weights for the animal's body weight. Among anthropoid primates, the New World species have the smallest brains, being less than 5 g in the pygmy marmoset (*Cebuella pygmaea*), a callitrichid. The biggest New World monkey brains, found in atelid woolly (*Lagothrix*) and spider monkeys (*Ateles*), are over 100 g. In contrast, Old World monkey brains are larger, ranging from 40 to more than 200 g.

The smaller size of platyrrhine brains is associated with smaller body sizes, and so the latter factor must be taken into account. A comparison of brain sizes after controlling for the effect of body weight is a comparison of *relative* brain sizes.

The equations that scientists have derived to describe associations between brain and body weights identify slopes and intercepts. The slope shows how much of an increase in brain weight occurs for every unit increase in body weight, and the intercept gives a theoretical brain weight at a particular body weight, usually zero. Typically the data points are the mean brain and body weights of species and are transformed logarithmically to even the variation and produce a straight alignment of the points.

Different slopes can be chosen a priori on the basis of statistical concerns or taxonomic considerations (see below). Using the least-squares equation assumes that the only variability of concern is with the dependent variable, in this case, brain weight. Other approaches, such as the major axis or reduced major axis, take into account variability in the independent feature, in this case body weight, as well. Slopes generated from the major axis or reduced major axis are typically steeper than ones generated from a linear regression, but with a strong correlation, little differences among slopes is observed (Martin, 1981). Because most of the previous research on New World monkey brains used the least-squares approach, we report observations based on that procedure here.

In studies that compare the sizes of adult brains from different taxonomic groups, the slope is less than 1 and brain–body associations are therefore considered to be examples of negative allometry (Armstrong, 1985b). That is, among animals of different sizes, body weights differ more than brain weights. If brain and body weights varied proportionately, the slope describing the association would be 1. Although it had previously been thought that the brain–body slope always decreased as taxonomic proximity increased (i.e., that body weights varied more than brain weights in closely related taxa), a recent analysis shows that this is not always the case. Some mammalian orders have as much variation in brain weights within a genus as within the order (Pagel and Harvey, 1989). The latter study underscores the importance of deriving an equation based on the taxonomic groups being studied rather than importing a slope from other animals.

Differences between the relative brain sizes of New and Old World

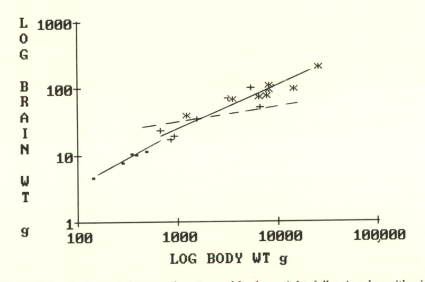

Figure 2.1. Brain weight as a function of body weight following logarithmic transformation of both variables. Callitrichids are dots, other New World monkeys are crosses, and Old World monkeys are stars. The callitrichid and noncallitrichid New World monkey slopes are drawn. The dashed line represents a hypothetical slope, like one of a carnivore family, that is lower than one set by the data in hand. Note that this would make the smaller bodied monkeys appear to have relatively small brains and the large bodied ones to have relatively big brains.

monkeys can be determined by measuring how far above or below their respective values is the slope describing the association between brain and body weights (Figure 2.1). Different equations alter the arrangement between the slope and the species' values, and thus change the distances of each species-specific point to the line.

Some investigators prefer to use an equation that is generated from brain and body weights of another taxonomic group (Stephan et al., 1972, 1981), from the encompassing group, such as all mammals (Jerison, 1973), or from a slope thought to be descriptive of a particular taxonomic distance, such as families (Hemmer, 1971). Another choice is to derive the equation from the data at hand (Armstrong, 1985a,b). The more the equations for standardizing brain size differ, the more the results and interpretations will differ as well (Holloway and Post, 1982; Radinsky, 1982).

The approach by Stephan and his colleagues is based on quantifying the difference between a particular animal's brain weight and that expected for a basal insectivore of equivalent body weight (Stephan et al., 1972, 1981). Basal insectivores are thought by Stephan and his colleagues to retain many attributes of the Cretaceous insectivores, including primitive cerebral patterns and brain–body relationships, which are thus interpreted as being ancestral (Stephan et al., 1972). The slope describing the basal insectivore brain–body association, 0.63, is then used to derive ratios between observed and expected weights for extant groups, like primates. The latter ratios are called encephalization indices, and those of New and Old World monkeys cannot be distinguished. That is, both groups have equally divergent encephalization indices from the basal insectivores (Stephan et al.'s data, 1981; $N = 23$; $t = 1.17$, n.s.). The relative size of callitrichid encephalization indices ($N = 5$), however, is smaller than those of other New World monkeys ($N = 8$; $t = 3.28$, $p <0.01$).

Similarly Jerison (1973) found New and Old World monkeys to have equivalent relative brain sizes and Callitrichidae to have significantly smaller ones. His study used a slope of 0.66, which described the brain–body relationship in his sample of extant mammals and is thus close to that used by Stephan et al. (1972, 1981).

Hemmer (1971) preferred to compare primates using a slope of 0.23, because this slope, originally determined within carnivores (Rohrs, 1959), was thought to describe intraordinal and intrafamilial relationships in all mammalian orders. The slope is significantly lower than those found among basal insectivores or all mammals, showing that among these carnivores, body weights vary much more than brain weights.[1] Using such a low slope gives New World monkeys a lower relative brain size than Old World monkeys, and callitrichids relatively smaller brains than either ceboids or catarrhines (Hemmer, 1971).

Slopes generated for different groupings of New and Old World monkeys, however, are steeper than the 0.23 slope Hemmer used. Values from Stephan et al.'s data (1981), for example, produce a slope of 0.77 for callitrichinae, 0.67 for other New World monkeys, and 0.48 for Cercopithecidae (the differences in slopes are discussed below). Using a slope of 0.23 when the data are arrayed more steeply exaggerates the distances below the line for small-bodied monkeys (in this case mostly New World species) and above the line for large-bodied primates (in this case mostly Old World monkeys). Consequently, using the slope derived from carnivores produces an artifact when it shows that New World monkeys' brains are relatively smaller than Old World monkeys, and Hemmer's (1971) interpretation that these taxa differ in terms of relative brain size is rejected.

Another approach is to determine the slope and intercept from the data. When Stephan et al.'s brain and body weight data (1981) are used in this manner, New World monkey brain-to-body slopes are close to the slopes given above (all Platyrrhine slopes = 0.77; all anthropoid slopes = 0.77). This demonstrates that for any given increase in body weight, closely related species of monkeys differ in brain weight more than carnivores do. It also corroborates the slopes derived by Pagel and Harvey (1989). The slopes generated from Stephan's data set are close to the basal insectivore slope used by Stephan and his colleagues and the panmammalian slope used by Jerison, and, consequently, so are the results. Once body size is taken into account, Old and New World monkeys have the same approximate relative brain sizes, whereas those of the Callitrichinae are smaller.

The relatively small brains of the callitrichids deserve further analysis. Other studies have shown that primates with brains that are relatively small for their body weights also have relatively low metabolic rates (Armstrong, 1983, 1985a,b, 1990). Metabolic activity plays an important role in the scaling between brain and body weights because of the critical importance in keeping the brain constantly supplied with oxygen and glucose.

Unlike muscle and many other organs, the brain does not use energy anaerobically and so does not store glycogen. The functioning of the brain and the survival of the animal depend on a constant supply of glucose and oxygen (Sokoloff, 1981). The major factors affecting the adequate supply of a brain depends on its absolute size, the size of the vessels supplying it (which is dependent on body size), and the rate by which the heart beats and glucose and oxygen are turned over (which is associated with metabolic activity). The need for glucose sets a minimal, but not a maximal, requirement for body–brain associations; excess serum glucose does not present with neurological symptoms the way hypoglycemia does (McDougal, 1981).

Consider comparing two primates of the same brain weight, one having a lower metabolic rate, perhaps to protect it from plant or insect toxins (Martin, 1983) or to be less careful in the regulation of body temperature (Muller and Jaksche, 1980) than the other. To supply the same sized brain in the same amount of time with the necessary amounts of glucose and oxygen, the primate with the lower metabolic rate (and slower heart rate) requires a bigger body. In a study of relative brain size, the former primate would be seen as having a relatively smaller brain than the latter (Armstrong, 1985a,b).[2] If this interpretation is correct, then the relatively small brained callitrichid species should have lower metabolic rates than expected for their body sizes.

Basal metabolic rates have been measured in three callitrichid genera: *Saguinus* sp., *Callithrix jacchus*, and *Cebuella*. The first has a typical mammalian resting metabolic rate (Scholander et al., 1950), whereas the latter two have lower rates than expected for their body weights (Morrison and Middleton, 1967; Rothwell and Stock, 1985; Petry et al., 1986). The relative brain sizes of these three species match their metabolic rates. The relative brain sizes and metabolic rates of *Cebuella* and *Callithrix* are lower than those of the tamarins. Furthermore, once the body weights are corrected for metabolism, the relative brain sizes of the Callitrichinae overlap those of other New and Old World monkeys.

Thus, these data corroborate a previous study that showed that differences in relative brain size between prosimians and anthropoids disappeared once metabolism was taken into account (Armstrong, 1985a). In other words, nonhuman primates, including callitrichids, have the same sized brains relative to their energy supply.[3]

The differences in slopes describing brain–body associations between Callitrichinae (0.77), the noncallitrichid New World monkeys (0.67), and Cercopithecidae (0.48), in which Old World monkeys have a lower slope than either New World group (see above), may also stem from the effect of metabolism. Among the Cercopithecidae only *Colobus*, an intermediate sized monkey, has a relatively low basal metabolic rate (Armstrong, 1985a). In contrast, among the New World monkeys, it is the small species, *Aotus*, *C. jacchus*, and *Cebuella*, that have lower basal metabolic rates than expected for their body weights. Consequently, these animals have bigger bodies for their brain sizes than other monkeys. Since the New World monkeys with relatively low metabolic rates are small bodied, they have the effect of tilting the brain–body slope to a steeper angle. *Colobus*, by having an intermediate sized body, does not have this influence.

Relative brain weights provide information concerning the integration of the brain into the total body unit, particularly in understanding how the brain is supported, but it tells us little about the behavior of any species. Absolute brain weight, when combined with an understanding of brain organization, on the other hand, is an important variable or standard when considering brain functions. Although it is not yet known how differences in absolute numbers translate into differences of behavior, the number of neurons involved in a particular activity is undoubtedly critical to its functioning (Williams and Herrup, 1988). It is well documented, for example, that density of neurons and sizes of different neural populations, like cortex or subcortical nuclei, are associated with the overall size of the brain (Armstrong, et al. 1986, 1989, 1990). Because absolute brain size can predict neuronal densities and sizes of neural populations, brain size helps standardize quantitative comparisons. At the same time, such comparisons are meaningful only if the neural organization is similar, and thus the predictions are strong only among closely related taxa (Armstrong, 1995).

Brain Organization

The brain consists of many different regions, all of which play critical roles in the perception of sensory stimuli, cognition, the organization and output of motor behavior, and the integration of information with memory, motivation, and other limbic functions. One primate specialization is that many of these neural functions are more highly dependent on cortical processing than is the case in other mammalian orders. The partial replacement of subcortical functions by the cortical ones has also been termed encephalization (Noback and Shriver, 1969). Although the overall level of encephalization is not known to differ between Old and New World monkeys, some small differences in connections and other anatomical features have been observed in the cortex. The cerebral cortex is by far the best studied portion of the primate brain and most of the New World–Old World differences have been identified here.

Cortical Folding

The cortex, the outer sheet of cellular material covering the cerebral hemispheres, has its neurons organized horizontally into layers and vertically into columns. Small brains have smooth cortical surfaces (are lissencephalic), but large ones are folded in a regular fashion. The constancy in the form and shapes of the folds (sulci and gyri) is thought to be the result of similar and complex growth factors throughout the brain (Todd 1982, 1986; Welker, 1990). The degree of cortical folding is associated with the volumetric proportions between the outer (layers I–III) and inner (layers IV–VI) cortical layers, such that a greater expansion of the outer layers relative to the inner ones results in increased cortical folding (Richman et al., 1975; Armstrong et al., 1991, 1995).

Analyses of cortical folding show that while larger primate brains are more convoluted than smaller ones and that the degree of folding is predictable on the basis of brain weight, two different associations between the degree of folding and brain weight exist among primates (Zilles et al., 1989). The degree of cortical folding in the prosimian[4] brains is less than that in anthropoids, so that for every unit increase in brain weight, anthropoid cortices buckle more than those of prosimians. A qualitative difference is also well known, namely, prosimian sulci tend to be longitudinal (the axes of the major sulci are in a rostral–caudal direction), whereas those of anthropoids run transversely.

It is not yet clear whether shifts in sulcal patterning are associated with altered degrees of folding, but New World monkey brains differ from those of Old World monkeys in sulcal patterns (Falk, 1981). Previously New World monkey brains were thought to be less convoluted than their Old World counterparts (Connolly, 1950), but a scaling analysis demonstrates that these differences in folding are significantly associated with brain size. After controlling for the latter, all anthropoids have an equivalent degree of cortical folding (Zilles et al., 1989). Thus, the difference in the degree of folding between Old and New World monkeys is the result of differences in brain size.

Differences between Cebinae (*Cebus* and *Saimiri*) and other New World monkeys have been observed in some neural systems (see below), but the degree of cortical folding does not separate Cebinae from other anthropoids. Although Clark (1959) thought that callitrichid brains were less convoluted than those of prosimians of an equivalent size, quantitative analyses show that the degree of folding present in callitrichid cortices fits both anthropoid and prosimian slopes (Zilles et al., 1989). As stated above, the degree of folding is associated with the amount of space in the outer cortical layers compared to that in the inner cortical layers (Armstrong et al., 1991, 1995), so differences in laminar proportions can help weaken or strengthen the inclusion of callitrichids on the prosimian or anthropoid slope.

The laminae in the posterior cingulate isocortex have been analyzed in such a way as to shed some light on this relationship. In this region, prosimians and anthropoids differ in the relative sizes of cortical layers, with anthropoids having relatively bigger outer (supragranular) layers than prosimians (Zilles et al., 1986). The volumetric proportions of the cortical layers of *Callithrix jacchus*, the only callitrichid analyzed in this study, resemble those of other anthropoids and not those of prosimians. This suggests that the folding index of the callitrichids is the result of its being a small anthropoid, and that the overlap with prosimians is a result of parallelism, not the retention of a primitive characteristic. Additional studies are needed to determine whether other callitrichids share the relatively enlarged supragranular layers with *Callithrix jacchus*.

The Visual System

Compared with other mammals, primate brains have disproportionately large visual structures. Only felids approach primates in this type of development (Allman, 1982). Furthermore, the visual system is the best studied of all the neural systems, and so the best documentation for differences between Old and New World primates is found in this circuitry.

The basic visual pathway (Figure 2.2) consists of rod and cone receptors in the retina projecting to bipolar cells, and the latter to retinal ganglion neurons. Axons of the ganglion cells leave the retina to end in either the superior colliculus or lateral geniculate body. Although the

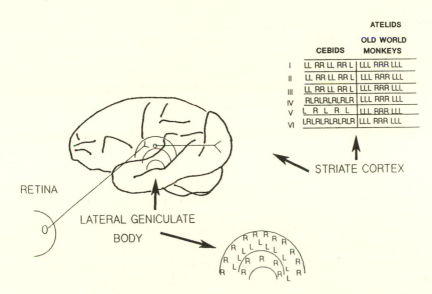

	CEBIDS	ATELIDS OLD WORLD MONKEYS
I	LL RR LL RR L	LLL RRR LLL
II	LL RR LL RR L	LLL RRR LLL
III	LL RR LL RR L	LLL RRR LLL
IV	RLRLRLRLRLR	LLL RRR LLL
V	L R L R L	LLL RRR LLL
VI	LRLRLRLRLRLR	LLL RRR LLL

STRIATE CORTEX

RETINA

LATERAL GENICULATE
BODY

Figure 2.2. Schematic drawing of the anthropoid visual system. LGB is the lateral geniculate body, R is the right ocular information, and L is monocular information from the left eye. The supragranular layers of cebid visual cortex have structural columns according to autoradiographic transport studies (shown here), but not physiological separation. See text for details.

superior colliculus is organized differently in primates than in other mammals, with the exception of variable projections from extrastriate cortex, no organizational differences between anthropoid taxa are known (Allman, 1982; Cusick, 1988).

In primates, the major input of visual information into the brain is from the retinal ganglion cells to the lateral geniculate body. The latter keeps input from each eye topographically separate (see Figure 2.2) and relays visual stimuli to the primary visual cortex (V1 or striate cortex) of the occipital lobe. Visual information is further elaborated here before being passed outward into a mosaic of specialized visual association cortices where more complex processing and perception occurs. Evolutionary changes are possible in every segment, but only a few have been documented separating Old from New World monkeys.

With the exception of *Aotus*, all monkeys are diurnal and have a retina with a well-defined fovea. *Aotus*, like other nocturnal mammals, uses rods for vision, but unlike most other nocturnal mammals, the retina of *Aotus* has cones and an area centralis, a region without visible blood vessels (Polyak, 1957; Ogden, 1975). In diurnal primates, the fovea of the area centralis contains a pit that is packed with cones, has very few rods,

and is specialized for accentuating discrimination of color. The area centralis of the owl monkey has an increased number of receptors, but the
cone–rod ratio is similar to that of the peripheral retina (Ogden, 1975).
The presence of an area centralis in *Aotus*, as determined by the lack of
blood vessels and the denser packing of rods and cones (Ogden 1975),
suggests that its recent ancestor was diurnal and possessed all attributes
of color vision (Clark, 1959).

Old World monkeys have excellent color vision that is on a par with
human color vision, but questions remain about subtle differences in the
color vision of New World monkeys (see Jacobs, this volume). Studies on
Ateles suggest that its color vision is similar to that of catarrhines. Analyses of *Saimiri* and *Cebus*, on the other hand, show that these animals
have certain characteristics that resemble color-blind people. The Cebinae, *Saimiri* and *Cebus*, can separate green or red from yellow, but compared with Old World monkeys (or people), they have a broader range
of red–greens they cannot separate from yellow. This is correlated with a
lessened ability to discriminate visual cues by hue and by a decreased
saturation of light necessary to separate colors from white light (De
Valois and Jacobs, 1968). Since *Saimiri* and *Cebus* may form a monophyletic group, Cebinae, this alteration in color vision may well be limited to
these two species and, as the data from the spider monkey attest, not be
characteristic of New World monkeys. Whether the Cebinae share their
visual characteristics with other members of the family Cebidae, *Aotus*,
and the callitrichids, is not yet clear.

Our knowledge of morphological correlates of these functional differences is limited, but some are known. The large visual relay center in the
thalamus, the lateral geniculate body (LGB), is a laminated structure in
primates. The degree of spatial lamination in the LGB is variably developed in primates, and New World monkeys may have fewer intercalated
cells and narrower interlaminar zones than Old World monkeys (Kaas et
al., 1978; Hendrickson et al., 1978; Diamond et al., 1985; Florence et al.,
1986), but a separate analysis of the Cebinae and the influence of brain
size has not been examined. The LGB, receiving information from both
eyes, but one visual field,[5] projects primarily to layer IV of the striate
cortex (Figure 2.2). In all simians, the information from one eye is kept
separate from the other in the LGB.

Monocular information is relayed to layer IV of the primary visual
(striate) cortex, where, in most anthropoids, it remains separate from
information arising from the other eye. The two sets merge into a functionally binocular system in other parts of the visual cortex.[6] The organization in the striate cortex differs between the Cebinae and other
anthropoids.

Crossing at right angles to the layers of cortex are repetitive functional

units, the ocular dominance columns, which contain neurons responsive to stimulation of one eye and which are themselves composed of
narrower orientation columns (Hubel and Wiesel, 1968). Although ocular dominance columns are physiological entities, structural correlates
are detected by transportation of radioactive substances across synapses
(autoradiographic transneuronal transport studies) and 2-deoxyglucose
studies in which radioactive material is carried from the blood stream
into metabolically active neural regions.

In Old World monkeys and *Ateles*, the ocular dominance columns are
seen in all layers of the primary visual cortex, layers I–VI (Hubel et al.,
1976; Florence et al., 1986; Hendrickson et al., 1978; Hendrickson and
Wilson, 1979; Tootell et al., 1988). *Saimiri* and *Cebus*, however, do not
segregate their input into ocular dominance columns in layer IV of their
visual cortex (Figure 2.2). Rather the Cebinae have a well-documented
binocular overlap in layer IV (Hendrickson et al., 1978). At the same
time, the Cebinae, like other simians, have periodic stripes of [^{14}C]de-
oxyglucose usually associated with ocular dominance columns in laminae I–III, whereas their lower layers (layers V and VI) are organized
differently from those of Old World monkeys (Hendrickson and Wilson,
1979; Tootell et al., 1988). In contrast, *Aotus*, the nocturnal third member
of Cebidae, has ocular dominance columns in layer IV (Tootell et al.,
1993), although they may be restricted to a small part of the striate cortex
(Diamond et al., 1985). *Callithrix*, less intensely studied, also appears to
lack ocular dominance columns in layer IV (Debruyn and Casagrande,
1981). If further work substantiates this, then the lack of ocular dominance columns may characterize diurnal members of the family Cebidae
(Cebinae + Callitrichinae). The presence of ocular dominance columns
in prosimians (Hubel et al., 1976), their variable development in New
World monkeys, and their structural variations among Cebidae all suggest a derived, not a primitive condition, for cebid striate cortex.

When stained, cytochrome oxidase, a mitochondrial enzyme associated with metabolic activity, shows alternating positive blobs and interblob spaces in primate striate cortex, particularly in layer III. Details
remain controversial, but the functional correlates of the blob–interblob
alteration are associated with receptive fields that are nonoriented and
color opponent (blobs) versus regions that are orientation specific and
noncolor opponent (nonblobs) and are topographically coextensive with
other markers of ocular dominance columns (Tootell et al., 1993).

Analyses of cytochrome oxidase blobs also suggest a lack of eye segregation in Cebidae striate cortex (Tootell et al., 1988; Rosa et al., 1991) or
a significant binocular overlap within monocular columns (Hess and
Edwards, 1988). The boundary between monocular and binocular fields
is associated with an increase in blob densities in Old World monkey

striate cortex, but no alterations in densities are observed in the Cebidae (Rosa et al., 1991). *Aotus* also has a unique cytochrome oxidase topography; blobs are present in all layers (Tootell et al., 1985).

Ateles, a non-Cebid New World monkey, shares the Old World pattern of monocular segregation of visual input in layer IV (Florence et al., 1986; Hubel and Wiesel, 1968). Because Galagos, Old World monkeys, and the diurnal Atelidae have continuous ocular dominance columns, the lack of ocular dominance columns in Cebinae striate cortex is most likely a derived, not primitive, condition.

Initially, fetal rhesus monkeys also lack ocular dominance columns in layer IV of the striate cortex: the binocular structure begins ending somewhat before embryonic day 130 (Rakic, 1976). The similarity between the adult Cebinae and immature rhesus monkey pattern suggests that the pattern in *Cebus* and *Saimiri* may have been attained through heterochrony, a change in timing in development. Thus evidence from the visual system corroborates evidence from nonneural tissue (Leutennegger, 1980; Ford, 1980) that evolution among several New World taxa may have occurred by altering growth rates or duration.

Visual perception is thought to occur in the visual association cortices, a mosaic of functionally and structurally discrete regions. The second visual association cortex, VII, surrounds the striate cortex from which it receives its major input. From there, information projects outward in a minimum of two major directions: one circuit goes toward the inferior temporal lobe where the form and shape of the object are perceived, while another branch carries visual information to the posterior parietal cortex where the spatial context and movement of the stimulus are perceived (Mishkin, 1972). Both the attention directed to visual stimuli and the control of motor responses to those stimuli are thought to occur in the posterior parietal cortex (Mountcastle et al., 1975).

All primates, including prosimians, have the dual pathways proceeding into separate target association cortex and the number of specialized visual association regions appears similar in Old and New World monkey brains (Tootell et al., 1993). Observed differences between primate taxa in these higher-order processing areas, such as the inferior temporal visual cortex of Cebinae (*Saimiri* and *Aotus*), but not *Macaca*, receiving information from a middle temporal region, MT (Weller and Steele, 1992), seem minor, but may have an additive effect. The organization of connections between visual and limbic regions separates cebids or possibly New World monkeys from Old World monkeys (Selzer and Pandya, 1978; Webster et al., 1991; Weller and Steele, 1992) more clearly than differences within association cortex, but kinds and degrees of homology remain unknown at this time.

The visual system, the best studied neural circuit in primate brains, appears conserved among New and Old World monkeys. The biggest differences segregate the subfamily Cebinae, or possibly the Cebidae, from other monkeys. Studies of other sensory systems, particularly in the somatosensory cortex, have also produced an analogous picture of structural similarity among anthropoids with the exception of the Cebinae.

The Somatosensory System

Information from cutaneous receptors is relayed to the parietal cortex where maps of the body's surface, "homunculi," are formed (Carpenter, 1991). Neurons in the somatosensory cortex respond to different receptors and are organized such that maps generated from cutaneous receptors (e.g., touch) in Brodmann area 3 are anterior to those for deep stimuli (Brodmann area 1) and joint movement (Brodmann area 2). Area 3 is further subdivided into regions 3a, which receives information from receptors that are deeper than 3b.

Although the precise nature of homologies is disputed, the organization of 3b is similar to that of the primary somatosensory cortex (SI) in nonprimates (Kaas, 1983). The representations of the tail cover a larger proportion of the somatosensory thalamus and cortex in monkeys with prehensile tails than in the others (Pubols and Pubols, 1971; Felleman et al., 1983). The relative enlargement supports the view that the cutaneous maps do not reflect the size of the structure, but rather the number of receptors and, therefore, its function.

Most anthropoids have two cortical cutaneous representations, areas 3b and 1. The callitrichids, *Saguinus*, and marmosets, like prosimians, have only a single map of their cutaneous receptors and this region is surrounded by cortex that either does not respond to touch or does so only following high levels of stimulation (Carlson et al., 1986; Krubitzer and Kaas, 1990). The presence of a single map in somatosensory cortex in callitrichids and prosimians suggests that the increased sets of somatosensory maps in New World and Old World monkeys arose through parallel evolution.

In both New and Old World monkeys, the two cutaneous maps in architectonic areas 3b and 1 of somatosensory cortex form mirror images with each other. However, the maps in the Cebinae, *Cebus* and *Saimiri*, contrast with the organization observed in *Aotus* and *Macaca*, namely, the maps of the trunk and portions of the limbs are reversed when the former group is compared with the latter (Felleman et al., 1983). Reversals in the topographic organization of digits versus palms also characterize the cuneate nucleus when the squirrel monkey is compared with the rhesus monkey (Florence et al., 1991). In contrast, the organization of facial maps is similar among *Saimiri, Saguinus, Aotus,* and Macaca (Cusick et al., 1986, 1989; Carlson et al., 1986). The reversals suggest parallel evolution in the somatosensory system of Old and New World monkeys.

A second group of somatosensory responsive neurons, on the superior bank of the lateral sulcus, comprises SII. Most of its features are equivalent in Old and New World monkeys, and differ from the organization present in the prosimian, *Galago* (Garraghty et al., 1991). The neurons in the monkey cortex are driven more strongly by input from somatosensory cortex than by a thalamic input.

The idea that parallel evolution is responsible for the double cutaneous maps in New and Old World monkeys receives support from data showing that the callitrichids share a single map with prosimians and that particular topographic details of the "homunculi" separate the Cebinae from the other Platyrrhini. Determining whether heterochronic mechanisms produced this divergence, as is possible in the visual system, requires further investigations.

The Motor System

The motor system is intimately connected with the somatosensory system, but less is known about taxonomic differences among anthropoids. The general patterns of cortical terminations in the spinal cord suggest great similarity. The fact that New World monkeys with prehensile tails have more direct corticospinal endings in coccygeal spinal cord levels, presumably allowing for better skilled movements of the tail (Petras, 1969), falls within the primate motor paradigm (Armstrong, 1989). The presence of direct corticospinal terminations at the coccygeal level and an enlarged somatosensory tail region leads to the expectation that the primary motor tail region is also enlarged in monkeys with prehensile tails.

A characteristic of the primate corticospinal tract is that cortical neurons project directly onto alpha motorneurons at all levels of the spinal cord. The axons of those motorneurons leave the CNS to synapse on skeletal muscles (Armstrong, 1989). The density of corticospinal synapses on the motorneurons innervating distal muscles of the extremities is less in squirrel monkeys than in rhesus monkeys and may be equivalent to that of a prosimian, *Galago* (Kuypers, 1983). These quantitative differences might reflect size rather than other aspects of phylogeny, however, because no shifts in density were noted between the other member of Cebinae, *Cebus*, or between *Ateles* and *Macaca* (Petras, 1969). Those differences would be expected if either Cebinae differed from other New World monkeys or platyrrhines differed from catarrhines.

Major connections between the motor cortex and motor thalamus appear similar in rhesus and owl monkeys, the two best studied species, but Old World monkey corticospinal tracts may have a greater input from the premotor cortex than New World one (Nudo and Masterton, 1990). New World monkey motor cortex may differ from that of catarrhines in that it also projects to the mediodorsal thalamic nucleus, an association and limbic region of the brain (Kunzle, 1976).

Association Cortex

The association cortices comprise a large proportion of the anthropoid cortex, but less is known about the details of size and organization of these cortices than for sensorimotor regions. Although originally association cortex was considered rather homogeneous in structure and function, more recent evidence shows it to be composed of distinct modules, each with a specific set of connections and functions.

In primates, the association cortex of each hemisphere is interconnected with homologous regions of the opposite side, the density of the projections perhaps reflecting cell density, itself a correlate of brain size. The interconnections are between anatomically identified cortical columns, vertical groups of cells that function together. Widths of anatomically identified columns have been measured in frontal association cortex and in sensory cortex and, in both types of cortex, squirrel monkeys have narrower columns than rhesus monkeys (Hendrickson and Wilson, 1979; Bugbee and Goldman-Rakic, 1983). Assuming a 1:1 ratio of columnar width and brain weight, the difference in columnar widths between the squirrel monkey and rhesus monkey is less than would be expected (Bugbee and Goldman-Rakic, 1983), but given that many morphometric features vary as a function of brain size in a nonlinear (allometric) fashion (Armstrong, 1990), the differences in cell column widths probably reflect brain size. More specimens are needed to determine whether the species differences in column widths are the result of a nonlinear scaling with brain weight.

The Limbic System

The limbic system has also been compared between different groups of primates. Detailed morphometric work has been done on Papez circuit, a portion of the brain that is concerned with memory, especially that of spatial relations, and emotions. The organization in noncortical regions does not differ between New and Old World monkeys (Armstrong, 1986), but parts of the subiculum and posterior cingulate cortex, regions thought to be important in memory and the direction of attention to various sensory stimuli and in motivation (Mesulem, 1981), have changed.

The subiculum can be divided into the subiculum proper and the pre- and parasubiculum. The parasubiculum, a small region wedged between the entorhinal cortex and presubiculum, is relatively enlarged in New World compared to Old World monkeys (Armstrong and Frost, 1987). This is one of the areas where New World, but not Old World, monkeys have connections from temporal visual association cortex (Webster et al., 1991; Selzer and Pandya, 1978).

One part of the allocortical posterior cingulate cortex, the agranular retrosplenial region (Area 29), has different laminar proportions in New World monkeys, thereby distinguishing them from other primates (Zilles et al., 1986; Armstrong et al., 1986). Here, New World monkey outer layers are relatively bigger than the inner ones. This cortex receives information from the anterior thalamus, the subiculum, and prefrontal cortex, thus putting it in a position to associate many different kinds of behaviors.

The development of a relatively bigger receptive and integrative zone of this cortex in New World monkeys than in other primates suggests a different function, which includes enhanced integration of cortical information (Zilles et al., 1986; Armstrong et al., 1986). The Cebidae that were examined, *Cebus*, *Saimiri*, and *Callithrix*, do not differ from the other New World monkey (*Alouatta*) in this regard. Because prosimians and catarrhines share a common pattern of having a relatively bigger output zone, it is most likely that the New World pattern is the derived pattern arising after the platyrrhines separated from the catarrhines and before the Cebidae diverged from the other New World monkeys. This distribution also suggests that changes in the limbic system preceded those of the cebid sensory systems.

Differences in central nervous system organization between New and Old World monkeys have begun to be identified. Despite the long separation time between these two taxa, the differences are small, requiring detailed observations to be seen. Differences in population size that cannot be accounted for by brain weight are too small to be manifest in volumetric studies (e.g., Stephan et al., 1981). To date, more features have been observed distinguishing the brains of the Cebidae from other New World monkeys than separating New from Old World monkeys. In particular, the Cebinae, *Saimiri* and *Cebus*, differ from other anthropoids in the organization of their sensory systems.

Discussion

Brains tend to be conservative in the amount and kinds of changes that occur in evolution, and this neural conservatism is shown in finding few differences separating Old and New World monkeys. Part of this conservatism reflects the fact that changes in peripheral structures, even those innervated by neurons from the central nervous system, do not greatly alter the latter. Taxa with prehensile tails, for example, have enlarged tail regions in sensory regions of cortex and thalamus, and altered corticospinal connections in coccygeal levels, but these alterations have not produced additional changes in organization or relative sizes in other parts of their brains.

Mechanisms of cell death allow the brain to adjust to changes in the size of peripheral structures (Hamburger, 1934; Hamburger and Yip, 1984). By this reasoning, the expanded size of tail regions in somatosensory areas of species with prehensile tails most likely stems from an increased survival of neurons rather than from more neurons being generated. That the enlargement of specialized areas has not led to additional changes throughout the brain suggests that evolutionary shifts driven by alterations in peripheral structures have limited effects on the central nervous system. That is, they do not serve as a basis for feedforward changes in the rest of the brain.

Sizes of neural populations are closely associated with the overall size of the brain, and thus regional comparisons must use an allometric analysis to detect differences between Old and New World monkeys. Despite the long separation, most neuronal regions have the same relative sizes in Old and New World primates. The sizes of neural populations scale according to brain weight among all anthropoids, suggesting that their major ontogenetic patterns use similar feedback mechanisms to link different components together.

Small differences between New and Old World monkey brains that are independent of brain size have been observed in the limbic region, the parasubiculum, and Brodmann area 29. The changed proportions suggest a relatively greater integration by these limbic areas, as does the presence of more projection pathways from higher order visual cortex to limbic cortex among Cebinae, but the precise functional significance of such a difference is not yet known. The findings show that detailed morphometric observations can elucidate differences between primates.

Organizational differences in the visual and somatosensory systems have been identified in the subfamily Cebinae, and possibly the family Cebidae (*Saimiri, Cebus, Aotus,* and the callitrichids). The Cebinae visual systems have been observed to differ both in higher order regions and in the receptive layer of the visual striate cortex where ocular dominance columns are lacking or have a great deal of binocular internal overlap. Whether the neuroanatomy correlates with the broad range of red–green spectra that Cebinae do not separate from yellow and the effects of a nocturnal existence for *Aotus* remain to be determined.

The different visual organizations could either have arisen as a more complex visual system evolved separately in New and Old World monkeys (parallel evolution) or be a derived (apomorphic) system unique to cebids. If New and Old World monkey visual systems evolved in parallel from a simpler prosimian configuration, the differences observed in one family, the Cebidae, may help identify what the organization was like in the common ancestor. On the other hand, the cebid visual system may have evolved separately after the platyrrhines and catarrhines separated. An examination of a larger number of noncebid New World species as well as prosimian and ontogenetic patterns is necessary to help determine which interpretation is more likely to be correct, but the presence of ocular dominance columns in *Galago* layer IV (Hubel et al., 1976) promotes the idea that the common ancestor of Old and New World monkeys most likely had ocular dominance columns as well.

Caution for a simple interpretation of cebid loss of previous neural organization is warranted. Computer simulations of the development of ocular dominance columns show that this organization occurs with asymmetric synaptic densities, and the latter can arise by many paths (Miller et al., 1989). An important question, then, is how similar synaptic development is among prosimians and anthropoids. The answer to this question is not yet in, but it is known that prosimian geniculate projections are denser and that the cortical layers within the prosimian striate cortex are less clearly defined than in cebids, atelids, or Old World monkeys (Diamond et al., 1985). If these architectural shifts are not merely the result of prosimian brains being smaller, they could suggest that the synaptic organization is not strictly homologous among prosimians and anthropoids. The separation of a specific effect versus one due to a scaling effect of brain size has not yet been made, however.

Until the kinds of homologies, particularly in forms and densities of connections in association cortex, are determined between New and Old World monkeys and between Cebids and other platyrrhines, questions about parallel evolution versus a derived status for Cebinae cannot be answered. Because dense projections from extrastriate cortex to striate cortex may play an important role in the structural organization of the latter during ontogeny (Dehay and Kennedy, 1993), analysis of association cortex may help clarify differences in structure within the primary visual cortex.

Ontogenetic studies will eventually help determine which interpretation, parallel evolution or a derived condition for cebids, is correct. Currently most prenatal developmental data are based solely on work in the rhesus monkey, where it has been determined that a binocular layer IV precedes a monocular one in the striate cortex (Rakic, 1976). If the developmental sequences are similar in New and Old World monkeys, and this is likely given the allometric similarity among the neural populations, then the cebid pattern may have arisen by heterochrony, a mechanism by which immature forms exist in adults (Gould, 1977).

The single somatosensory maps in callitrichids and prosimians and the dissimilarities in maps between the Cebinae and other anthropoids suggest that the increase in numbers of maps arose in parallel among New World and Old World monkeys. Although the shifts may have arisen from selective pressures acting on precise aspects of the visual and somatosensory systems, a pleiotropic effect from a changed developmental strategy would alter several systems at the same time (Finlay et al., 1987). The latter is especially likely given that many nonneural systems manifest parallelisms among atelids and Old World monkeys (Kinzey, 1971, 1986). If the changes in the cebid sensory systems were not the result of selection for specific adult behavioral attributes, but for an altered development, one prediction is that Cebinae would manifest other neural differences. Data do not exist that can answer this question at this time, but research on Cebinae ontogeny may elucidate insightful contrasts to the patterns seen in other anthropoids.

The brains of New World primates are smaller than those of Old World monkeys, but controlling for metabolic reserves (body weight and specific basal metabolic rate) shows that relative brain weight is very stable in the two groups. Although callitrichids have smaller brains for their body sizes than do other platyrrhines, once body size is corrected by the specific metabolic rate, no significant differences exist in the relative sizes of callitrichids (or any other platyrrhine) brain and those of nonhuman catarrhines. Analyses of relative brain size need take into account the rate that oxygen and glucose are delivered as well as body size.

Platyrrhines have had a separate evolutionary history probably since the lower Oligocene, during which time they maintained an anthropoid form of neural organization that differs only in small details from the brains of Old World monkeys. The largest structural differences in the brain are found between the subfamily Cebinae (or possibly the cebid family) and other anthropoids, and may have arisen by selecting aspects of developmental rates or duration. The otherwise remarkable conservation of the central nervous system suggests that, in primates, neural organization and size are tightly constrained.

Acknowledgments

We would like to thank S. Juliano, T. Preuss, and S. L. Florence for critically reading an earlier version of this manuscript.

Notes

1. The more recent work by Pagel and Harvey (1989) suggests that carnivore families have a slope of 0.63. Unlike the earlier Rohrs (1959) study, the Pagel and Harvey (1989) family slope is derived by using mean values of the genera, rather than of the species.

2. The role of metabolism in understanding differences in relative brain size is not restricted to primates, but has been observed for mammals in general (Armstrong, 1983), birds (Armstrong and Bergeron, 1985), and dietary groups (Armstrong, 1983).

3. Only by taking into account metabolism do the relative brain size studies account for the findings by neurochemists and physiologists that primates use a higher percentage of their energy reserves for their brains than do other mammals (Armstrong, 1983, 1990).

4. *Tarsius* has a degree of cortical folding similar to that of prosimians. In many quantitative features *Tarsius* clusters with prosimians, although in qualitative features it sometimes resembles anthropoids (Armstrong et al., 1986). Because of the morphometric similarity of *Tarsius* with lorises and lemurs, the term prosimian and not strepsirhine will be used in this chapter.

5. Because visual information from one eye partially crosses in the optic chiasma (fibers from the nasal half of the retina cross the midline), information from the left visual field comes from both the left and right retinas to the right side of the brain. The information of the same visual field from both eyes remains on one side of the brain in the lateral geniculate body and in layer IV of the striate cortex, but usually in alternating monocular modules.

6. Although individual neurons in the supra- and infragranular layers of the striate cortex are potentially binocular, they respond more easily to a particular visual stimulus from one eye.

3

Color Vision Polymorphisms in New World Monkeys: Implications for the Evolution of Primate Trichromacy

GERALD H. JACOBS

Introduction

A key to the success of the primates is superb vision. One need look no further than the structure of the primate nervous system for evidence of the importance of vision. The darting eyes of a macaque monkey, for instance, contain upward of 130 million photoreceptors (Packer et al., 1989). The outputs from the eyes are transmitted along about 3 million fiber pathways (Perry and Cowey, 1985) and the most intricately organized portion of the primate brain, the neocortex, is dominated by tissue dedicated to visual analysis. The macaque primary visual cortex comprises some 15% of the total area of the neocortex and there are at least 20 additional, distinct neocortical regions involved in the processing and elaboration of visual information (van Essen, 1985). These impressive brain substrates provide the basis for the hallmark properties of primate vision—high spatial acuity, excellent discrimination of depth and distance, and keen color vision.

This chapter provides a review of the status of our understanding of one of these visual attributes—color vision—among New World monkeys. In doing so, I first provide a brief tutorial on some of the basic features of color vision, next note the history of studies of color vision among New World monkeys and document the current understanding, and, finally, discuss how the variations in color vision among these monkeys may provide some insights into the evolution and ecology of primate color vision.

Some Basic Features of Color Vision

Color in the Laboratory

Much of what we know about the nature of color vision comes from laboratory studies in which the visual world consists of monochromatic lights, is spatially homogeneous or is sharply divided into homogeneous regions, and is abruptly turned on and off. Under these (very) unnatural conditions the stimulus can be specified with respect to its spectral energy distribution (radiance per wavelength interval). In turn, the resulting chromatic experience can be described according to its color (hue), the extent of its coloredness (saturation), and its overall effectiveness (brightness). Each of these response dimensions can be accurately measured by using appropriate behavioral responses. It has long been known that most humans have trichromatic color vision. This conclusion derives principally from color matching experiments in which a subject establishes a perceptual identity (a match) between a single light (a test light) and mixtures of other lights (primaries), each of which is of fixed spectral content. In practice, the test and primary lights are often presented as two halves of a bipartite field that is imaged directly onto the retina. Although there are some restrictions that must be satisfied, under these conditions most observers can precisely match the appearance of any test light (be it monochromatic or a mixture of wavelengths) by varying the relative proportion of the three primaries (Figure 3.1). Since only these three values need be specified to describe this behavior, normal human color vision is termed trichromatic.

The facts of trichromacy require the presence in the visual system of

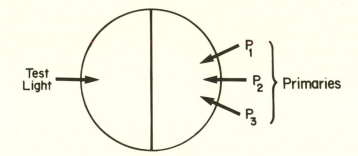

Figure 3.1. Stimulus arrangement in a typical color-matching experiment. The test light is projected onto the left half of the field. The primary lights are superimposed in the right half of the field. The latter are fixed in spectral content. The task of the subject is to adjust the relative proportions of the primaries until the two halves of the field match in appearance.

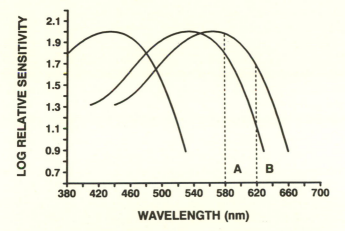

Figure 3.2. Absorption curves for the three classes of cone photopigment found in normal humans. Note that particular spectral stimuli (locations indicated by the vertical lines labeled A and B) will differentially affect the pigments and, thus, a signal based on the ratio of the responses of the two cone types will be consistently different for the two stimuli irrespective of their relative intensities.

three independent mechanisms. The first-stage mechanisms to accomplish this task are three types of cone receptors, each containing a spectrally distinct photopigment. Over the past 20 years increasingly precise measurements have been made of the absorption characteristics of these pigments. Figure 3.2 shows the spectra for the three cone types found in normal humans. These have peak absorption (λ_{max}) values of about 430, 530, and 560 nm (viewed in isolation these wavelengths would appear as blue, green, and yellowish-green, respectively). To initiate vision, light must activate molecules of one or more of these pigment types. The spectral positioning of the three pigment types of Figure 3.2 limits the perception of color to wavelengths lying between about 400 and 750 nm. No other primate species is able to exceed the human spectral range. Each of these pigment types responds univariantly in that its response is proportional to the total quantal capture. As a result, in the responses of any single cone class the effects of wavelength and intensity change are confounded, i.e., the response does not contain unique information about either of these dimensions. However, since the absorption curves of different pigments overlap, a particular spectral stimulus will be differentially effective in producing responses in cones containing different pigment types (see Figure 3.2) and thus a signal based on the ratio of the responses of two cone types can be used as a basis for an effective color signal. These ratios are computed by cell–cell interactions that occur in the visual nervous system.

It has been known for at least 200 years that our species does not have uniform color vision. The variations from the normal, the color vision defects and color vision anomalies, are of several types. In a color-matching task of the type illustrated in Figure 3.1, individuals having variant color vision produce aberrant color matches. Some individuals require only two primaries instead of the standard three to make color matches. These people have dichromatic color vision and three subtypes (protanopia, deuteranopia, tritanopia) are recognized based on the details of the color matches they make. Others require the use of the three primaries to complete all color matches, but they select distinctly abnormal proportions of the three. These subjects are categorized as having anomalous trichromatic color vision and, again, depending on the details of their matches, they fall into one of three different groups (protanomalous, deuteranomalous, tritanomalous). On very rare occasions an individual is found to be able to match the appearance of any test light to that of any single primary by simply adjusting the relative brightnesses of the two. Such subjects are monochromats; they constitute the only individuals for whom the designation "color blind" is strictly appropriate.

Note in Figure 3.2 that the three cone pigments are not evenly spaced across the spectrum. One result of this uneven spacing is that the short wavelength-sensitive cones have very low sensitivity in the longer wavelengths. A consequence is that although three primaries are required by a trichromatic subject to complete matches for an exhaustive range of test lights, only two primaries are required if the test light is restricted to wavelengths longer than about 540 nm. Since it is easier to instrument color-matching tests that require two primaries rather than three, this peculiarity of trichromatic color vision is often exploited in the design of color vision tests. An example of this is a standard screening device for color vision called an anomaloscope. In this case the test light is a monochromatic yellowish light and the two primaries are green and red (the Rayleigh match). Normal trichromats all require about the same proportion of red to green light to match the yellow, whereas the predominant types of anomalous trichromat select either much more red light (protanomalous) or much more green light (deuteranomalous) in the mixture and they thus can be accurately diagnosed according to their matches. Whereas trichromats become dichromatic for long-wavelength test lights, those who are "normally" dichromatic become monochromatic. This means that in a Rayleigh match such subjects will accept all mixtures of red and green light as matching the yellow test light.

The incidence of significant color variation among humans is substantial. A total of about 4% of the population of Western Europe and the United States have a congenital color defect or a color anomaly. One of the categories of anomalous trichromacy, deuteranomaly, alone accounts for about 50% of this total. The mode of inheritance of congenital color vision defects and anomalies has attracted much study since Horner first noted in 1876 that, like hemophilia, color blindness affects predominantly men. The basis for this disparity is now known. The middle- and long-wavelength cone pigments (λ_{max} = 530 and 560 nm) arise from the activity of genes located on the X-chromosome, and thus any loss or alteration of these genes, which in turn results in loss or alteration of a photopigment type, will cause direct color vision changes in males. Unless a female has these changes on both X-chromosomes (i.e., is homozygous), there is typically little observable change in her color vision. The short-wavelength cone pigment is linked to the activity of an autosomal gene, so that the rare color vision defects involving the short-wavelength cone are equally likely among males and females.

Color-matching tests are most useful for determining the nature of color vision a particular subject might have, but they are less informative about the quality of their color vision. For assessments of the acuteness of the color vision, scientists have frequently turned to discrimination experiments. These experiments are designed to determine the magnitude of the change required in the stimulus to yield a discrimination step along one or more dimensions of chromatic experience. Results from two such common tests are illustrated in Figure 3.3. At the top are results from a wavelength discrimination experiment in which the subject is tested to determine how much wavelength must be changed ($\Delta\lambda$) at various spectral locations for the two wavelengths to produce discriminably different color experiences. It can be seen (1) that the normal trichromat shows an acute ability to discriminate wavelength differences, with thresholds of 1 nm or less at some spectral locations, and (2) that ability is not equally good at all parts of the spectrum being most acute around 500 and 570 nm. By contrast, the dichromatic subject shows only a single region of acute wavelength discrimination and, most strikingly, becomes effectively blind to color differences in the middle and long-wavelengths. The bottom panel of Figure 3.3 shows corresponding results from trichromatic and dichromatic humans in a test designed to evaluate the extent of the saturation of spectral lights. In this case the dependent measure is the amount of white light that had to be added to each of the spectral lights for the two lights to be indiscriminable. Note that for the trichromatic subject, wavelengths at the ends of the spectrum appear highly saturated, with saturation declining to a minimum at a wavelength of about 570 nm. For the comparably tested dichromat, short-wavelengths appear highly saturated; however, a wavelength of about 500 nm is completely unsaturated (i.e., appears colorless) and all other wavelengths in the middle and long-wavelengths have low saturation values. Numerous other tests of these kinds have

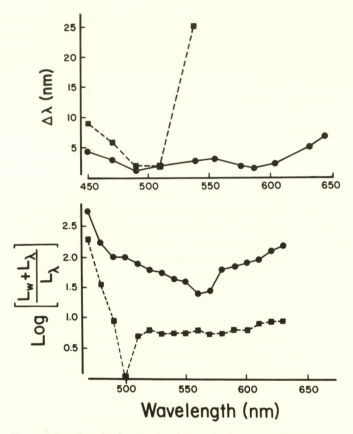

Figure 3.3.   Results from color discrimination experiments conducted on trichromatic (circles and continuous lines) and
dichromatic (squares and dashed lines) subjects. Top:
Wavelength discrimination. The dependent measure is the
change in wavelength ($\Delta\lambda$) required at each wavelength for
the subject to just detect a difference in color appearance.
Bottom: Spectral purity discrimination. The dependent
measure is the amount of achromatic (white) light that had
to be added to a monochromatic light to make it indistinguishable from an equally bright white light.

been used to probe the quality of color experience (see, for example,
Boynton, 1979; Pokorny et al., 1979).

Color vision categorization and color vision acuity tests of the types described have been routinely performed on both human and nonhuman
subjects. A procedural problem stems from the fact that in such investigations the experimenter must separate the use of color and brightness as
possible cues for discrimination; that is, in all explicit tests of color vision

procedures must be employed to eliminate, or to make irrelevant, brightness differences between the stimuli being evaluated. For tests of human subjects this presents no great challenge. In the case of a color matching test, for instance, one can simply instruct the subject to adjust first the relative intensities of the test and mixture lights so they appear equally bright and then to adjust the proportions of the latter to complete the match. It is quite another matter in tests of animals. In these cases the usual approach has been to require the animal to discriminate between stimuli whose features have been selected to probe some aspects of color vision—for instance two lights having different wavelength contents. If there are consistent brightness differences between these stimuli, successful discrimination may reflect that fact alone rather than anything about the animal's color vision. It is hard to overestimate the cleverness of a motivated monkey subject in picking up slight brightness cues and thus misinforming the experimenter as to what is and what is not a true color discrimination. Various strategies to solve the "brightness problem" in studies of animal color vision have been discussed elsewhere (Jacobs, 1981).

Color in the Natural World

Color vision in the natural world is a more complex affair than that studied in artificial situations of the kinds described above. Partly this reflects the fact that color is a property of objects in normal scenes (so-called surface colors), whereas in most laboratory investigations, including those described above, color is not associated with an object (and such colors are thus called aperture colors.) The colors of objects can have complex perceptual properties (e.g., "transparency," "luster," "glow") that cannot be completely encompassed by the color dimensions (hue, saturation, brightness) found sufficient to characterize aperture colors (Evans, 1974). These properties are attributable to the varied surface characteristics of the objects themselves, and to the varied arrangements of lighting in the scene.

A major source of the difference between aperture and surface colors is that the latter are typically viewed in a spatially and temporally variegated context, while the former are not. These spatial and temporal variations lead to powerful interaction effects, simultaneous and successive contrasts, between regions in the field of view. As a result, it is generally quite misleading to try to evaluate the color properties of objects without considering the context in which these objects appear.

As stimuli for color vision, objects can also be specified with respect to their spectral energy distributions. Unlike the monochromatic lights typically used in laboratory studies of color, the surface reflectance properties of objects (i.e., the proportion of light reflected at each spectral wavelength) usually involve much of the spectrum. For instance, Figure

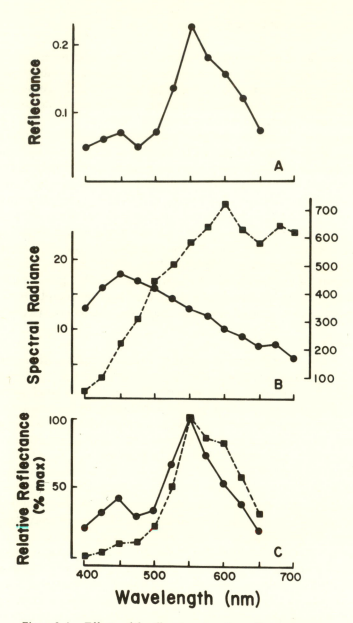

Figure 3.4. Effects of the illuminant on the reflectance properties of objects. (A) Surface reflectance function for
grass. (B) Spectral radiance functions recorded for two
daylight conditions: direct sunlight (squares and right
abscissa) and scattered light (circles and left abscissa)
coming from a direction 120° separated from the solar
axis. The radiance units are photons/cm²/nm × 10¹².
Note that the two functions have different scaling. (C)
Relative reflectance spectra for grass under the two
illuminants of B (circles and squares, respectively).

3.4A shows the reflectance function for a common object, grass. Note that to a greater or lesser extent all wavelengths in the visible spectrum are reflected by this object. If the illuminants in the world were all identical, then reflectance functions like that shown in Figure 3.4A would be sufficient to specify the object as a stimulus for color vision. Since that is not the case, then an adequate specification must include both the reflection properties of the object and the spectral composition of its illuminant. Sources of illumination, like the sun, are themselves subject to variation through the joint processes of absorption and scatter, and thus the spectral composition of the illuminant of an object can also be highly varied.

The implications of variations in the spectrum of the illuminant are illustrated in Figure 3.4B. Shown there are the spectral compositions of two illuminants recorded at the same time on a cloudless day: direct sunlight (dotted line) and scattered light (solid line) coming from the sky at a point 120° separated from the axis of the sun. In addition to having considerably higher overall radiance, direct sunlight has a much higher relative representation of energies from the long-wavelengths. Since, as noted, the light reflected from an object is a joint property of the reflectivity of the object and the spectrum of the illuminant, an object separately illuminated by these two sources can present two diffferent reflectance spectra. The bottom panel of Figure 3.4 illustrates this fact by showing the reflectance spectrum of grass as illuminated either by direct or indirect sunlight. Even though the locations of peak reflectance are identical, the differences in the two cases are clearly noticeable.

If perceived color slavishly followed the spectral reflectivity of an object, then one might expect the color of grass to be drastically different when viewed under the two illuminants of Figure 3.4B. Of course, it is not. The fact that an object reasonably maintains its characteristic color in the face of changes in the spectrum of the illuminant is a remarkable (and highly useful) feature of color vision called color constancy. The presence of color constancy in effect implies that the visual system must in some way acquire information about the spectral distribution of illuminants, and use that information in the analysis of color signals. Just how that process works remains mysterious, although there has been much recent research and theorizing on the topic (for a recent review, see Jameson and Hurvich, 1989).

Color Vision in New World Monkeys

Early Studies

In 1939 Grether published a comparative study of color vision in several species of Old and New World monkeys. The major conclusion from this pioneering investigation was that the color vision of the Old World monkeys (macaques and baboons were tested) appeared similar to that of comparably tested normal human trichromats, whereas the color vision of a common New World monkey, *Cebus*, seemed to be much like that of a human protanopic dichromat. Subsequent measurements on an additional *Cebus* monkey appeared to substantiate this outcome (Malmo and Grether, 1947). The results from several studies of color vision in squirrel monkeys (*Saimiri*) led to a similar conclusion—color vision in this genus also appeared to be more similar to that of human color defectives than to that of the normal human (Miles, 1958a; Jacobs, 1963; De Valois and Morgan, 1974). Tests run on marmosets (*Callithrix*) yielded essentially the same result (Miles, 1958b). What was less certain was the correct diagnosis of the severity of the color loss in these New World monkeys—were they protanopes (i.e., dichromats) or merely severely protanomalous (i.e., anomalous trichromats)?

Whatever the correct diagnosis might be, the generalization that emerged from these early studies was that Old World monkeys have color vision much like that of most humans, whereas New World monkeys have color vision that is more similar to some of the forms of human color defective vision. Research done over the past few years makes clear that the picture of color vision in New World monkeys is both more complex and more fascinating than these early studies indicated.

Recent Studies on Squirrel Monkeys

A large fraction of the recent research on color vision in New World monkeys concerns the squirrel monkey (*Saimiri* sp.).[1] The reevaluation of color vision in this monkey was triggered by the discovery in electrophysiological experiments that there are substantial variations among individual animals in the relative proportions of different cell types in the primary visual pathways (Jacobs, 1983a). These variations were soon shown to have behavioral importance in an experiment that sought to determine the sensitivity of individual squirrel monkeys to two different test wavelengths, 540 and 640 nm (Jacobs, 1983b). Sensitivity to these two wavelengths showed significant variation from animal to animal. The ratios of sensitivity to the two lights covered a total range of about 1.5 log units in more than 50 subjects. By comparison, analogous ratios for a large sample of normal human trichromats spanned a range of only about 0.7 log units. These variations strongly implied that there must be variation in the complements of cone pigments found in different squirrel monkeys and, by extension, also variation in their color vision. That possibility was directly evaluated in studies of color vision.

Several different color vision tests of the types summarized above (e.g., color matching, wavelength discrimination) were run on about 36 squirrel monkeys. As was predictable from the results of the electrophysiological investigation and the sensitivity tests, this work revealed that there are striking individual variations in color vision among squirrel monkeys (Jacobs, 1984; Jacobs and Blakeslee, 1984). These variations were such as to identify six discretely different forms of color vision in the genus *Saimiri*. Three of these were types of dichromacy; these three could be separated by the types of color matches that individual animals made. Likewise, three alternate forms of trichromacy were found to be present among squirrel monkeys and these too could be distinguished according to the results from color-matching tests. Of particular interest is the fact that several of these subvarieties of color vision in squirrel monkeys are similar in character to the various categories of normal, anomalous, and dichromatic human color vision (Jacobs, 1984).

The nature of color vision variation among squirrel monkeys is illustrated in Figure 3.5. That figure shows the results from a color-matching experiment in which three animals were required to discriminate various mixtures of red and green light from a standard yellow light. One of the monkeys fails to discriminate any red/green mixture from yellow. As noted earlier, this behavior is characteristic for a dichromat. The other two animals, whose results are summarized in Figure 3.5, could discriminate some red/green mixtures from yellow and, thus, they have trichromatic color vision. Note, however, that the red/green mixtures that match (that is, are indiscriminable from) the yellow are different for the two animals, so there must be different forms of trichromacy represented in these monkeys. Several other tests of color vision were used to validate these distinctions (Jacobs, 1984).

On the basis of earlier studies of color vision in New World monkeys, the results of this investigation of squirrel monkey color vision were quite surprising. They showed that like the human, an individual squirrel monkey may have any of several polymorphic forms of color vision. However, unlike the human, where alternate forms of color vision are found in only a small fraction of the total population, each of the alternate forms of color vision appears to be abundantly represented among squirrel monkeys. Indeed, in the original sample of 36 monkeys that had been subjects for color vision testing, it was unclear which, if any, of the polymorphic forms were more frequently represented than others.

Photopigment Polymorphism in Squirrel Monkeys

Individual variations in color-matching behavior can often be attributed to individual variations in the complement of cone photopig-

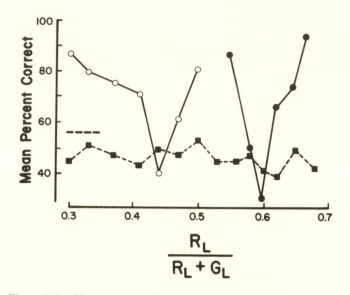

$$\frac{R_L}{R_L + G_L}$$

Figure 3.5. Variations in color matching among three squirrel monkeys (results for the three are indicated by different symbols). Each data point shows the performance of one monkey in a task where the animals were required to discriminate various mixtures of red and green light (ordinate values) from a standard yellow light. The horizontal line to the left indicates the performance level required to demonstrate nonchance discrimination. Note that two animals fail the discrimination over small (and different) parts of the mixture range. The third animal (squares) fails to discriminate any red/green mixture from yellow. Data taken from Jacobs (1984).

ments. Since the six types of color vision in squirrel monkeys were defined by differences in color matching, it seemed likely that these animals would differ in their photopigment complements. That expectation turned out to be correct. Photopigment measurements were made on squirrel monkeys whose color vision had been previously determined. These pigment measurements were made first by direct microspectrophotometric measurements (Jacobs et al., 1981; Mollon et al., 1984; Bowmaker et al., 1985, 1987), and then, later, using a noninvasive electrophysiological technique (Jacobs and Neitz, 1987a). The results from the two approaches are essentially identical.

Squirrel monkeys were found to have four different classes of cone photopigment. The spectra of these pigments are illustrated in Figure 3.6. All individuals have small numbers of cones containing a short-wavelength pigment having peak absorbance of about 433 nm.[2] The

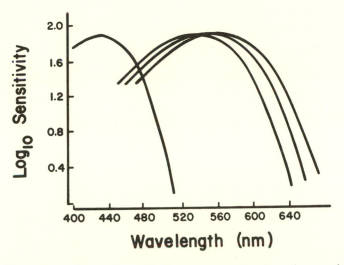

Figure 3.6. Spectral sensitivity curves for the four classes of cone pigment found in squirrel monkeys.

additional three pigments (λ_{max} = ca. 538, 550, and 562 nm) are distributed differently among individuals, and this distribution corresponds with the established color vision of the monkey. Monkeys found to have dichromatic color vision have any one of the three middle- to long-wavelength pigments; trichromatic individuals have any pair. Several other direct comparisons of pigment complement and behavior were offered in the papers cited above, and these show, in sum, that there is a direct correspondence between type of color vision and cone pigment complement in the squirrel monkey. The six variant types of color vision among these monkeys thus seem completely explained on the basis of variations in the cone photopigments.

Studies on Other Cebinae

Squirrel monkeys have significant cone pigment and color vision polymorphism. Is this a unique condition among New World monkeys or is this indicative of a larger trend? Although it is not yet possible to know how common color vision polymorphism might be among these animals, it is already certain that it is not unique to the squirrel monkey. Some years ago, Gunter et al. (1965) reported that *Cebus* monkeys show individual variations in color vision. That report was effectively ignored by subsequent writers, but its conclusion turned out to be correct. Photopigment measurements have now been made on tufted capuchins (*Cebus apella*) and white-faced capuchins (*Cebus capucinus*) (Bowmaker and Mollon, 1980; Jacobs and Neitz, 1987b; Jacobs and Deegan, unpublished measurements). Although the pigment measurements are much less extensive than they are for the squirrel monkey (a total of 22 animals have been examined), and although there have been no correlative measurements of their color vision, the outcome appears to be much the same as for the squirrel monkey, i.e., individual *Cebus* monkeys have different combinations of three middle- to long-wavelength cone pigments. These pigments seem to be spectrally similar, and very probably identical, to those found in the squirrel monkey (Table 3.1).

Callitrichid Monkeys

Recent studies of cone pigments and color vision have been reported for two species from the Callitrichinae—the saddle-backed tamarin (*Saguinus fuscicollis*) and the common marmoset (*Callithrix jacchus*). In behavioral tests of color vision in five tamarins, both dichromatic and trichromatic individuals were identified (Jacobs et al., 1987). Correlative measurements of the cone photopigments in these animals, as well as in several additional tamarins, revealed a picture qualitatively similar to that described above for the squirrel monkey. In addition to a short-wavelength cone pigment (λ_{max} of ca. 433 nm), three middle- to long-wavelength pigments were discovered (λ_{max} = 545, 557, and 562 nm). As for the squirrel monkeys, individual animals had any one of these three pigments (and dichromatic color vision) or any pair of these pigments leading to trichromatic color vision.

Microspectrophotometric measurements of cone pigments in the marmoset also provide evidence for polymorphism. In addition to finding small numbers of short-wavelength sensitive cones (λ_{max} = ca. 424 nm), believed common to all individuals, Travis et al. (1988) identified three types of middle- to long-wavelength cone pigments. These had average λ_{max} values of 545, 558, and 565 nm, and they occurred either singly or in pairs in individual animals. Although the color vision of these animals was not tested, the former would be expected to have dichromatic color vision; the latter would be trichromatic. Just such a correlation between photopigment complement and behavioral tests of vision has now been reported for common marmosets (Tovee et al., 1992).

These recent studies indicate that there are both similarities and differences in color vision between Cebinae and Callitrichinae monkeys. The similarities are (1) species drawn from both groups have polymorphic color vision, (2) some individuals are dichromats while others are trichromats, (3) subtypes can be found within each of these two categories, and (4) the polymorphic variants appear to be of high frequency. The most obvious difference between cebinae and callitrichinae species is in the spectral positioning of their cone pigments. The results of several measure-

Table 3.1. Estimates of the λ_{max} Values for the Cone Pigments of Several Genera of New World Monkeys

Genus	Method[a]	Spectral peak (nm)				Reference
Saimiri	MSP	433[b]	536	549	564	Mollon et al. (1984)
Saimiri	ERG	c	538	551	561	Jacobs and Neitz (1987a)
Cebus	MSP	c	534	c	c	Bowmaker and Mollon (1980)
Cebus	ERG	429	537	550	562	Jacobs and Neitz (1987b); Jacobs and Deegan (unpublished)
Callicebus	ERG	c		549	562	Jacobs and Neitz (1987b)
Saguinus	ERG	433	545	557	565	Jacobs et al. (1987)
Callithrix	MSP	424	545	558	565	Travis et al. (1988)
Aotus	ERG		542		563	Jacobs et al. (1993)
Ateles	ERG	c		550	555	Jacobs and Deegan (unpublished)
Leontopithecus	ERG	c	544		555	Jacobs and Deegan (unpublished)
Old World monkeys	MSP and ERG	430	530		560	Concensus

[a] MSP, microspectrophotometry; ERG, electroretinography.

[b] Entries in this column probably represent more than one type of photopigment.

[c] Indicates the probable presence of a cone type that was not directly measured in the experiment.

ments of cone pigments in these animals are summarized in Table 3.1. Note that although the three middle- to long-wavelength pigments seem common to members of each group, they differ between groups. For the cebinae species these pigments have λ_{max} values of about 538, 550, and 562 nm; for the callitrichinae the corresponding values are about 544, 557, and 562 nm. Note also that of the five types of middle- to long-wavelength photopigments found in these New World monkeys, only one type (λ_{max} = ca. 562 nm) is common to all of the species.

Atelid Monkeys

There is some evidence about photopigments and color vision for three genera from this group. First, photopigment measurements have been made on eight titi monkeys (*Callicebus moloch*) (Jacobs and Neitz, 1987a). Subject to limitations imposed by a relatively small sample size, the story for these animals seems very similar to that documented above for *Cebus* and *Saimiri*: all these genera have the same three types of middle- to long-wavelength cone pigments (Table 3.1) and the polymorphic variation among individuals seems the same. With respect to the proposed genetic explanation for plathyrrhine photopigment polymorphism (see below), I note here that all of the individual males of these genera have been found to be dichromatic.

The situation with regard to the spider monkey (*Ateles*) has been less clear. Grether (1939) tested color vision in a single spider monkey (identified as *Ateles ater*). This animal had good wavelength discrimination abilities, in fact, strikingly more acute color vision that that of *Cebus* monkeys and at least as good as that of several Old World monkeys. More recently, color vision tests were performed on two spider monkeys (Blakeslee and Jacobs, 1982). In confirmation of the Grether result, these monkeys seemed to have keen color vision as evidenced by good abilities to discriminate wavelength differences. Both of these animals also had trichromatic color vision, but color-matching tests of the kind illustrated in Figure 3.5 suggest that the two did not have the same complement of middle- to long-wavelength photopigments. The latter result would imply that the polymorphism documented for other New World monkeys might also be characteristic of spider monkeys. Based partly on what seemed unusually keen color discrimination for platyrrhines, and partly on differences in their lifestyles, there was the expectation that further study of spider monkey color vision might be particularly revealing. We have recently had the opportunity to measure the photopigment complements in a number of spider monkeys (Jacobs and Deegan, unpublished results). Some of these animals had only a single type of middle- to long-wavelength pigment (and thus would be presumed dichromatic). The actual pigments measured varied from animal to animal and had spectra very similar to those previously characterized for *Cebus*, *Callicebus*, and *Saimiri* (Table 3.1). Other spider monkeys had two middle- to long-wavelength pigments and should have trichromatic color vision. Although much remains to be understood about the color vision of spider monkeys, they seem at present less likely to be unique than they once did.

The third genus in this group that has received some study is the owl monkey, *Aotus. Aotus* is unique as the only living nocturnal anthropoid. In accord with its lifestyle, the retina of *Aotus* contains many fewer cones than that of any other monkey. This monkey is of interest from the viewpoint of cone-based vision because it is believed that the nocturnality of *Aotus* was secondarily acquired, that its ancestors were diurnal (Martin, 1990). Some years ago I studied color vision in *Aotus* monkeys, and from the results of several discrimination tests concluded that this animal has a weak color vision capacity (Jacobs, 1977). From that behavioral study it was not possible to determine the nature of the *Aotus* cone pigments. We have recently reexamined the cone vision of *Aotus* (Jacobs et al., 1993a). There are two striking results. First, it seems clear that *Aotus* has only a single class of cone photopigment (mean = ca. 542 nm) and, importantly, there was no polymorphism among the six animals we studied—each had the identical cone pigment. Unlike all other monkeys, both Old and New World (Table 3.1), there is no evidence that this monkey has a short-wavelength sensitive photopigment. These electrophysiological results are in accord with a recent immunocytochemical study of *Aotus* photoreceptors that also failed to find evidence for any cones containing a short-wavelength-sensitive photopigment (Wikler and Rakic, 1990). From a technical point of view *Aotus* appears, thus, monochromatic. The weak color discriminations documented in the earlier study probably reflected the ability of this monkey to jointly utilize signals from rods and the single class of cone. Second, our new study gives some hint as to why *Aotus* is monochromatic. A few years ago the genes that encode the opsins required to make human cone pigments were isolated and sequenced (Nathans et al., 1986). Since the genes for all photopigments are greatly similar, it is possible to use the human genes as probes to search for the presence of particular pigment genes in other species. We followed this procedure for *Aotus* and discovered that this monkey has (at least) some portion of the gene required to code for a short-wavelength-sensitive photopigment. Since its retina does not contain such a pigment, this implies that the monochromacy of *Aotus* is due not to the lack of an appropriate photopigment gene, but rather to some defect in the pathway between gene and pigment product. In turn, one might interpret this picture to suggest that the ancestors of contemporary *Aotus* may well have had the two classes of cone photopigment that seem to minimally characterize all other monkeys.

Genetics of Color Vision in New World Monkeys

The behavioral studies that first revealed the presence of color vision polymorphism among squirrel monkeys also showed that the variation in color vision in this species has a striking gender-related component: although both dichromatic and trichromatic females were discovered, no males were found to possess trichromatic color vision (Jacobs, 1984). If that difference held for all squirrel monkeys, a simple mechanism could be conceived to account for the inheritance of color vision in these animals. That mechanism would include the feature that, in contrast to the human (see above), there must be only a single X-chromosome photopigment locus in the squirrel monkey (Mollon et al., 1984; Jacobs and Neitz, 1985b). This locus was hypothesized to have three alleles, each of which specifies the opsin for one of the three classes of middle- to long-wavelength cone pigment. Since males have only a single X-chromosome, this means that each would have one of the three possible middle- to long-wavelength pigments and, in conjunction with the invariant short-wavelength pigment, one of the three forms of dichromatic color vision. With two X-chromosomes, female squirrel monkeys would have the possibility of being heterozygous at the pigment locus, hence producing two middle- to long-wavelength photopigments. Random X-chromosome inactivation (Lyon, 1962) would ensure that each cone receives the product of only one of the alleles and thus the heterozygote achieves trichromatic color vision. Homozygous females would, of course, be restricted to dichromatic color vision.

We were able to evaluate this "single locus model" for the inheritance of color vision in squirrel monkeys by examining the photopigment complements in a large sample of squirrel monkeys and by determining pigment pedigrees (Jacobs and Neitz, 1987a). The model requires that no male should have trichromatic color vision, and that proved to be true for a sample of more than 30 male monkeys. If one knows the frequency of the three allelic genes in the population, it is also possible to predict the relative frequency of female dichromacy and trichromacy. Examination of a sample of more than 70 squirrel monkeys suggested the three genes to be about equifrequent. If so, the single locus model predicts about 33% of all females to be dichromatic. That value was close to the 38% figure actually observed. Pedigree analysis of 11 squirrel monkey families was similarly in accord with the model. Finally, in recent years it has become possible to use molecular genetic techniques to examine directly the X-chromosome pigment genes in the squirrel monkey. Such an examination showed that male squirrel monkeys have only a single type of cone pigment gene on their X-chromosome (Neitz et al., 1991; Jacobs et al., 1993b). In sum, the single locus model seems to economically account for the inheritance of color vision among squirrel monkeys.

It has not yet been possible to ascertain if the single-locus model also accounts for the inheritance of color vision in other New World monkeys. It is the case, however, that among all of platyrrhine species mentioned above (except *Aotus*), all of the males had pigment complements that would restrict them to dichromacy while, at the same time, several of the females similarly examined had two middle- to long-wavelength cone pigments and trichromatic color vision. Although for the present the conclusion must be held provisional (and, for instance, Tovee et al., 1992 have suggested this model may be inadequate for *Callithrix*), it seems a reasonable guess that the genetic mechanism that seems to account for the inheritance of color vision among squirrel monkeys may be appropriate for several genera of New World monkeys.

Unknowns and Uncertainties

The picture of color vision among New World monkeys is far from complete and, thus, the brief survey presented above is as important for what it does not say as for what it does say. Two points need to be made explicit.

First, and most important, we still lack much information. The comments made to this point concern representative species from only 7 of the 16 genera of New World monkeys. There is a particular lack of information about pigments and color vision among the atelid monkeys. Beyond that, it is by no means certain that even within these genera the pattern of color vision is constant across all species (see below). Clearly, more work needs to be done.

A second reason why our understanding of color vision among New World monkeys is incomplete stems from the existence of some results that are contradictory to the view summarized above. Savage et al., (1987) recently examined color discrimination in four cotton-top tamarins (*S. oedipus*). On the basis of results from other callitrichinae one might have expected them to find color vision polymorphism and a gender-linked variation in color vision among cotton-top tamarins, but they found neither. The results of this study could be interpreted to mean that the nature of color vision and its mechanisms in this species are different from those of these other species. However, it might also be that the experiments were not adequate to reveal color vision polymorphism. The latter possibility has been addressed elsewhere (Jacobs, 1991). We recently were able to measure the photopigment complements in a sample of cotton-top tamarins (Jacobs and Deegan, 1994). We found clear evidence for a total of four types of cone photopigment among the members of this species. In addition to a short-wavelength-sensitive photopigment, there were three classes of middle- to long-wavelengths photopigments having peak sensitivities of about 545, 556, and 562 nm. These latter three pigments were polymorphic in the same apparent fashion as described above for *Saimiri* and for several other genera of New World monkeys. These results suggest, contrary to the conclusions of Savage and her colleagues, that the cotton-top tamarin does not constitute an exception to the developing picture.

Evolution of Primate Color Vision

There is no hard information about the actual course of evolution of color vision, either among the primates or for any other group of vertebrates. A standard view is that promoted by Gordon Walls (1942). Walls suggested that the considerable morphological variation in cones among contemporary vertebrates known to have color vision indicates that the biological substrates of this capacity must have evolved separately many times. He was particularly confident that this must be the case for primates concluding, "the color vision of primates is assuredly a law unto itself" principally because, "primates originated as a nocturnal group" (Walls, 1942, p. 520). Whether or not this is the case is no longer so clear. Paleontological evidence suggests that there may have been both nocturnal and diurnal forms among early primate-like mammals and among early prosimians (Fleagle, 1988). Even if the primates did not evolve color vision *de novo*, it seems reasonable to believe that they at least elaborated the capacity considerably. Certainly compared to all other mammalian groups, color vision among the primates is a superior capacity, quantitatively and qualititively (Jacobs, 1981).

The discovery of color vision polymorphism among some New World monkeys suggests what the sequence of evolution of color vision among the primates might have been. To make that suggestion explicit, it will be first necessary to note briefly how the picture of color vision among the New World monkeys described above differs from that believed to be appropriate for Old World monkeys.

Color Vision among Old World Monkeys

One would like to be able to answer two questions about color vision in Old World monkeys: What is its nature? Is there polymorphism of the kind that appears to characterize New World monkeys? Far too few studies have been done to allow completely confident answers to these questions. Most of the direct studies of color vision among Old World monkeys have been done on four species of macaques (*Macaca mulatta, M. nemstrina, M. fascicularis, M. fuscata*). The consensus conclusion is that these monkeys have trichromatic color vision, very similar in nature to that of normal human trichromats (Grether, 1939; De Valois et al., 1974; Oyama et al., 1986). Although it is clear that macaques are trichromats, what is undecided is whether macaque color vision is identical to that of normal human trichromats or merely highly similar. That issue has been recently discussed elsewhere (Jacobs, 1991). Sufficient numbers of macaques have been studied in visual discrimination experiments to indicate that not only are they trichromatic, but that trichromacy is routine, i.e., there is no obvious polymorphism (Jacobs and Harwerth, 1989).

Beyond the macaques, not too much is known of color vision among Old World monkeys. An early behavioral study, noted previously, suggested that baboons are trichromats (Grether, 1939). More recently, measurements of cone pigments in baboons (*Papio papio*) by microspectrophotometry (Bowmaker et al., 1983) and spectral sensitivity measurements made electrophysiologically in vervets (*Cercopithecus aethiops*) in our laboratory (Jacobs et al., 1991) indicate the strong likelihood that members of both of these species are routine trichromats. Measurements of cone pigments in individuals from several additional species (*Cercopithecus diana, C. petaurista, C. cephus, C. talapoin, Erythrocebus patas*) lead to the same conclusion (Bowmaker et al., 1991). The concensus values for the cone photopigments for all of these Old World monkeys are entered in Table 3.1.

On the basis of these several studies the most reasonable projection is that Old World monkeys are likely to be routine trichromats, most having similar if not identical color vision to that of normal human trichromats. If color vision polymorphism exists among these animals it seems apt to be of the low-frequency kind seen among humans, rather than the high-frequency type characteristic of New World monkeys.

Steps to Routine Trichromacy

What is minimally required for color vision is a retina that contains at least two spectrally distinct classes of receptor and a nervous system that is able to make use of these differences in spectral absorption. Recent work indicates that the molecules that provide the spectral selectivity of receptors, the photopigments, are greatly similar in structure, suggesting that all arose from a common ancestor. For instance, the deduced amino acid sequences of the rod photopigments of bovine, mouse, and human are about 90% identical (Pak and O'Tousa, 1988) and the pigment genes of even very distantly related organisms have homologous DNA segments (Martin et al., 1986). The implication of these results is that the photopigment molecule is conservative. Thus we would expect that photopigments would share many commonalities of operation.

There is little rational ground for supposing what receptor types and photopigments might have characterized the early anthropoids. Although the data are scanty, one can defend the views that among contemporary mammals outside of the primates, whether diurnal, nocturnal, or crepuscular in habit, (1) most species have some cones and at least one type of cone photopigment, and (perhaps less certainly) (2) no species seems to have achieved better than two classes of cone pigment and dichromatic color vision. On these grounds alone its seems likely that early primates had some cones and at least one type of cone pigment. In addition, comparison of the DNA sequences for human cone pigments suggests that the divergence of short-wavelength cone pigments from middle- and long-wavelength cone pigments may have taken place 530–670 million years ago (Yokoyama and Yokoyama, 1989). If so, it seems likely that the early primates had already acquired two cone pigment types.

Because the absorption bandwidths of photopigments vary only modestly among photopigment types (note Figures 3.2 and 3.6), the only available mechanism for broadening spectral sensitivity is to add new pigment types to the retina. Since the long-wavelength limbs of photopigment absorption curves fall off steeply (about 0.3 log unit/10 nm), the addition of a new pigment only slightly displaced in peak sensitivity toward the long wavelengths will allow the capture of substantial additional long-wavelength energies. The neural differencing networks that allow for the extraction of chromatic contrasts are generically the same as those employed to extract brightness contrasts. Consequently, crude color vision could emerge following the addition of a second cone pigment without any concurrent redesign of the nervous system.

However the early evolution of color vision among primates might have proceeded, the color vision polymorphism of contemporary New World monkeys may encapsule one of the stages on the path to routine trichromacy. At that stage there would be an autosomal gene specifying a short-wavelength cone pigment and a single X-chromosome gene producing a middle- to long-wavelength cone pigment. Addition of an allele at the X-chromosome site would allow for individual variations in spectral sensitivity with possible consequent advantage or disadvantage. Most importantly, however, the presence of two alleles would make some female monkeys heterozygous at the photopigment locus. For the first time these females would have two different classes of middle- to long-wavelength cone pigments and, with the proper neural interconnections, trichromatic color vision. Assuming that trichromatic color vision is, on balance, superior to dichromatic color vision, this would provide a heterozygous advantage and the frequency of the new allele would rise in the population. At equilibrium this arrangement would allow only 50% of the females to have trichromatic color vision. To increase trichromatic representation requires more alleles. In the case of the squirrel monkey, and apparently several other New World species, there is a third allele, so that the proportion of trichromatic females probably reaches about two-thirds. Although adding still more alleles could increase the representation of female trichromats, males could never share in the color vision spoils with this arrangement. To give males trichromacy requires a gene duplication so that a second pigment locus is established on the X-chromosome. At that stage trichromacy can become a species characteristic.

Trichromacy is very likely routine among all Old World monkeys and apes. This suggests that the gene duplication needed to yield two (or more) X-chromosome pigment loci occurred at some point after the divergence of Old and New World primates, that is, about 50 million years ago. Whether an X-chromosome pigment gene duplication may also have subsequently occurred in some of the lines of New World monkeys remains an open question.

Although the scenario just outlined seems plausible, Mollon (1991) recently pointed out that there are other possibilities. One is that pigment gene duplication could precede as well as follow the divergence of pigment genes; if so, Old and New World monkeys might have followed separate routes to their forms of trichromacy. The other is that New World monkeys may have lost a second X-chromosome pigment gene locus, and thus abandoned routine trichromacy at some stage in their evolution. Those who spend their days studying the elegance of human trichromacy would surely find this idea unpalatable, and the case of *Aotus* outlined above might be used to argue against this possibility. At any rate, it seems likely that the rapid increase in information about the cone pigment genes in a wide variety of primates will shortly provide a much better idea of which (if any) of these schemes is the more likely correct.

Ecology of Color Vision

The discussion in the previous section provides a scenario for the mechanical steps involved in the evolution of routine trichromacy among primates. However plausible or not that may be, nothing has yet been said about the much more slippery issue of why color vision would evolve in this way. As a way of approaching this, one might want to consider two questions: (1) How does the presence of color vision enhance survival? (2) How do the alternate forms of color vision found among some New World monkeys map into their normal visual behaviors? The reader is forewarned, likely unnecessarily, that there are no clear answers to either of these questions (for additional discussion of this issue, see Jacobs, 1993).

Utility of Color Vision

Of what use is color vision? Walls (1942) was perhaps among the first to offer one answer: that color vision promotes the perception of contrast between objects and surrounding regions, and hence enhances the visibility of the former. Walls argued that animals that can make use of local differences in spectral reflectivity (color cues) in addition to using differences in total reflectivity (lightness cues) have an advantage over those competitors that cannot do both, and that advantage is sufficient to impel the evolution of a color capacity. Consider the extreme: if all objects in a scene were equally light, then, obviously, the only way to discriminate among the objects would be to use color cues. In such an environment the animal without color vision would be blind. Scenes in the real world are probably never equally light, but the example can serve to illustrate this argument about the utility of color vision.

While there is little doubt that this visibility argument is plausible, recent authors have suggested additional reasons why color vision is a useful aid to discrimination. One arises from the fact that scenes in the real world are frequently illuminated by some combination of direct and indirect lighting, and this creates a situation in which there is much shadowing and substantial local lightness variations. Although there are great lightness variations in such a situation, the color variations introduced by shadowing are substantially less variant. The discrimination of objects should thus be more accurate and reliable if these discriminations can be based on color cues than if they must be based on lightness cues alone. The point is that the variable distribution of light intensity produced by shadowing may serve as a hindrance to detecting the form of an object; the less variable distribution of spectral energy could serve to disambiguate the situation and thus make the discrimination possible (De Valois and De Valois, 1988).

An additional advantage of color vision is that color can be used as a property for perceptual segregation. The idea is that separable components of an object (for instance, a predator viewed through a screen of dense vegetation) that differ in lightness might be linked together by their shared color property so as to allow object recognition (Mollon,

1989). Familiar examples of this include the pseudoisochromatic plates often used as quick tests of color vision. In these plates a viewer who has appropriate color vision perceives a figure even though that figure is defined only by spatially separated colored elements in a scene that otherwise varies randomly in lightness.

All of these arguments show how color vision may be a highly useful capacity in the discrimination of objects. Quite apart from the discrimination aspect, a further utility of color vision is that it allows an independent source of information about objects. This has been referred to as the signal significance of color (Jacobs, 1981, 1990a). For instance, with the appropriate color vision a fruit might be rendered more discriminable from leafy surrounding if it were any of several colors. At the same time, however, only one of these colors might convey information about the ripeness of the fruit. In this case color vision allows the perceiver to learn something specific about contents of an article through an analysis of its particular color. Finding the fruit might depend on using color as an aid to discrimination, but deciding whether or not it should be eaten may require the ability to recognize and categorize specific colors. There is evidence both from neurophysiological studies of color vision and from the study of human color vision pathologies that these two aspects of color vision may have separable neural bases (Jacobs, 1981).

The Feeding Hypothesis

On the grounds that there can be mutual benefit for both, it has long been believed that there must be an intimate relationship between the evolution of plant pigmentation and the color vision of birds and animals (e.g., Polyak, 1957). The argument suggests that under this arrangement the forager with appropriate color vision is guided to a food source and then, in turn, may act as an agent for seed dispersal. Couched in the present context, it might be that one should examine the spectral properties of food sources to understand monkey color vision and its evolution.

Snodderly (1979) made a promising start in this direction by attempting to interrelate the color vision of South American monkeys and the spectral reflectance properties of various plants that form part of the monkey diet. To do so he measured the diffuse reflectance properties of a number of food plant samples from a site in northern Peru that constituted the home range of a group of titi monkeys (*Callicebus torquatus*). The major food sources of these monkeys are fruits. Although there was much individual variation in the spectra recorded from different species of fruits, most appeared (to the human with normal color vision) as either green, purple, red, yellow, orange, or brown. Snodderly notes that the critical factor for discrimination of these fruits is probably their color versus the predominantly green background of the arboreal habitat of these monkeys. Given this, he concluded that the color discriminations these animals would be likely to have to make would involve discriminations among various shades of green, or between green and the several other colors listed above. He further concluded that these monkeys would not be required to make any subtle color discriminations among colors in the long-wavelength part of the spectrum and, thus, an animal with impaired long-wavelength color vision would not seem to be disadvantaged.

As we have seen, at the time Snodderly did his investigation it was believed that South American monkeys routinely had some sort of protan color defect (i.e., characterized by difficulty in making discriminations in the middle to long wavelengths), and thus his results seemed to fit with the idea that there might be an intimate link between food selection and color vision. Although the general conclusion is still attractive, it is clear that much more work will be required to make it compelling. Three things seem crucial.

First, spectral reflectance measurements of objects and surroundings should be made *in situ* so that the effects of variations in illumination can be taken into account. As noted above, variations in the spectrum of the illuminant affect the spectral distribution of energies reflected from an object. Such effects can often be substantial. This might be particularly important in a rainforest environment where sunlight is frequently filtered through a dense chlorophyl canopy.

Second, it is clear that most New World monkeys have a broad diet, feeding not only on fruits and plants, but also on insects and plant exudates (Terborgh, 1983; Sussman and Kinzey, 1984). Many of these will provide potential color discriminations of a type quite different from those examined by Snodderly. Beyond the issue of the range of food items and their spectral characteristics, particular attention might profitably be paid to the items of diet selected at times when food is scarce (W. Kinzey, personal communication). Those species that continue at such times to use color as a principal source of information for food selection may well have been subject to strong selection for those mechanisms that promote better color vision.

Third, since we now know that color vision varies strikingly from animal to animal (at least among some monkey species), it becomes crucial to understand what discriminations animals having particular color vision phenotypes make. Thus some way must be found to identify individuals with different color vision phenotypes and establish the sorts of discriminations that each type makes. Although this will not be an easy task, only in this way will it be possible to evaluate the feeding hypothesis, to see how closely color vision capacity is matched to the crucial discriminations required to obtain food.

The Fit between Color Vision and Visual Behavior

Laboratory investigations suggest that within any troop of squirrel monkeys one may expect to find wide individual variation in color vision abilities. How does this variation impact normal behavior? One possibility is that the variation is without consequence for normal behavior, that the significant differences in discrimination capacity seen in the laboratory are unimportant in the natural environment. If this were the case there would be no selection for particular color vision phenotypes and the documented polymorphism would have little consequence for understanding the evolution of color vision. Although this possibility cannot be rejected out of hand, it seems an unlikely explanation given (1) the fact that the three allelic cone pigment genes appear to be equifrequent in the population, suggesting that they have been subject to selection and, as noted above, (2) the fact that the variations in pigments required to yield color vision polymorphism will have significant behavioral impact, on color vision surely, but also on all other features of visual discrimination.

If the color vision variation of New World monkeys is not selectively neutral, then one must seek an explanation for it. Three general explanations for occurrence of polymorphism would seem potentially applicable to this particular case. One is frequency-dependent selection (Clarke, 1979). This supposes that there is some survival advantage that accrues directly to the presence of diverse phenotypes within a group. Applied to this situation it might be that monkeys having different kinds of color vision are able to exploit preferentially different food resources, e.g., some individuals might specialize in fruit harvest while others are more likely to seek a diet of insects. A second explanation is multiple-niche polymorphism (Levene, 1953). This assumes that the environment is composed of visually distinct niches and that different phenotypes are able to make advantageous use of these local variations. For instance, for the monkeys these niches might be defined by differences in the average level of illumination or in the dominant spectral characteristics of the illuminant or both. It is noteworthy that both of these factors will vary significantly as a function of depth from the top of the canopy of the rainforest. The third explanation, heterozygous advantage (Ford, 1971), has already been mentioned. This assumes that trichromats (heterozygotes) have a visual advantage over dichromats, and thus the variation in genotype in the species allows a mechanism for the maximization of trichromacy.

There are few facts to guide a choice among these possible explanations for the polymorphism of color vision among New World monkeys and, indeed, the three contain common elements. On balance, however, heterozygous advantage may offer the most likely explanation for this polymorphism. Perhaps most telling, it is hard to find evidence that any visual discriminations can be more efficiently made by dichromats than by trichromats as one might expect under the explanations offered by either frequency-dependent selection or multiple niche polymorphism. There have been persistent suggestions that human dichromats have advantages over trichromats in certain visual tasks (e.g., movement or brightness discriminations) and, indeed, there is some evidence to show that dichromats can penetrate some camouflages that defeat trichromats (as well as vice versa) (Morgan et al., 1989). However, a recent comparison of the capacities of dichromatic and trichromatic conspecific monkeys revealed no clear differences in ability of individuals to make either flicker or brightness discriminations (Jacobs, 1990b). In that same study there was some indication that for tasks in which both brightness and color cues are present, certain discriminations can be more efficiently made by dichromatic monkeys than by trichromatic monkeys. Further investigation of this possibility will be required. For the present, however, there seem to be few disadvantages associated with trichromatic versus dichromatic color vision; and this is particularly true as weighed against the very substantials advantages that the trichromat holds in making a variety of visual discriminations.

Finally, the following observation: In examining the visual capacities of monkeys that have the several different color vision phenotypes, clearly the most striking differences are those between the dichromat and the trichromat, irrespective of the particular type within either of these groupings. There are much less impressive differences among the various dichromacies or among the various trichromacies. For example, Figure 3.7 shows wavelength discrimination functions obtained from two squirrel monkeys having different trichromatic phenotypes (one has cone pigments having λ_{max} of 430, 550, and 562 nm; the pigments of the other animal are 430, 538, and 550 nm). Even though these laboratory tests probably tend to maximize any differences among the phenotypes (by using narrow-band lights and by removing all accessory cues to discrimination), it can be seen that there are only small differences between these two different types of trichromat. A similar conclusion tends to hold for comparisons among the dichromatic subtypes. The point is that there would seem to be a great advantage to be gained by being trichromatic, but perhaps only relatively smaller advantage (or disadvantage) to having any particular trichromatic phenotype. If that is the case, the thrust of the process would seem to be to maximize the representation of trichromacy per se and so one might not expect to find any very perfect fit between a specific color vision phenotype and real world visual behavior. Whatever the case may be, it is already clear that some important general questions about the evolution and ecology of

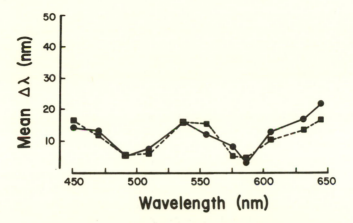

Figure 3.7. Wavelength discrimination functions obtained from two types of trichromatic squirrel monkey. The λ_{max} values for the cone pigments of the animals were (squares) 430, 550, and 562 nm; (circles) 430, 538, and 550 nm. Note how similar the two wavelength discrimination functions are despite the difference in cone complement. Data from Jacobs (1984).

color vision may be profitably approached through studies of New World monkeys.

Acknowledgments

I thank Jay Neitz for helpful comments on an early version of this manuscript, Christine Allen for assistance with the illustrations, and the National Science Foundation for research support (IBN93-18770).

Notes

1. For years a variable number of squirrel monkey subgroups have been identified and, even today, the systematics of *Saimiri* remain contentious (Hershkovitz, 1984; Thorington, 1985). Since a direct comparison of four squirrel monkey subgroups revealed no differences between the groups with respect to their color vision (Jacobs, 1984; Jacobs and Blakeslee, 1984), this issue does not seem important for present purposes.

2. Although photopigment absorption spectra are broadly peaked (see Figure 3.2), it is traditional to specify them by giving a peak wavelength. To accomplish this, a theoretical curve is best fit to the array of spectral sensitivity values.

Since (1) these measurements are often somewhat noisy, and (2) there are a variety of different procedures that have been used to fit these absorption curves, separate measurements of the same type of photopigment may lead to slightly different estimates of the absorption peak. This presents a problem for the reader in that it is hard to appreciate what are real differences in pigment peaks and what simply reflect these uncertainties. No general rule can be offered, but since it seems evident that the minimum peak separation between any two different types of nonhuman photopigment is about 6 nm (Jacobs and Neitz, 1985a), then variations of, say, 2–4 nm in estimates are unlikely to be biologically relevant variations. For a variety of technical reasons the short-wavelength-sensitive pigments are harder to specify accurately so for these pigments that range of tolerance should probably be expanded. My current best guess is that all of the measurements collected into each column in Table 3.1 represent the same type of photopigment. As measurement techniques become more precise, these annoying uncertainties should disappear.

4

Is Speech Special? Lessons from New World Primates

CHARLES T. SNOWDON

Introduction

Speech and language have long been thought to be among the phenomena that distinguish human from nonhuman primates. In recent years studies of chimpanzees and bonobos trained on various language analogues have chipped away at the notion of language being distinctively human. These animals can learn to use arbitrary symbols as referents for real objects, they show mastery of word order or rudimentary syntax, and they respond appropriately upon the first presentation of novel utterances (Savage-Rumbaugh, 1988).

The success of some great apes demonstrating a degree of linguistic competence came only with the abandonment of attempts to teach these species to speak. All of the successful language analogues are nonverbal. Thus, the evolution of speech must have a different course from the evolution of language. [Although recent studies of "babbling" in deaf children (Petito and Marentette, 1991) suggest that "speech" need not be vocal.] If one accepts the premise that it is as important to hunt for nonhuman precursors to speech as it is to hunt for nonhuman precursors of language, then where should one look for precursors? The traditional model in biological anthropology has been to seek parallels in those species that are most closely related to human beings either phylogenetically (and thus the study of chimpanzees, bonobos, and other apes) or ecologically (and thus the study of savannah-dwelling species from East Africa—baboons). Yet great apes are remarkably silent compared with human apes. Their vocal repertoire appears impoverished compared to those of other species (Marler and Tenaza, 1977). Little work has been done on the vocal skills of baboons (see Waser, 1982). All of these species appear to utilize visual communication to a much greater degree than vocal communication.

I argue that the New World primates offer greater promise for the understanding of speech and vocal language origins than the great apes and baboons. All New World primate species are arboreal, and this fact of ecology places severe constraints on the use of visual communication. Not only are visual signals difficult to perceive through the dense vegetation of a tropical forest, but arboreal monkeys need their limbs for locomotion and support and cannot readily free them for visual gesturing. The variety of complex social systems displayed by neotropical primates coupled with the arboreal environment has led to the evolution of complex vocal communication systems. These complex vocal systems are likely to provide us with insights concerning the origins of speech-like phenomena.

But why should we give any credence to models developed from species so phylogenetically distant from human origins? Similar traits or processes have appeared in several distinct phylogenetic lines, presumably in response to similar environmental pressures. Thus color vision has evolved in several distinctive lines (bees, fishes, birds, mammals). (See also Jacobs, this volume, for the importance of color vision in neotropical primates.) Complex vocal communication has also developed in several distinct lines. If we are interested in the precursors of the human vocal communication system, then it makes sense to study species that have complex vocal communication, regardless of their phylogenetic relationship to human beings. Marler (1970) has made a persuasive argument for the utility of studying bird song, based on several parallels between bird song and human speech. This argument is also valid with respect to the vocal communication of New World primates for the reasons given above. Of all primates (with the possible exception of gibbons) New World primates appear to be the most vocal.

Given that New World primates might be the most relevant species in which to study the origins of human vocal communication, how should we study these species? Should we bring them into captivity and attempt to teach them to speak? There are two distinct approaches to the study of linguistic origins. Comparative psychologists have isolated animals from their normal social environments and provided human companions as social surrogates. Animals are given extensive training on a symbolic system that is analogous to English. The linguistic ethologist seeks to understand the language of animals rather than to make the animals understand language. The linguistic ethologist is forced to play detective to decode the utterances of another species and must develop clever observational techniques to understand the complexity of the animals' utterances. While the comparative psychologist can tell us something about the ultimate capacities of animals when subjected to rigorous training, the linguistic ethologist can tell us about the natural precursors of linguistic phenomena. What environmental conditions might lead to symbolic communication? Under what circumstances do animal species develop a rudimentary syntax? How are phonology and usage acquired within the ontogeny of an individual?

Although I am a comparative psychologist by training, I am more comfortable with the position taken by the linguistic ethologists. However, rather than simply document the complexity of utterances of other species, I am motivated to look for parallels between animal communication systems and human speech and language phenomena. Several phenomena that were once thought to be exclusive to human speech and language have been documented in New World primates: phonetic variability, categorical perception of phonetic differences, variation of call structure with respect to location of intended recipients, syntax, conversations and turn-taking, metacommunication, and the ontogeny of vocal communication. For each of these phenomena, New World primates present some of the most compelling examples of parallels to human speech and language.

Phonetic Variation

One characteristic of a complex communication system is a large vocabulary size. An organism must be able to communicate about subtle differences in social context or environment with distinct signals. Several years ago it was argued that animals had a very limited repertoire of vocal signals (Moynihan, 1970a). In recent years many ethologists have demonstrated in birds and primates that vocalizations previously treated as unitary call types actually were quite variable and that these variations correlated with subtle differences in the context in which the variants were given. Furthermore playback studies indicated that at least some of these variants were perceived by the animals themselves and truly represented an increase in vocabulary.

In several bird species there are variations in song structure that appear to correlate with different contexts (Ficken and Ficken, 1967; Morse, 1967, Lein, 1978, Smith et al., 1978, Kroodsma, 1981). One song variant typically is used at the periphery of a territory, presumably in defense against intruders, while other variants are used at the center of the territory, near the nest, possibly to communicate with one's mate.

The first study on nonhuman primates to demonstrate subtle call variants within a previously unitary call type was Green's (1975) study on Japanese macaques (*Macaca fuscata*). Green described seven variants of "coo" calls that were used in 10 different contexts, but the distribution of call type and context were highly correlated. Green hypothesized an underlying motivational continuum of increased arousal to account for these variant calls. Subsequently Lillehei and Snowdon (1978) found two of the same coo variants in young stumptail macaques (*Macaca arctoides*). Zoloth et al. (1979) trained Japanese macaques in a discrimination paradigm and found that they would learn to discriminate between two of the coo variants more rapidly than other primate species that were tested.

The use of a discrimination paradigm to test whether the monkeys actually perceive the differences in call variants is a necessary step. An observer may be too enthusiastic in attributing variation in call structure to separate call types with their own meanings. Call structure can also vary as a function of individual variation, of geographic, age, or subspecies, or population variation (Snowdon, 1989). Thus, before we can be confident in assigning a larger vocabulary size to a species, we must test the animals in an operant discrimination paradigm as did Zoloth et al. (1979) or in a playback paradigm (see below) to determine that the monkeys discriminate call variants in a functionally significant way.

There are a few examples of call variants indicating different contexts or functions in New World primates. Cleveland and Snowdon (1982) described four variants of long calls in the cotton-top tamarin (*Saguinus oedipus*) similar in function to bird song. One was used at the start of territorial encounters, a second variant was used later in a territorial encounter at the peak of arousal, and a third form appeared to be used for group cohesion and communicating with separated animals. The final form was highly variable and appeared to be characteristic of reproductively immature animals, much as Marler and Peters (1982) have characterized subsong and plastic song as immature stages of song for various species of sparrows.

A playback study was done with two of these tamarin long call variants, the Normal Long Call given at the start of a territorial encounter and the Quiet Long Call used in intragroup cohesion (Snowdon et al., 1983). Playbacks of long calls of each type from familiar and unfamiliar tamarins elicited very different results. Tamarins responded only to the calls of unfamiliar tamarins and ignored playbacks of long calls of their own group. They displayed increased arousal and agonistic behavior to the territorial form (Normal Long Call) and showed different behavioral responses to the Quiet Long Call. These long call variants are perceptually distinct to the tamarins and elicit different responses.

Cleveland and Snowdon (1982) also described eight variants of chirp vocalizations, which are short, frequency modulated calls. Each of these variants was associated with a very different context. One was a mobbing call, another an alarm call, a third given as animals approached food, a fourth given as animals took food and moved away, a fifth given when animals first heard the calls of an unfamiliar group, and so on. In contrast to Green's report that the coo variants of Japanese macaques could be mapped onto a single motivational continuum, the chirps of cotton-top tamarins represent a wide variety of putative motivational states. In a subsequent study Bauers and Snowdon (1990) selected two chirps that were most similar in acoustic structure and most different in context (alerting to a strange group of animals versus maintaining vocal contact within an unaroused group). Examples of each type of chirp were played back through speakers to eight groups of tamarins. All of the animals alerted to both types of chirp playbacks, but they showed increased rates of vocalization, gave increased numbers of territorial calls, and showed increased piloerection and locomotion only in response to the playback of the territorial chirp. The monkeys not only discriminated between the two chirp types, but they gave contextually appropriate responses to each.

In the pygmy marmoset (*Cebuella pygmaea*) Pola and Snowdon (1975) observed four types of trill-like vocalizations. These are sine wave frequency-modulated calls. Two of the trills differed only in duration (Closed Mouth Trills, mean = 176 msec; Open Mouth Trills, mean = 334 msec). A third variant differed from the Closed Mouth Trill only in frequency range (Closed Mouth Trill mean = 4.0 kHz range; Quiet Trill mean = 1.5 kHz). The fourth form (J-call) was highly distinctive from the other three trills. It consisted of an interrupted series of notes representing just the upsweep of the sinusoidal modulation. The frequency range was large (>5 kHz) and the duration was significantly longer than those of the other trills (>700 msec).

The Closed Mouth Trill was used in calm, relaxed conditions as animals moved through the environment while the Open Mouth Trill preceded aggressive or agonistic encounters. No differences were observed in the captive pygmy marmosets between the contexts where Closed Mouth Trills, Quiet Trills, and J-Calls were used, despite the large number of acoustic differences among these call types (but see Locational Cues below). How do pygmy marmosets discriminate between Open Mouth and Closed Mouth Trills?

Categorical Perception

To many students of speech perception, one of the major features that makes speech special or different is the mechanism by which it is perceived (Liberman et al., 1967, Liberman, 1982). When human subjects are tested with synthesized versions of different phonemes, they display categorical perception in that different sounds (phones) that are given the same label (allophonic) cannot be discriminated from one another. For example, if we are presented with a synthesized continuum of sounds varying in voice onset time (the time between the forcing of air through the lips and the activation of the vocal cords) that mimics the sounds /ba/ and /pa/, we do not hear each individual sound as different. Instead we classify the different sounds as either /ba/ or /pa/. Instead of perceiving continuous variation, we hear a sharp boundary between two of the stimuli. This dichotomization of a continuum into discrete categories is categorical perception. Do monkeys perceive their sounds in a similar categorical fashion?

In Goeldi's monkeys (*Callimico goeldii*) there are two types of predator alarm calls. One elicits mobbing and the other elicits freezing responses (Masataka, 1983). To determine how the monkeys perceive these calls, Masataka synthesized each type of call as well as several intermediate types of calls on each of several acoustic continua that separated the two call types. Using playback studies he found that frequency range was the main dimension that separated the two types of calls. When monkeys were presented with synthesized calls varying in frequency range from 1.6 to 2.4 kHz, the animals responded with antiphonal calling and increased locomotion typical of a mobbing response. When the frequency range was increased from 2.4 to 2.6 kHz or greater, the animals responded with a freezing response. Thus, calls varying from 1.6 to 2.4 kHz produced mobbing responses and calls ranging from 2.6 to 5.2 kHz elicited freezing, demonstrating a sharp category boundary in frequency range between the two call types. Goeldi's monkeys appear to ignore much of the variation within each call category and attend only to differences at the boundaries between the two call categories.

Several years before Masataka's study Snowdon and Pola (1978) synthesized the trill vocalizations of the pygmy marmoset. Synthetic trills were created varying on each of four dimensions: duration, frequency range, center frequency, and rate of frequency modulation. Since the Closed Mouth Trill and Open Mouth Trill are given in very different contexts and differ only in duration, this became the dimension of greatest interest. Synthetic calls that ranged in duration from 176 to 248 msec elicited antiphonal responses from pygmy marmosets, while calls ranging from 257 to 334 msec elicited no responses. Thus a difference of only 9 msec (the temporal resolution of the synthesizer) in duration between 248 and 257 msec sufficed to produce different responses, while trills varying from 176 to 248 msec produced similar responses. This appears to be a case of categorical labeling of trills by pygmy marmosets similar to the categorical labeling of speech sounds shown by human subjects. When the same synthesized calls were played back to human listeners, they were able to discriminate between each of the synthesized trills (Snowdon and Pola, 1978). The human subjects showed no evidence of categorization. This last finding raises a paradox. Why should human subjects be more adept at distinguishing between pygmy marmoset trills than pygmy marmosets themselves? Why aren't captive pygmy marmosets able to make fine distinctions between their trills?

We need to consider the functional significance of categorical perception to resolve this paradox. Most research on human speech perception has ignored the social components of speech. Human subjects are tested with synthesized speech sounds that can be well-controlled in terms of acoustic structure, but that do not sound like any familiar human speaker. There are at least two levels of processing that are involved in attending to speech (Snowdon, 1987). First, we must attend to the phonetic structure to interpret the phonemes, words, and phrases we hear. But we also attend to many nonlinguistic aspects of speech. Intonation contours tells us if a speaker is making a statement or asking a question. We can identify whether a speaker is young or old, male or female, familiar or unfamiliar. If we are sitting in a dark movie theater and we hear someone yell "Fire", it is essential that we attend only to the linguistic input. If, on the other hand, we hear a voice whispering: "Kiss me, I love you," our response depends not only on our interpretation of the linguistic content, but also on whose voice we hear and the location we hear it from. We need to interpret a variety of social information incorporated in this utterance to decide how or whether to respond.

So what does this have to do with why pygmy marmosets categorically label Closed Mouth and Open Mouth Trills? Like the human subjects in speech perception studies, the pygmy marmosets were presented with a synthesized version of a trill that represented the average characteristics of a pygmy marmoset trill but represented no single individual. Furthermore, the nature of the task given to the marmosets only required them to label the two types of trills. They were not given the opportunity to show if they could discriminate between different types of trills. To see if social variables and a different testing technique would affect their categorical perception of trills, we performed another experiment (Snowdon, 1987). We recorded the calls of several individuals in our colony and determined the features that made one individual's call different from others. We then created synthesized versions of trills that matched the frequency, frequency range, and rate of frequency modulation for three individual monkeys, and we systematically varied duration.

We also altered our testing procedure to focus on whether marmosets could attribute these synthesized calls to particular individuals. We had found previously (Snowdon and Cleveland, 1980) that we could test for individual recognition with a very simple paradigm. If we played animal X's calls through a speaker hidden in X's cage, other animals in the colony would respond with an increased rate of antiphonal trilling. If we played X's trills through a speaker in animal Z's cage, however, we found no antiphonal response. Thus, our pygmy marmosets appear to connect the sound of an individual with the location where that animal typically is found. In our new study using synthesized trills we made use of this technique, playing synthesized trills representing an individual through a speaker in that individual's home cage and equally frequently through speakers located in other parts of the colony. If there was a significant increase in antiphonal calling to calls played back from the caller's own cage compared with the same call played back from another animal's cage, then a particular synthesized call was considered to be perceived as produced by that individual.

We varied individual calls over six durations and presented each call type 10 times from the animal's own cage and 10 times from another location. The results were quite different than those we reported earlier (Snowdon and Pola, 1978). For example, one animal typically had very short Closed Mouth Trills. When synthesized versions of these trills were played, other animals responded only to the shortest versions of the trills. With another animal whose calls were typically longer in duration, the other animals responded only to those in the typical range for that animal. Thus, in contrast to the results of Snowdon and Pola (1978), pygmy marmosets can discriminate between calls within a category, if the task demands are arranged to facilitate this discrimination and if features of socially familiar individuals are present in the calls presented. There are at least two levels of acoustic processing displayed by pygmy marmosets. Calls of unfamiliar animals are labeled according to a set of acoustically broad categories. However, once an individual's calls are familiar, a listener creates much more precise categories specific to each individual. No similar study has ever been done with human subjects, but I would not be surprised if human perception of speech was similar to this two-level processing found in pygmy marmosets perceiving their vocalizations.

Pygmy marmosets can show categorical perception and they can show discrimination of call types within a category depending on the nature of the stimuli that are used and on the task presented to the marmosets. That marmosets can make within-category discriminations when calls of familiar individuals are synthesized demonstrates that these responses are not due to a biologically fixed feature detector as was suggested after the results of the first study (Snowdon and Pola, 1978). Perceptual responding is flexible and dependent upon acquiring knowledge of the call structure of familiar individuals. Although there is little evidence of learning being involved in call production (see Ontogeny below), these results imply the importance of perceptual learning.

Locational Cues

The preceding sections demonstrate that there are clear functional and perceptual differences between Closed Mouth and Open Mouth Trills. However, we found no differences in our captive animals in responses to Closed Mouth Trills, Quiet Trills, and J-calls. Why should these three calls with quite diverse call structures be given under similar circumstances and elicit similar responses? Again let us speculate on the social function of these contact calls. It would be important for pygmy marmosets in the field to be able not only to identify who is calling but also to obtain information about the location of the caller. However, providing information about location can be dangerous since potential predators as well as members of one's group could use the information. A compromise solution to this problem would be to use ventriloqual calls when one is close to other members of the group and to use highly locatable calls only when one is relatively far from other group members. The three trill types can be ranked according to the acoustic features each provides for sound localization.

The Quiet Trill has no sharp onsets and offsets, and has a narrow frequency range. Brown (1982) showed that monkeys can significantly improve their sound localization in a three-dimensional test (similar to the task of an arboreal pygmy marmoset) when at least 2 kHz of frequency modulation was added. All Quiet Trills we have recorded have a frequency range of 1.0 to 1.5 kHz well below the critical value for three-dimensional sound localization. These calls should be cryptic and difficult to locate.

The J-call, on the other hand, has several features allowing for precise sound localization. The frequency range is 5–6 kHz, and the call consists of a series of brief notes, providing both temporal and frequency cues for sound localization. The J-call should be very easy to localize, and the Closed Mouth Trill should be intermediate in localizability.

If pygmy marmosets make use of these acoustic features for sound localization in an adaptive way, then we would predict that in the field they should use the Quiet Trill at close distances to other group members, the Closed Mouth Trill at intermediate distances, and the J-call only when animals are far apart. To test this hypotheses we followed a group of pygmy marmosets in the Peruvian Amazon (Snowdon and Hodun, 1981), recording each of the three trill types when they occurred and noting the location of the caller and the nearest other group member we could find on a three-dimensional map of the home range. We calculated the distance between the two individuals and plotted the distribution of each call type against distance between group members. The results clearly supported our hypothesis. Over 50% of all Quiet Trills were given when animals were less than 1 m apart, and we could always find another pygmy marmoset within 5 m when a Quiet Trill was given. In contrast, the only call given when animals were more than 20 m apart was the J-call. As predicted, the Closed Mouth Trill was used at intermediate distances. Thus, pygmy marmosets can alter the structure of their contact trills according to how far away they perceive themselves to be from other group members.

A similar finding has been reported for squirrel monkeys (*Saimiri sciureus*) by Masataka and Symmes (1986). They systematically varied the distance between separated infants and members of their natal group. As distance increased both infants and adults used calls with increasing duration, and they prolonged the high-frequency components of their calls. Snowdon and Hodun (1981) had noted the presence of a "window" above 8 kHz where there was little ambient noise. Thus, squirrel monkeys, like pygmy marmosets, can adjust call structure to improve localization according to how far they are from other group members.

Localization cues of a different nature have been described by Whitehead (1987, 1989) for mantled howler monkeys (*Alouatta p. palliata*). In a tropical rainforest the reflection of sound waves from leaves, twigs, and branches produces increasing reverberation to sound heard at a distance from the caller. Calls that are emitted close to the listener have little reverberation, while calls emitted at a distance have considerable reverberation by the time they reach the listener. Whitehead (1987) created a clever playback experiment to test whether howler monkeys could make use of these acoustic differences. He created a playback tape that had a pure howl followed by a reverberated howl, which would be the pattern heard if a monkey were moving away from the listener. Whitehead created a second tape that had a reverberated howl followed by a pure howl, the pattern that would be heard if the monkey were approaching the listener. Male howler monkeys approached the playback speaker when the sound pattern was similar to that of an approaching intruder, and the same males retreated from the speaker when the sound pattern was similar to that of a retreating intruder. Thus, howler monkeys also can use information in call structure about distance and direction.

I am unaware of any studies on human beings altering speech patterns to increase localization or to make their location more cryptic, yet we know of three New World primate species that can alter localization cues and respond to localization cues. These examples show an area where the study of animal communication can suggest new ideas for the study of human speech. The study of vocal communication in neotropical primates is useful not only for understanding possible parallels to speech and language, but also for developing new directions for speech research.

Syntax

One of the major criteria for language proposed by linguists and psycholinguists is the presence of syntax (Chomsky, 1957). Much of the debate on whether chimpanzees and bonobos have an understanding of language has been based on whether these linguistically trained animals have the capacity for syntax. Terrace et al. (1979), based on their study of Nim Chimpsky and a review of the studies of other signing chimps, concluded that chimpanzees do not have an understanding of syntax. Recently, Savage-Rumbaugh's (1988) work with the bonobo, Kanzi, has shown that Kanzi is capable of understanding syntax in simple English declarative sentences. Is there any evidence of naturally occurring syntactic structures in primate communication? Yes, but the only examples are found in the New World primates.

Syntax can be examined at both structural and functional levels. Does a syntax allow for the potential creation of an infinite number of sequences as is the case in human language or is the syntax limited to the production of a limited number of sequences? Even if a syntax does not allow for an infinite number of utterances to be produced, a simpler syntax can still be informative from the perspective of precursors to human speech and language. An animal might have a very complex syntax but not make functional use of it, or an animal might be able to use a relatively simple syntax to communicate in more complex ways than it could without a syntax. Syntax can be used to create sequences of sounds that have entirely different meanings from the meanings of the individual components or syntax can be used to produce sequences of calls that preserve the meanings of individual components.

Marler (1977) distinguished two different types of syntax, a phonetic syntax that is equivalent to the formation of different words from phonemes and a lexical syntax that is equivalent to the formation of phrases or sentences from different words. In phonetic syntax, the units combining into a sequence do not maintain the functions or meanings of the individual components. However, with lexical syntax the sequence that results maintains the meaning of the individual constituents. Based on a review of information available at the time, Marler (1977) concluded that phonetic syntax occurred quite commonly in nonhuman animals, but that lexical syntax was unknown in nonhuman animals and would be extremely rare if it occurred.

The most complicated syntax yet shown for a nonhuman animal is not found in a primate, but in the chickadee (*Parus atricapillus*). Hailman et al. (1985) found four distinct types of notes in the eponymous "chick-a-dee" call. The arrangement of these notes followed a predictable sequence; out of 3500 calls recorded there were 362 different sequences. All but 11 of these sequences followed a simple rule system. If the first note is A, it can be repeated any number of times, and then it is followed by note D, which can be repeated any number of times, and note D, is followed by silence. Or a call can begin with note type B, which can be repeated *ad libitum*, which is then followed by note C, which can be repeated, and then is followed by note D, which is then repeated, and then silence. English and Chickadee are both generative grammars in that both can generate an infinite number of sequences. The main problem with the chickadee example to date is that there is no evidence that the birds can make use of the potential creativity of their grammar.

The only other examples of syntactic complexity come from New World primates. Robinson (1979b) described the vocal repertoire of titi monkeys (*Callicebus moloch*), which give complex sequences of calls in response to the approach of intruders. These sequences were organized hierarchically into phrases that were then organized into more complex sequences. Six different types of sequences were used in different circumstances. Robinson constructed playback stimuli with notes sequenced into both normal and abnormal sequence types. The titi monkeys responded appropriately to the normal sequences, but showed high levels of "disturbance behavior" to playbacks of the syntactically rearranged sequences, indicating that the monkeys perceived not only the individual components of sequences, but were also sensitive to the structural arrangement or syntax of the sequences.

Cleveland and Snowdon (1982) found a general structural syntax for the call sequences used by cotton-top tamarins. The vocal repertoire consisted of several types of short, frequency-modulated calls (chirps) and long unmodulated calls (whistles). In the formation of sequences, a chirp or a whistle could be repeated several times. In a sequence involving both chirps and whistles, all chirps preceded all whistles. Within a sequence each note had a lower frequency than the note that preceded it.

In a study of the wedge-capped capuchin monkey (*Cebus olivaceus*), Robinson (1984a) found several examples of lexical syntax. Several call types were used independently, but they could also be combined to create compound calls. These compound calls constituted 38% of the total sample that Robinson analyzed. The situations in which multiple calls were given were intermediate to the situations in which each call would have been used alone, suggesting that the combined calls conveyed the combined meaning of the individual call types. Many of the sequences appeared to represent the additive properties of the individual components.

Many of the combined calls of cotton-top tamarins were repetitions of the same note, which served to intensify the meaning of an individual note, or they represented new meanings that did not relate to the meanings of the individual components (Cleveland and Snowdon, 1982). However, two examples of lexical syntax were found. One was a combination of an alarm call with a whistled note used in calm, relaxed situations. This combination was typically emitted after animals had already responded to an alarm call by freezing, just prior to when they began to move about again. The combined call appeared to serve as an "all clear" call following an alarm. The second example was a combination of a territorial defense call given mainly by males with a territorial call given mainly by females. The combination of both territorial calls was given with equal frequency by both sexes and occurred at the time of peak arousal in a simulated territorial conflict (McConnell and Snowdon, 1986).

Newman et al. (1978) described rules for the sequencing of components that make up the twitter vocalizations of squirrel monkeys. Different individuals use different combinations of structural components and different combinations of units were associated with different social contexts. Pola and Snowdon (1975) described two types of calls in pygmy marmosets that appeared to be made of identical note structures, but the notes were given at different intervals. The J-call, which is used for contact between animals that are far apart, consists of a series of upwardly rising frequency modulated sweeps that occur with a period of about 50 msec between successive notes. Typically 10–20 notes constitute a J-call sequence. The alarm call that elicits freezing and quiet in pygmy marmosets appears to be made of an identical note structure, but the notes are given approximately 500 msec apart. The rate of repetition of the notes or the number of notes may be used to indicate the context of alarm or contact rather than the structure of the note itself. While these examples of syntax do not come close to the syntax of human language, the finding of syntactic structures and lexical syntax in New World primates indicates the possible origins of a complex syntax from the natural vocal signals of New World primates.

Turn Taking

Examples of turn taking can be found in the patterns of calling between animals. Turn taking is an interesting phenomenon, since, in human children, it is considered to indicate a significant advance in cognitive ability as the child passes from egocentricity to the ability to understand another individual's perspective. Turn-taking conventions can also be viewed as a between-individual syntax. Antiphonal calling is quite common among bird species, and Deputte (1982) has shown complex turn-taking behavior in the songs of white-cheeked gibbons (*Hylobates concolor leucogenys*). There are several examples of turn-taking behavior and rule-governed calling between individual animals in New World primates. Duets are quite common in howler monkeys (*Alouatta* sp.), in situations similar to those in which gibbons use duets (Sekulic, 1982b, 1983b). Duetting is also common among titi monkeys (Robinson, 1979a; Kinzey and Robinson, 1983) and Maurus et al. (1985) described "dialogues" in squirrel monkeys. Each of these examples requires a close coordination between two animals in both the timing of vocalizations and in the structure of the vocalizations emitted.

Snowdon and Cleveland (1984) recorded the sequence of trill vocalizations from a group of three pygmy marmosets and found clear evidence of turn taking. The animals did not call at random, but a significant number of the sequences of three calls recorded included a call from each of the members of the group. There are two possible sequences of turn taking among three animals (123, 231, 312 or 132, 321, 213). This group of pygmy marmosets used the first sequence of calling almost twice as frequently as the second sequence. Both of these orders occurred significantly more often, while sequences in which one animal called twice before all animals called once were significantly less frequent. If trills are simply an expression of emotional state, then calls should be uttered at random rather than in predictable sequences.

Another example of calling between animals following some sort of rule system emerged from a study by McConnell and Snowdon (1986) simulating territorial encounters between groups of cotton-top tamarins. We found an increase in the intensity of vocalizations from unfamiliar groups led paired tamarins to alert and begin a complex vocal interaction with the unfamiliar groups. Many of the behaviors observed in these tests were similar to those Neyman (1977) observed during territorial encounters between groups of cotton-top tamarins in the wild. There was an apparent syntax in the sequencing of vocal interactions (McConnell and Snowdon, 1986). The response of a tamarin depended on the type of call heard most recently, and on whether that preceding call came from the individual's own group or the other group. Animals who called immediately after one of their own group members had called tended to escalate the encounter by using a more intense form of territorial call. Animals who called immediately after a call was heard from the unfamiliar groups responded with the same type of call as the unfamiliar animal had given. There were also differences in the types of calls given by males and females.

There are a variety of rule systems that various New World primate species have developed to regulate calling between mates, between group members, and between different groups in territorial encounters. Turn-taking behavior appears to be quite common among New World primates.

Metacommunication

Metacommunication is defined as the ability to communicate about communication or to modify the usual message of a signal with the use of an additional signal. Metacommunicative signals are functionally more complex signals than signals that have simple referents. Two examples of possible metacommunicative signals have emerged from studies of squirrel monkeys. Smith et al. (1982a) described the chuck calls exchanged between female squirrel monkeys who were preferred partners or friends. The chucks did not appear to elicit contact or approach between females nor were they used among the nearest neighbors in a cage. But the exchange of chuck notes occurred mainly between females who had a close relationship based on total time spent near one another, grooming, etc. Smith et al. (1982b) also found statistically significant individual differences in the structure of chuck calls, indicating that squirrel monkeys could identify specific individuals from these calls.

Biben and Symmes (1986) described play vocalizations in squirrel monkeys. The rate of play vocalization varied with the nature of play and the duration of play bouts, with longer bouts having longer and more complex calls. Biben and Symmes hypothesized that play vocalizations have a metacommunicative function, telling other group members that what is about to occur is not serious, even though components of aggressive and sexual behavior will occur. The function of special calls to indicate that the signals that follow are not to be taken seriously appears to fit the definition of metacommunication.

Ontogeny

Marler (1970), in arguing for parallels between bird song and human speech, focused exclusively on parallels in ontogeny. The search for ontogenetic parallels between New World primate vocalizations and human speech has so far been elusive. Although there are a few suggestive results indicating parallels, the bulk of the data to data suggest that most of vocal development is under genetic control.

Squirrel monkeys have received the greatest attention with respect to ontogeny of vocalizations. Lieblich et al. (1980) reported that individual and population-specific features of squirrel monkey isolation peeps were virtually unchanged from the first day of life through the end of the second year. Although vocalizations were quite short at birth and became progressively longer over development, there was no change in the basic structure of the call. Studies of deafened monkeys found few changes in vocal structure (Winter et al., 1973; Talmage-Riggs et al., 1972) However, there has not been a longitudinal study following a deafened monkey from infancy through adulthood. Herzog and Hopf (1984) reared squirrel monkeys in social isolation, and found that when isolation-reared monkeys were first presented with the alarm call used to avian predators, they ran to their mother surrogate showing appropriate behavioral response on the very first hearing of the call.

Newman and Symmes (1982) have examined the heritability of population-specific features of isolation calls in squirrel monkeys. Two forms of squirrel monkeys (the Roman Arch and Gothic Arch forms) representing different species and subspecies (Hershkovitz, 1984) have distinctively different isolation peep structures (Symmes et al., 1979). When Roman Arch × Gothic Arch hybrids were examined, the hybrids with Roman Arch mothers inherited the maternal facial pattern and maternal form of isolation peep. This close parallel did not emerge with hybrids whose mothers were from Gothic Arch populations. Only 46% of the isolation peeps of these hybrids were classified as Gothic Arch in form. These results from all of the studies on squirrel monkeys suggest that early auditory experience and early social interactions are of little importance in vocal development.

In marmoset and tamarin species the data are more promising for vocal ontogeny, but the data available at present are far from conclusive. Hodun et al. (1981) recorded long calls from four subspecies of saddleback tamarins (*Saguinus fuscicollis*) in the Peruvian Amazon. There were structural differences in the notes of the long calls that distinguished among each of the subspecies. However, one animal that was physically similar to *Saguinus f. nigrifrons* was found on the boundary of the range between *S. f. nigrifrons* and *S. f. illigeri* and had a long call structure that was an amalgam of the structures of both subspecies. Three possible explanations could account for this amalgam. First, the monkey might have been a hybrid and its call reflects components of both parents. Second, if a troop of the other subspecies was in close proximity during development, the animal may have learned components of both subspecies. Third, the calling pattern may follow an ontogenetic trajectory that includes a juvenile stage with components of long calls of adults of both subspecies. To partially test among these hypotheses, Hodun et al. (1981) recorded long calls from captive tamarins that were known to be hybrids of *S. f. nigrifrons* × *S. f. illigeri*. These hybrid monkeys failed to show the amalgam structure of the animal recorded in the field. Rather these hybrids included components of a third subspecies (*S. f. fuscicollis*) that was the most numerous subspecies housed in the same colony room. This suggests, but does not prove, that some vocal learning might be possible.

A third strategy for studying vocal development is to follow individuals longitudinally to see if any consistent developmental trajectories appear. We have recently completed a longitudinal study of vocal development in pygmy marmosets over the first 3 years of life (Elowson et al., 1992). Contrary to our preliminary report on the first infants (Snowdon, 1988), there were no consistent changes in vocal development across different individuals in vocal structure. However, there were clear developmental changes in the usage of trill vocalizations. In infancy, pygmy marmosets use trills indiscriminantly in conjunction with a wide array of contexts and with a wide variety of vocal types. We have previously described an infant "babbling" phase (Snowdon, 1988) where infants juxtapose a wide variety of call types together in extensive vocal bouts. Over time there is a progressive restriction of trill vocalizations to situations where trills are appropriate. As animals develop they cease using trills in conjunction with other types of vocalizations.

A reverse phenomenon has been observed in the ontogeny of predator alarm calls in cotton-top tamarins (Hayes and Snowdon, 1990). Captive-born cotton-top tamarins were presented with a live boa constrictor. Adult tamarins failed to show the full array of vocal and behavioral responses that has been described from the field. Furthermore, the animals gave identical responses to a boa constrictor and to a laboratory rat, suggesting a lack of specificity of response to a predatory versus nonpredatory species. Animals born in our colony that were known never to have seen a snake before showed much less intense responses than tamarins born in facilities with outdoor cages. However, young animals in each group did give large numbers of alarm calls, and showed great agitation in the presence of the snake. This suggests that appropriate responses to predators might be well-developed in infant monkeys, but may need social reinforcement throughout development to maintain the intensity and specificity of the response displayed by free-ranging tamarins.

So far the evidence on vocal development suggests that learning and experience do not seem to be necessary for vocal production. However, experience does appear to be critical to the perception of vocalizations, as indicated by the study on pygmy marmosets responding to individual specific features of trill vocalization, and to the proper usage of vocalizations as indicated by the ontogenetic study on trill usage in pygmy marmosets and the responses of naive cotton-top tamarins to predators. However, it is in vocal production that the strong parallels between bird song and human speech exist. The failure to find a similar parallel in vocal development in New World primates is a major gap in developing parallels to human speech and language. However, there are also significant ecological differences between birds and nonhuman primates that may explain these differences (Snowdon and Elowson, 1992).

Summary

The vocal communication systems of New World primates are quite complex, due most likely to the arboreal adaptations that minimize the usefulness of visual communication. New World monkeys exhibit great structural variability in calls. They display subtle variations in call structure that correlate with differences in social context or in function. They display a human-like categorization of their speech sounds, but they are able to show within-category discrimination of calls that belong to familiar conspecifics. New World primates display an ability to alter the acoustic parameters of calls to increase or decrease the potential for sound localization according to how far they perceive themselves to be from other group members. They can combine calls into complex sequences according to syntactic rules, and they display both phonetic and lexical syntax. These monkeys can alternate calls in turn taking behavior in both friendly and aggressive encounters, and there appears to be some evidence of metacommunication. In contrast to birds and humans, there is no strong evidence of a role for learning in vocal production. However, experience seems likely to be important in perception of calls and in developing appropriate usage. There are a large number of parallels between New World primate vocal communication and human speech, suggesting that speech is not special, but that its origins can be traced in New World primates. Furthermore, close study of the social and ecological significance of vocal communication in New World Primates has suggested new directions for research on the structure and function of human speech and language. The study of complex vocal communication in arboreal neotropical primates can be as useful to understanding the origins of speech and language as studies on great species using language analogues.

Acknowledgments

The author's research and the preparation of this chapter were supported by USPHS Grant MH 29,775 and a NIMH Research Scientist Award MH 00,177. I am grateful to Karen B. Strier and Warren G. Kinzey for their critiques of earlier versions of this chapter. The revised chapter was completed in 1991.

5

Sex Differences in the Family Life of Cotton-Top Tamarins: Socioecological Validity in the Laboratory?

W.C. McGREW

Introduction

When Terborgh and Goldizen (1985)reported that wild saddle-backed tamarins (*Saguinus fuscicollis*) in Peru were *not* primarily monogamous, but instead were predominantly polyandrous, the implications reverberated throughout primatology. Long-cherished but possibly simplistic assumptions about the mating system of the Callitrichidae therefore need to be reexamined. For example, *all* of the sociosexual data from Neyman's (1980) pioneering field study of cotton-top tamarins [*S. (o.) oedipus*] could be interpreted as being monogamous or otherwise.

One of the implications had to do with differences in behavior between the sexes. So long as marmosets and tamarins were thought to be monogamous and monomorphic, there was no reason to expect sexual diethism. But nonmonogamy means different mating strategies and life-history trajectories for females and males, and these are likely to express themselves in many aspects of social life, not just in reproduction. The aim of this paper is to explore sex differences in the social lives of a colony of cotton-top tamarins (see Figure 5.1) and to compare these captive subjects with their natural counterparts.

Both field (Neyman, 1977, 1980) and laboratory (McGrew and McLuckie, 1986) studies suggest that cotton-top tamarins may show sex differences in dispersal. In Columbia, Neyman found that transients between wild groups tended to be females, while membership of breeding groups was biased toward males (see Table 5.1). (Neither of these trends reached statistical significance, however.) McGrew and McLuckie (1986) found that eldest daughters were the keenest "prospectors" in

Figure 5.1. Cotton-top tamarins at Stirling. Individual on right is older
 sibling carrying younger twin siblings.

exploring unoccupied "territories" in the author's laboratory colony at
the University of Stirling.

Thus, the most likely hypothesis regarding dispersal is that in nature,
all daughters leave the natal family (that is, emigrate), while at least
some sons seek to stay (that is, are philopatric), though most sons leave
too. The former is likely to result from fierce competition to acquire or to
inherit the family's territory.

However, published findings on behavioral sex differences in groups
of cotton-top tamarins are inconclusive. Cleveland and Snowdon (1984)
found *no* sex differences in infants up to the age of 20 weeks and few in
their caretakers. French and Snowdon (1981) found sex differences on
some measures, but not on others, between members of breeding pairs
in their responses to intruders. Wolters (1978) found many sex differ-

Table 5.1. Neyman's (1980, pp. 89–90) Data on Wild Cotton-Top Tamarins

	Males	Females	Sex unknown
Groups (*N* = 6, July 1975)	15	7	1
Transients (total sum)	11	16	4
Overall population	39	33	18

ences in another colony, including some behavioral patterns done by *only* one sex, such as food-presenting by males. It is likely that variables other than sex have sometimes confused results, so more controlled comparisons are needed. For example, if on average sons stay longer in their natal families than do daughters, then two independent variables, sex and age, may be confounded.

Mating System

To what extent do cotton-top tamarins depart from monogamy, given Terborgh and Goldizen's (1985) report of varied mating systems in the congeneric *S. fuscicollis*, and the inconclusive nature of Neyman's (1980) tentative data? Price and McGrew (1991) sought to investigate this among captive colonies by a survey: 163 questionnaires were sent world-wide to all places known to breed cotton-top tamarins. Seventy-five (46%) were returned, and 15 of these returns reported departures from monogamy (DFM). These DFMs totaled 23 usable cases, when Stirling's five cases were included.

Table 5.2 summarizes the results, which must be considered as hypothetical trends, given the small numbers. While DFMs deliberately set up by human caretakers were evenly distributed among polygyny, polyandry, and polygynandry, those DFMs that emerged spontaneously from the monkeys tended toward polyandry. Similarly, polyandry tended to be the most stable mating arrangement, though breeding success was more equivocal. Overall for DFMs, polyandry seems preferred, followed by polygyny, with polygynandry a distant third choice. This mirrors the general picture from the wild saddle-back tamarins (Goldizen, 1988). To move from this general picture to the specific means focusing on a single, longitudinally studied population, and the rest of this report does so.

Table 5.2 Summary of Main Results of Departures from Monogamy in Captive Cotton-Top Tamarins

Type of group[a]	Number of cases	Origin		Stability	
		Deliberate	*Spontaneous*	*Stable*	*Unstable*
PG	8	3	5	0	6
PA	11	3	8	6	4
PGA	4	3	1	1	3
Totals	23	9	14	7	13

[a]PG, polygyny; PA, polyandry; PGA, polygynandry.

Table 5.3. Composition of Families of Cotton-Top Tamarins Compared in Nature (Neyman, 1980) and in Captivity (Stirling, this study)

	Neyman	*Stirling*
Family size (range)	3–13	4–14
Sex ratio (male/female)		
Population (mean)	1.18	1.37
Breeding groups (mean)	1.87	1.45
Adults/group (mean)	4.2	4.0
(range)	3–6	2–7

Methods

In the Primate Unit of the University of Stirling, each family of cotton-top tamarins lived in a room (33–40 m³) or in a cage (4–8 m³) with adjacent outdoor exercise space (26–32 m³), in visual (but not auditory or olfactory) isolation from neighbors (see Price and McGrew, 1990 for details). Indoor housing had tropical-like temperature (20–25°C), humidity (ca. 50%), and light–dark cycles (adjusted quarterly to mimic the latitude in the wild), plus natural lighting from skylights. Rooms and cages had extensive furnishings, mostly of natural materials, including living plants. The monkeys were fed three times daily, with a varying menu of natural (e.g., fruit) and artificial (e.g., monkey chow) foods. Husbandry was minimally disruptive, and all research was noninvasive. Each individual was identified by a colored disc worn on a necklace or by a dye-mark on the topknot, and was well used to daily observation.

Table 5.3 shows the results of efforts made to maintain socioecological validity, that is, to keep captive monkeys in the same types of groups as found in the wild. Neyman's (1980) findings from the most comprehensive study of the species yet published were compared with the colony at Stirling. The range of family sizes, the mean sex ratio both overall and for breeding groups, and the mean number and range of adults per group were all much like groups found in nature.

Eviction

In an earlier study, McGrew and McLuckie (1986) found that daughters were more often forced to leave their natal families than were sons. However, the sample size was small (*N* = 6), and so the results were

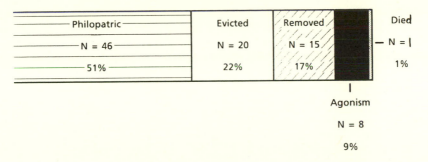

Figure 5.2. Status of 90 offspring after the neonatal stage.

inconclusive. Later we accumulated more cases, and 20 are presented here. Similarly, McGrew and McLuckie (1986) presented data on the status of 28 offspring, with regard to aggression as well as eviction. This dataset also grew much larger ($N = 90$), and is updated here. However, in both cases the data remain tentative, and so are termed *early* data, as few individuals were followed long enough for life-history strategies to have emerged.

Ninety offspring were born to nine females (range: 1–22 offspring per female) and survived to more than 30 days of age at Stirling. Figure 5.2 gives their status in one of five categories:

Just over half (46 or 51%) remained in their natal families at the time of analysis; they ranged from 2 to 61 months old.

The other 44 (49%) were no longer in their families, for one of four reasons:

One (1%) 2-month-old male died suddenly of unknown causes.

Fifteen (17%) offspring were removed for management reasons, *not* prompted by agonism in the family, which in most cases means that they were needed for setting up new breeding pairs.

Eight (9%) offspring were removed for varied reasons related to agonism in the family: Three "bullied" younger siblings, two responded aggressively to the entry of a stepmother, and three died from a bloody fight.

Twenty (22%) offspring were "evicted," that is, victimized by one or more family members to the extent of having to be removed to avoid fatality.

The last two types require further description: In evictions, there is a

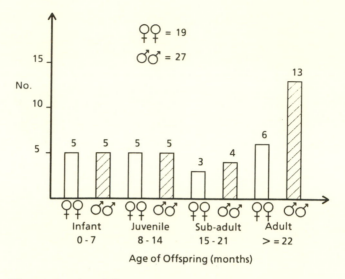

Figure 5.3. Sex differences in philopatry according to age of
offspring. *N* = 46.

progressive, predictable sequence (see Figure 5.3). One family member
targets another, and relations between them worsen. Sometimes the
agonism wanes, and peaceful relations are resumed, but there is a point
of no return beyond which the victim is likely to die if not removed. If
such processes occur in nature, presumably the targeted individual is
temporarily peripheralized or forced to emigrate before being seriously
injured. If so, then fatality in captivity is likely to be an artifact of confinement. (See Hampton et al., 1966, for the first published description
of a fatality from eviction in a group of cotton-top tamarins.)

The eight cases of agonism-related removal were unpredictable and
varied (and sometimes confusing to us). The three cases of bullying
might have ended in evictions if they had been allowed to proceed.
However, we removed the aggressor and not the victim, usually because
we judged the victim to be too young or inexperienced to emigrate.
Thus, human intervention may have cut short an incipient eviction, but
this cannot be known, as the cases were not allowed to continue.

Similarly, when two adult offspring in a family resisted the entry of a
stepmother to replace their recently deceased mother, we removed
them. They might have repelled her, or she them, but for management
purposes the quarreling was stopped by removing the offspring. Finally,
the fatal fighting flared up over a weekend when human presence was
minimal, so no details are known. It may have been an eviction that

Table 5.4. Three Ways of Calculating a Null Hypothesis of No Difference between the Sexes

Type	Males (%)	Females (%)	Definition
a. Secondary sex ratio	50	50	Expected sex ratio at birth
b. Tertiary sex ratio	53	47	Actual sex ratio of offspring surviving to 30+ days
c. Tenure	56	44	Actual cumulative residence in natal family

went drastically wrong, but this case of three deaths in less than 48 hours was unique in our colony and remained unexplained.

To test for sex differences requires a null hypothesis of no difference between the sexes. Often this is assumed to be a 50:50 probability based on the expected evolutionarily stable strategy for a sexually reproducing organism (see type a in Table 5.4). However, there was an actual (though statistically nonsignificant) departure from a balanced tertiary sex ratio (b) in our population of post-neonatal survivors: Of the 90, 48 were males and 42 were females. Further, there was an even greater actual (though still statistically nonsignificant) departure in terms of the two sexes' overall accumulated residence before departure (c). The 48 males were resident for a total of 1,242 (56%) "monkey-months." The last is the most specific probability allowing the most sensitive null hypothesis for this population.

There was no sign of departure from randomness for the 15 cases of removal for management purposes, for the 8 cases of agonism-related removal, or for the 46 individuals still living in their natal families (i.e., philopatric). All observed versus expected proportions are close to identical (see Table 5.5).

Table 5.5 Testing for Sex Differences According to Four Categories of Residence Status of 89 Offspring

	N	Males		Females	
		Expected	Observed	Expected	Observed
Philopatric	46	26	27	20	19
Removal	15	8	7	7	8
Agonism	8	5	5	3	3
Eviction	20	11	8	9	12

Table 5.6. Testing for Age Differences between the Sexes According to Four Categories of Residence Status

		Age (months)					
		Males			Females		
	N	*Mean*	*(SE)*	*Range*	*Mean*	*(SE)*	*Range*
Philopatric	46	24	(3.5)	2–61	16	(3.4)	2–61
Removal	15	35	(3.8)	21–52	36	(3.5)	20–47
Agonism	8	30	(6.2)	20–53	30	(1.3)	27–31
Evicted	20	27	(3.7)	12–40	23	(2.2)	16–38

The exception to this occurs in the 20 cases of eviction: There is a statistically nonsignificant trend for daughters to be evicted more than sons ($\chi^2 = 1.64$, $df = 1$, $p > 0.10$, one-tailed).

The same data were examined for age differences (see Table 5.6). It can be seen by inspection that there were no differences in the ages at which offspring were removed, either for management reasons or as a result of various types of agonism.

Similarly, there was no real difference in the ages at which sons and daughters were evicted. (One aberrant eviction of a juvenile son aged 12 months prevented this sex difference from being a strong trend. If excluded, then $n_1 = 7$ males, $n_2 = 12$ females, $U = 22$, $p < 0.10$, one-tailed, Mann–Whitney U test.)

Finally, it appears from the mean values in Table 5.6 that philopatric sons were older than were daughters. Figure 5.3 shows that this applied only to adult offspring, as there was no difference between the sexes in numbers of infants, juveniles, and subadults still in their natal families. However, this apparent sex difference among adult offspring proved to be only a nonsignificant trend. There were no more adult sons left at homethan adult daughters, and the philopatric adult sons were no older than their philopatric sisters. (One aberrantly philopatric daughter 61 months old prevented the latter sex difference from being a statistically significant one. If excluded, then $n_1 = 5$ females, $n_2 = 13$ males, $U = 12.5$, $p < 0.05$, one-tailed.)

Who did the evicting? The numbers of cases are too few for statistical testing, but some patterns emerge. First, fathers took no part in evictions, and, on the contrary, they studiously seemed to avoid quarreling as if it did not concern them.

Eviction

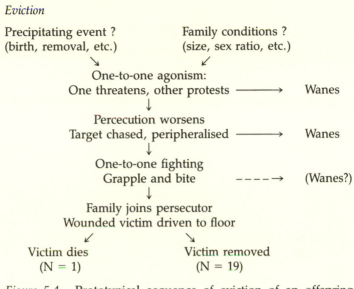

Figure 5.4. Prototypical sequence of eviction of an offspring from a family. *N* = 20.

Second, with one exception, mothers took part in the evictions of daughters but not sons. (The one exception, when a mother alone clearly expelled her 12-month-old son, was unique in our colony. It was thus doubly bizarre.)

Third, sons were more active than daughters in evicting their siblings. The only daughter to have been a principal persecutor was also the only daughter to remain beyond 4 years in her natal family. Well past the age of 5 years, she behaved in some ways like a son and not like a daughter.

Fourth, when offspring were evicted by their siblings, it seemed more likely than one would expect by chance that the instigator was their twin. It seemed to be more likely with twin brothers than with twin sisters or with twin sister and brother.

Finally, in many cases, by the time the process of eviction was completed, two or more family members led the persecution (see Figure 5.4). Thus, individually initiated actions became familial ones, and this was true both of maternally and fraternally led evictions.

Overall, there were consistent trends for families to be biased toward male membership and the (forced) emigration to be biased toward females. These trends mirror those found by Neyman (1980) in the wild for familial constituence and transients.

Social Life of Offspring

Given the likely sex differences in philopatry and dispersal, corresponding sex differences were expected in offspring before departure from their natal families. As the more philopatric sex, males were expected to be more intimately involved in family life than females. Thus, sons were predicted to show friendly behavioral patterns more often. Conversely, being less philopatric on average, daughters were expected to stay at greater distances from other family members than did sons.

To test this in a more controlled way than in previous studies, pairs of sister–brother twins were compared. Thus, almost all other variables such as age, family number and composition, cage size and features, birth order, and parental parity were held constant except for the variable at issue, which was sex.

Ten pairs of female–male twins from seven nuclear families were watched; the twins ranged in age from 14 to 217 weeks (that is, from infancy to adulthood). Each of the 20 subjects was the focus of two observational sessions within a fortnight. Each session lasted until 50 minutes of unobscured observation was obtained. For social behavior, one/zero scores were noted through each minute; for spacing, point-samples were taken on the minute. The resulting data were pooled to give a set of 100 data points for each subject.

Four categories of social behavior were noted: *play* (= wrestle, grapple, mouth, tug at, chase, etc.); *huddle* (= rest without moving, in contact with another monkey); *groom* (= fingers or mouth repetitively touch the body surface of another monkey); and *groomed* (= is groomed by another monkey). Though nonexhaustive, these patterns are the ones most often seen in social relations, and the definitions match those of other studies (e.g., Cleveland and Snowdon, 1984).

Five inclusive categories of "nearest-neighbor" spacing were recorded in terms of distance in centimeters from the focal subject to the closest other member of the family. The categories were zero (= touching), < = 10 cm, < = 50 cm, < = 100 cm, > 100 cm. (These are absolute rather than relative measures of distance, and the monkeys lived in different-sized rooms or cages. However, all comparisons are *within* pairs of twins rather than *across* pairs, so interval level data are justified.)

As predicted, sons showed more friendly social relations than did daughters (see Table 5.7). For all four categories of behavior, the brothers' mean scores exceeded those of their twin sisters. However, only one of these differences, in *play*, reached statistical significance.

Figure 5.5 gives the results for the data on proximity, and shows that sons showed more intimate spacing than did daughters, who stayed further away from others. The overall distribution of nearest-neighbor

Table 5.7. Sex Differences in Four Patterns of Affiliation
in 10 Pairs of Female–Male Twin Offspring

	Males		Females		
	$\bar{X}^a$	SE	$\bar{X}$	SE	p^b
Play	32.0	(5.4)	23.6	(5.4)	<0.025
Huddle	13.1	(3.0)	7.9	(3.4)	>0.05
Groom	6.6	(2.3)	4.3	(2.0)	>0.05
Groomed	5.8	(2.1)	3.7	(1.1)	>0.05

[a]Mean one–zero scores.
[b]Wilcoxon matched-pairs, signed-ranks test, $N = 10$, one-tailed.

distances was tested statistically by using the differences between scores
for each pair of twins at each category of distance. This gave a statistically significant sex difference (Friedman two-way analysis of variance,
$N = 10$, $k = 5$, $\chi^2 = 14.44$, $p < 0.01$). To specify the sex differences, pair-
wise comparisons were done for each of the five categories of differences: Males were more often in contact with others than were females

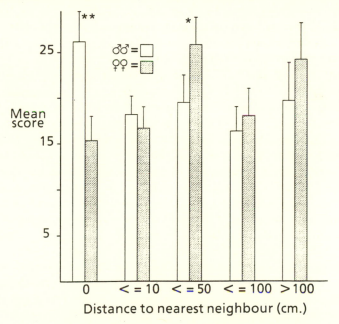

Figure 5.5. Sex differences in distance to "nearest-
neighbor" for 10 pairs of opposite-sexed twins.
Mean and standard error. $^*p < 0.05$; $^{**}p > 0.01$.

(Wilcoxon test, $N = 10$, $T = 1$, $p < 0.005$, one-tailed), while females were more often between 10 and 50 cm from others than were males (Wilcoxon, $N = 10$, $T = 9$, $p < 0.05$, one-tailed). For the other three categories, the apparent sex differences were statistically nonsignificant.

The two types of measures are congruent, in that males spent more time in contact, engaged in friendly social interaction, while females more often were further from others and showed less social involvement. It is tempting to interpret this as sons investing in social bonds that, in some cases at least, might benefit them over a lifetime. What is not clear from these data is whether this social investment is directed to parents or to siblings. Further study was conducted at Stirling (Moore, unpublished data) of the extent and direction of this socializing, but unfortunately there are no comparable data from wild cotton-top tamarins yet available.

Conclusions

Both types of data reported here, however preliminary, notably fail to reject the hypothesis of sex differences in the social lives of cotton-top tamarins. However, many questions remain to be tackled. For example, it seems likely that age (or sibling birth-rank, as the two variables of age and birth order are hard to disentangle) is important, given the complexities of life-history trajectories. A simple prediction is that the differences between the social lives of opposite-sexed twins will increase as they grow older, especially after sexual maturity, when (at least in principle) daughters could leave to start their families and (dominant?) sons could usurp or inherit the role of breeding male at home.

These results reinforce the view that generalizations about social behavior across the callitrichid radiation of about 20 species are likely to founder. Within the same genus as the cotton-top tamarin, *S. fuscicollis* in nature shows a tendency for male-biased emigration and a similar significant sex difference for immigration (Goldizen and Terborgh, 1986). A sibling species to the cotton-top tamarin, *S. (o.) geoffroyi*, studied in Panama, shows no sex difference in dispersal (Dawson, 1977). On the other hand, tamarins of another genus, *Leontopithecus rosalia*, show patterns of familial aggression and eviction virtually identical to cotton-top tamarins (Inglett et al., 1989). Such cross-taxa differences in cooperatively breeding mammals should come as no surprise, given that cooperatively breeding birds of similar taxonomic relationships show both across- and within-species variation in behavior (e.g., Strahl and Brown, 1987).

Finally, to what extent can the question posed in the title be answered? Ecologically, a campus colony in a Scottish university can never emulate a tropical forest in Colombia. At best one can hope to simulate some key features of the physical and biotic environment (Price and McGrew, 1990). Socially, if the cotton-top tamarins are kept in largely self-regulated groups that approximate in size and age–sex composition those found in nature, the monkeys behave naturalistically. That is, they show the full social repertoire, patterned in ways that both make sense in terms of evolutionary theory and lead to impressive survivorship, longevity, reproductive success, and active and stimulating social lives.

Acknowledgments

I thank E. Halloren and I. Rodgerson for their careful husbandry of the monkeys, and undergraduates C. Beattie, L. Begg, L. Beveridge, P. Bors, J. Brown, E. Coburn, S. Downie, A. Fletcher, L. Frazer, P. Graham, L. Hurwitz, G. Kirkwood, J. Kuhn, W. Lightfoot, S. McCabe, and C. McInnes for help in data collection. S. Evans supplied extra data on evictions, and she, A. Ensminger, W. Kinzey, K. Moore, and E. Price made helpful comments on an earlier version of the manuscript, as did K. Valley, who made the figures.

6

Subtle Cues of Social Relations in Male Muriqui Monkeys (Brachyteles arachnoides)

KAREN B. STRIER

Introduction

Primate social relationships are often described in terms of their aggressive and affiliative components. Such reductionist approaches are useful in interpreting the functional significance of agonistic and affiliative interactions, but they tend to oversimplify the complex compromises that social relationships reflect (de Waal, 1986). Agonistic interactions are generally attributed to competition over access to limited resources. Evolutionary theory predicts that primates will compete with one another for resources, such as food and mates, that directly affect reproductive success (Wrangham, 1979a). In many cases, the outcomes of such competitive contests are so consistent that it is possible to construct linear dominance hierarchies. In the majority of Neotropical primates, however, agonistically determined dominance relationships occur less frequently than among the better-studied Old World species (Strier, 1990a), and dominance relationships may be difficult to detect or nonexistent.

Affiliative interactions are also assumed to enhance the fitness of the participants. A number of studies have shown that individuals who maintain close spatial proximity are also more likely to groom one another and come to one anothers' aid in agonistic conflicts with other group members (Seyfarth, 1977; Smuts, 1985). The expression of affiliative bonds is most striking among closely related individuals who may also be one anothers' greatest competitors. In the matrifocal Japanese macaque, aggressive threats occur most often between close female kin. Yet, these same females spend more time in proximity, have the highest grooming rates, and are most likely to support one another against nonkin (Kurland, 1977). Presumably the advantages to inclusive fitness (Hamilton, 1964) outweigh the costs that greater proximity imposes.

A similar compromise between aggressive competition and affiliative behavior is observed among male chimpanzees. The patrifocal society of chimpanzees is believed to facilitate cooperation between males in patrols of a community range that encompasses the core areas of independently foraging females (Wrangham, 1979b). Related males compete agonistically for copulations with estrous females (Tutin, 1979), and can be ranked by dominance status (Bygott, 1979). However, males also have strong affiliative relationships that are evident in their close spatial proximity and frequent grooming bouts (Goodall, 1986). In one captive colony, the frequency of grooming between males increased in the presence of an estrous female. As de Waal (1986, p. 470) notes, "male–male grooming thus appeared to be the 'price' of an undisturbed mating session."

Although male baboons, who are unrelated to one another, form affiliative coalitions (Packer, 1979; Smuts and Watanabe, 1990), they do not exhibit a comparable repertory of affiliative behaviors that reduces tension in competitive situations. Kinship, it appears, provides an important impetus for the evolution of behaviors that mediate aggressive tendencies, and may thus be fundamental to considerations of primate social relationships.

The lower frequency of occurrence of both affiliative interactions such as grooming and agonistically determined dominance hierarchies among Neotropical monkeys may reflect the fact that in many of these species, both males and females disperse from their natal groups, and are thus unrelated to other adults (Strier, 1990a). In the Brazilian woolly spider monkey, or muriqui (*Brachyteles arachnoides*), males remain in their natal groups while females migrate (Strier, 1987a, 1991a). Kinship among male muriquis may inhibit the expression of aggression, and facilitate the formation of strong affiliative bonds that resemble those found among males in other patrilineal nonhuman and human primate societies.

During over 1,200 contact hours with a group of 23–26 muriquis, only 31 agonistic interactions between group members were observed (Strier, 1990a, 1992a). These interactions never occurred over access to obvious resources such as food or mates, and the same pair of individuals was never observed to interact more than once. While food abundance may account for the lack of direct feeding competition, it is difficult to account for the absence of overt contests between males for access to sexually receptive females.

Such low levels of aggression among muriquis are particularly noteworthy because muriquis spend over 30% of their time within 1 m, and over 50% of their time within 5 m proximity to at least one other individual (Strier, 1990a). Hypotheses about the constraints on muriqui aggression are presented elsewhere (Strier, 1992a). Here I examine the dynamics of muriqui spatial relationships for insights into the social relationships between males.

Study Site and Methods

The data presented in this paper were collected during a 14-month study of one muriqui group from June 1983 through July 1984 at Fazenda Montes Claros, a privately owned forest located at 19° 50'S, 41° 50'W in Minas Gerais, Brazil. During this period, the group grew from 23 to 26 individuals due to two immigrations by subadult females and one birth (Strier, 1987a). Muriquis at this site have distinct facial markings that made it possible to recognize them individually in most instances.

Systematic data on muriqui spatial relationships were obtained from scan samples and focal animal samples of recognized individuals (Altmann, 1974). The scan samples were conducted at 15-min intervals throughout the study. During each scan sample, the identity, activity, and nearest neighbors were recorded for all individuals in view (see Strier, 1987b for details on activities).

Nearest neighbors were distinguished by their proximity to the scanned individual within the following categories, in contact, <1 m, and <5 m. Only nearest neighbors within the closest proximity category were recorded, such that if A and B were in contact, and C was within 5 m of B, A and B would be scored as one another's nearest neighbors, while B would be scored as C's nearest neighbor. In these analyses, nearest neighbors from different proximity categories are assigned equal value (Strier, 1986a), and are considered to be in proximity if they were within 5 m of one another.

Focal animal samples of adult males were conducted on preselected days from August 1983 through June 1984. Focal subjects were chosen in sequence from a predetermined list; however, if the individual slotted could not be located within 5 minutes, the next individual on the list was sought. Focal animal samples varied in duration from 2 to 247 minutes (mean = 50 minutes), and were terminated whenever the focal male was lost to view for 5 continuous minutes. At 1-minute intervals, the activity of the focal subject, the distance between him and his nearest neighbors, and the identity of his nearest neighbors were recorded as described above. I also kept track of which individual in each dyad involving the focal male was responsible for approaching and terminating proximity.

Focal animal samples were strongly biased against traveling individuals because poor visibility and steep terrain made it difficult to maintain continuous visual contact at these times. Focal samples are used, nevertheless, to provide quantitative data on the dynamics of muriqui spatial relationships. For these purposes, and because males always associate in the same group, proximity data are assumed to be independent of the focal samples in which they were obtained. Comparisons are based on the proportion of time each male was observed to control for differences in the number and duration of focal samples for each male.

Using Hinde's (1977) method for analyzing partner preferences, it was possible to determine which individual in each nearest neighbor dyad in the focal samples was responsible for maintaining the relationship. The total number of approaches by male *A* was divided by the total approaches between both *A* and *B*. The proportion of departures from proximity, calculated in a similar way, was subtracted from the proportion of approaches. In this example, a positive score indicates that *A* was responsible for maintaining proximity with *B*; a negative score indicates that proximity was maintained by *B*.

Results

Spatial Relationships

Muriqui spatial relationships were examined from nearest-neighbor data obtained from over 4,300 scan samples involving more than 17,000 individual observations and from 3,644 male focal minutes obtained from 73 focal male samples. In general, adult males and females did not differ from one another in the proportion of time spent in proximity to members of their own age–sex classes. Adult males were one anothers' nearest neighbors in 69% of the male proximity scan samples ($n = 3,564$). Although adult females were one anothers' nearest neighbors in only 41% of the cases ($n = 3,523$), when scan samples involving semidependent offspring as nearest neighbors are excluded ($n = 1,343$), adult females were one anothers' nearest neighbors in 69.5% of the remaining 2,180 cases.

The nearest neighbors of each adult male were neither randomly nor evenly distributed across the group. Rather, each male was observed in proximity to only a few other individuals during a majority of the samples. Other adult males accounted for at least 10% of each male's spatial associates, although the identities of these "top-ranking" nearest neighbors reveal interesting patterns.

Table 6.1. Male Popularity Scores

Adult male	Proximity/total minutes[a]	Popularity
SC	718/1283	56
CL	246/343	72
MR	408/378	108
IV	1016/967	105
MK	83/156	53
PR	58/213	27

[a]Because the proximity data were broken down into their dyadic components, it is possible for an individual who had more than one nearest neighbor at any time during a focal sample to obtain a popularity score that exceeds 100.

Individual Variation in Sociality

Differences in the proportion of time individual males were observed in proximity are evident from "popularity scores," calculated by dividing the number of minutes each focal male was observed in proximity by the total number of focal minutes for that male. Two adult males, IV and MR, had exceptionally high popularity scores (Table 6.1), but only IV was a top ranking nearest neighbor of all five of the other males. Differences in male popularity do not appear to be related to the number of focal minutes different males were observed (Spearman $r_s = 0.54$, $n = 6$, $p > 0.05$).

Individual Variation in Maintaining Proximity

There were also differences in the efforts individuals made to initiate or avoid proximity with others. "Initiation scores" were calculated by summing all approaches and departures by each adult male to other adult males over all approaches and departures involving that male and others. The difference, $A\% - D\%$, provides an overall index of responsibility between males (Table 6.2).

Male popularity scores and initiation scores were not correlated (Spearman $r_s = 0.49$, $n = 6$, $p > 0.05$). Of the two most popular males, IV had the second lowest initiation score while MR had the highest. IV appears to have been popular largely through the efforts of others, whereas MR was responsible for his popularity. Conversely, the male who was third ranking in popularity, SC, had the fifth ranking initiation

Table 6.2.   Male Initiation Scores[a]

Adult male	Initiation score
SC	+0.21
CL	+0.01
MR	+0.22
IV	−0.22
MK	+0.13
PR	−0.15

[a]See text for method of calculation.

score. This male, it appears, took an active role in initiating spatial proximity with males which avoided him.

Reciprocity and Responsibility

The proximity data from the scan samples indicate few reciprocal relationships in which two individuals were among one anothers' top ranking nearest neighbors. For nearly all of these nearest neighbor dyads, responsibility was strongly asymmetrical (Table 6.3). Maintaining proximity with others was also highly specific for any particular male. For example, MR was among the top ranking nearest neighbors of both IV and CL, yet while his relationship with IV was due to his own efforts, his relationship with CL was due to CL.

Table 6.3.   Responsibility in Reciprocal Relationships

Reciprocal dyads	$A\% - D\%$ [a]
SC–IV	0.29
SC–CL	0.08
CL–MR	0.18
MR–IV	1.00
IV–CL	1.00

[a]See text for method of calculation.

Discussion

Differentiated Relationships

Although the spatial relationships of adult male muriquis consistently favored other adult males, individual males did not distribute their associations evenly with one another. Rather, each male associated preferentially with only a few other males. Individual males were among one another's top ranking nearest neighbors in only five of the 15 possible adult male dyads. The majority of spatial associations were not reciprocal, and the responsibility for maintaining proximity was asymmetrical.

These data suggest that social relationships among male muriquis are strongly differentiated. Some males showed more initiative in maintaining proximity, while other males appeared to be more preferred spatial associates. It is difficult, nonetheless, to interpret the significance of such variation among males, particularly since the near absence of intragroup agonistic interactions among muriquis precludes the application of more traditional measures of social relationships.

Differences in male initiative and popularity correspond to other differences in male behavior, which may provide indirect insights into their social relationships. The male who received the highest score for initiating proximity (SC) also engaged in twice as many embraces as any other male (Strier, 1992a). His light coloring and less robust build, compared to his present (1991) appearance and that of other males at the time, suggest that he was the youngest adult male during the study period. Perhaps his greater initiative and participation in affiliative interactions are characteristic of young males seeking to integrate into the adult social network.

The spatial relationships of both subadult males (SO and CY) are consistent with the suggestion that young males seek proximity to adult males (Strier, 1992a). However, whenever either subadult was among the top ranking nearest neighbors of an adult male, it was the subadult male who was responsible for maintaining the association. During an expanded study between 1986 and 1987, when both of these males were classified as young adults, they were still almost entirely responsible for initiating their affiliative contacts with the other adult males (Mendes, 1990). By the time both of these males were fully adult, they had become the targets of proximity by younger resident males as well as other adults.

Further systematic data are clearly needed before conclusions about the ontogeny of social relationships among male muriquis can be advanced. However, similar efforts by immature males to associate with adult males have also been reported in chimpanzees (Goodall, 1986).

It is tempting to infer that the males with the highest popularity scores, IV and MR, also had higher dominance status. Together these two males accounted for over 56% of the copulations observed during this study period. Although males never interrupted the copulations of one another, at least one of these two males was a potential father of the four infants conceived during this time (Strier, 1987d, 1992a).

Both popularity and reproductive success have been correlated with dominance rank in other primates that exhibit more conventional ago-nistic behavior. Dominant individuals are groomed more than subordi-nates in both male-bonded (Simpson, 1973) and female-bonded species (Seyfarth, 1977). Dominant males may also copulate with sexually recep-tive females more than subordinates (see review in Gray, 1985). Because muriquis do not compete overtly for access to receptive females (Milton, 1985b; Strier, 1986b, 1992a), independent assessments of dominance rank and reproductive opportunities cannot be made. It is striking, nonetheless, that the males who were the most preferred spatial associ-ates were also the most sexually active.

Other males may maintain proximity with more experienced or suc-cessful males to monitor the reproductive activities of the latter (Strier, 1992a). Further research is clearly needed to evaluate whether male pop-ularity and experience are a function of age or individual competitive ability, and whether relationships between male muriquis are stable over time (Mendes, 1990; Strier et al., 1993).

In a systematic study conducted 3 years after the present one, Men-des (1990) found some changes in the popularity and initiative shown by the same adult males described here. Although IV was still the most popular male and still received the highest proportion of affiliative con-tacts, by Mendes' study period, MR had dropped to middle-ranking positions in his frequency as a nearest neighbor of other males and in the proportion of affiliative contacts he initiated. Thus, there is evidence that male relationships change over time.

Importance of Male Associations

The strong affiliative associations between male muriquis are consis-tent with demographic evidence, accumulated over a 9½-year period, that males are closely related to one another. During this time, natal male residency has remained constant, while all but one natal female emigrated before reaching reproductive maturity (Strier, 1991a). Male philopatry has been attributed to the importance of cooperation in re-source competition with other groups of related males (Wrangham,

1979a, 1980). Kin are assumed to be more reliable allies than nonkin because of their shared genetic interests (Hamilton, 1964).

Because muriquis during this study period associated in a cohesive group (Strier, 1989) and occupied a home range that overlapped extensively with that of a neighboring group (Strier, 1987d), male cooperation and kinship do not appear to be related to defense of an area. However, cooperation among related males may be important in preventing unrelated males from associating with group females. Intergroup encounters have changed dramatically over the years, from active avoidance by group males (Strier, 1992a) and prolonged, vocally mediated confrontations (Strier, 1986a), to temporary "takeovers" in which males from the other muriqui group in the forest associate exclusively with group females (Strier et al., 1993). Although the reasons for these changes are still unclear, the *size* of adult male kin groups is likely to be as important in determining the outcome of intergroup encounters in muriquis as it has proven to be in chimpanzees (Bygott, 1979; Nishida et al., 1985). Indeed, the frequency of incursions by males from an adjacent group increased following the disappearances and presumed deaths of two adult males, MK and MR, in 1988 (Strier et al., 1993). If the size of male kin groups is positively related to successful monopolization of females, then adult male muriquis may tolerate their younger male relatives because their future participation in intergroup competition for association with females is essential.

Relevance to Humans

The strong associations among related male muriquis resemble those observed among human, as well as nonhuman primate, males living in patrilineal societies. With the exception of their wives, the closest associates of Efe pygmy men are other closely related men (Bailey and Aunger, 1989). The highest frequencies of affiliative interactions occur among Ye'kwana men who are related to one another by at least one fourth (Hames, 1979a). Yanomamo men preferentially support close male kin in both intra- and intervillage conflicts (Chagnon and Bugos, 1979; Chagnon, 1981, 1988). Among Grand Anse villagers, male kinship is associated with lower frequencies of agonistic interactions over mates (Flinn, 1988).

Human kinship may be one way that men can assure themselves of reliable allies who will assist them in mate competition (Irons, 1981). Male kin aid one another in the recruitment of wives and in protecting females from harassment or capture by other groups of related men (Chagnon, 1988; Bailey and Aunger, 1989). According to Hames (1979a), residence is the proximate mechanism that explains how related males can interact preferentially.

The apparent parallels in both the relationships of male muriquis, chimpanzees, and humans, and the explanations for why such cooperative alliances are important in intergroup competition over females, emphasize a continuum in male behavior. Due to a historical research bias focusing on the semiterrestrial Old World monkeys (Southwick and Smith, 1986; Anderson, this volume; Strier, 1994a), in which males are generally unrelated and highly competitive (Wrangham, 1980), such continuums involving affiliations among related males have often been overlooked. However, when we broaden our comparative base to include the New World monkeys, it is the traditional perceptions of aggressive male relationships that appear to be exceptional (Strier, 1990a, 1994b).

Male philopatry and affiliative relationships between related males are not as unusual as was previously thought (Moore, 1984; Strier, 1994a). Male philopatry occurs in other New World primates such as spider monkeys (Symington, 1987a) and squirrel monkeys, where it is also coincidental with nonaggression between males (Boinski, 1987a). Extrapolating from muriquis (or any other nonhuman primate) to humans is as problematic as any referential model of behavior (sensu Tooby and DeVore, 1987). Nevertheless, such comparisons support the utility of common evolutionary principles to understanding social relationships across nonhuman and human primates.

Acknowledgments

I am grateful to the Brazilian government and CNPq for permission to conduct research in Brazil, and to Dr. Celio Valle for his essential support. The research described here was financed by NSF Grants BNS 8305322, BNS 8619442, and BNS 8959298, the Fulbright Foundation, Grant 213 from the Joseph Henry Fund of NAS, the L.S.B. Leakey Foundation, and the World Wildlife Fund. E. Veado, F.D. Mendes, and J. Rimoli contributed to the long-term demographic data. K. Bassler and M. Steele assisted in manuscript preparation. A version of this paper was presented at the 87th annual meeting of the American Anthropological Association, and I thank W. Kinzey for inviting me to participate in his symposium, and for his and C. Snowdon's valuable comments on an earlier draft of this manuscript.

7

The Influence of New World Mating and Rearing Systems on Theories about Old World Primates

CONNIE M. ANDERSON

Introduction

Fifteen years ago I went to Africa to study baboons because I wanted to shed some light on the evolution of hominid mating and rearing systems. My foreparents, the founders of anthropological field studies of primates, Sherry Washburn, Irv De Vore, and Ron Hall (e.g., Washburn and De Vore, 1961; Hall, 1965), studied the African terrestrial, savannah-dwelling primates as models for early humans, and I followed in their footsteps both literally and figuratively. After 8 years in Africa, however, I began to feel that I should have gone to South America instead, because most of the studies relevant to the problems on which I was working seemed to be coming from there.

I had thought that baboons and chimpanzees were the two most valuable species for early hominid reconstructions, baboons because of their australopithecine habitat and ecological niche (Washburn and De Vore, 1961), and chimpanzees because of their genetic similarity to ourselves (Hasegawa, 1990). The more we learn about primate variability, though, the more we realize that we need to look at as many species under as many different conditions as possible before we can draw conclusions of general applicability (see Kinzey, 1987, for further discussion of this point). I will address the relationship of the New World studies to understanding the population of baboons I have studied, and to understanding australopithecines, for whom I have been using baboons as a surrogate.

Among evolutionary biologists, primatologists are in the enviable position of being able to work with an order uniquely valuable not only for its similarity to ourselves, but also for the large number of different

119

species, different adaptive radiations, and different social types it contains (Crockett, 1987). When we include the New World species, we can concentrate on ecological determinants or constraints on behavior across the order, because the confounding effect of phylogenetic closeness among Old World species is alleviated (see also Strier, this volume, for effects of ecological constraints on inhibiting aggression in woolly spider monkeys).

I am surprised at how often I still hear the phrase "Baboons just don't *do* that," or, especially, "Female/male baboons don't do that," since it is now well-established that variability within the genus *Papio* approaches that within the genus *Homo*. Primatologists working in the New World have repeatedly demonstrated the importance of primate adaptability, where variation due to environmental and/or social conditions is frequently greater within a single species than it is between two different species (e.g., Crockett, 1987; Kinzey, 1986).

Research in the New World

The New World primates have sometimes been described as more primitive or less intelligent or adaptable than Old World species (e.g., Bramblett, 1976; Fleagle, 1988; Szalay and Delson, 1979), largely due to their unspecialized dentition and features of their cranial structure. Intraspecific variation in New World species is at least as great as and probably greater than that in Old World species. Even though variation among the Old World species is increased by the addition of terrestrial species, whereas all New World species are arboreal, the greater seasonal variation in New World than in Old World tropical forests (Terborgh, 1983) apparently results in as much or more New World variation as the arboreal/terrestrial continuum produces in the Old World.

The first successful field study of any wild primate focused on a New World species. The howler monkeys of Barro Colorado Island were studied by Carpenter in 1931 and 1932. Despite his important observations, the New World primates were considered largely irrelevant by most anthropologists for the next 40 years. The New World species were too difficult to study in their arboreal habitat, in addition to being phylogenetically remote, physically diminutive, and environmentally too restricted to be judged of any importance for understanding Old World species.

New World primates were also felt to be unimportant initially for some of the very reasons they are considered especially *interesting* today: several species are monogamous, others are even polyandrous; males often invest in young; and many species have a very restricted distribution. In the academic climate of the 1950s and 1960s these qualities made them seem irrelevant, since humans were believed to have begun as a polygynous species with irresponsible fathers (e.g., Coon, 1954), much like the current model that sociobiologists began to popularize in the late 1970s. Now we see that the picture is much more complicated, partly because of persistent efforts by researchers in the New World to document the wide range of variability in all primate species and partly because of the frequent occurrence of male investment observed throughout the primate order.

Many of us have begun to feel a sense of urgency about New World studies by now, since many of them are highly endangered (Mittermeier et al., 1989). This is probably because they are more arboreal than most Old World species (Robinson and Janson, 1987); a tropical rainforest is incredibly difficult to duplicate in a zoo. Baboons will probably be available for study as long as there are surviving primatologists, while a sort of "salvage primatology" sadly seems appropriate for the New World. Those of us who have graduate students should encourage them to go out and study some New World species before the chance to understand its social behavior and ecology is gone forever.

Although I, for one, had not bothered to read many articles on New World species except for historical interest, things changed when I began to write my dissertation. Then I found that most of the observations of several behaviors I needed to explain at my baboon field site had already been reported and explained among New World species. Some of the exciting hypotheses that have come out of New World studies and are applicable to the study of both *Papio* and *Australopithecus* follow.

It would seem that among New World primates the environment is a major determinant of group size, and group size in turn largely determines the social system (e.g., Kinzey, 1986; Terborgh, 1986). If this is true, we should eventually be able to predict the basics of australopithecine social organization from preserved environmental and demographic parameters, the original "promise of primatology" articulated by Washburn in the 1950s (Washburn, 1951, 1973). Terborgh (1986) has shown that this principle applies to baboons and other Old World species, where only one social system is observed for each group size. These causal connections have been very useful to me in understanding why the social organization of baboons at my site, Suikerbosrand, differs from those of populations in environments that select for larger or less divisible groups.

The close correlations among food, locomotion, and social behavior that have been found for the New World species (Kinzey, 1986) have especial promise for anthropologists. We can recover much information on both diet and locomotion from fossil forms, including the australopithecines. When we know where the terrestrial, bipedal omnivores should be plugged into this continuum, we have hope of determining at least some of their basic social behavior.

The effect of competition on individual success in feeding has periodically been touted by some Old World primatologists as an important determinant for a particular species under study (e.g., Wrangham, 1979b), while in the New World this has been shown to be important over and over again (Terborgh, 1986). Another consistent determinant of New World, and by extension all, primate social organization is the effect of predation (Terborgh, 1986). The weight of this independent evidence has helped to resolve an Old World controversy about the influence of predation (e.g., Anderson, 1986; Cheney et al., 1988). At Suikerbosrand where predation was absent, the baboons foraged in small family units averaging 12 individuals in the winter (Anderson, 1981). As Terborgh (1983) demonstrated consistently for New World species, individuals forage alone or in small family groups where predation is low or absent.

Furthermore, the New World studies have provided another demonstration that the largest of the arboreal species repeatedly divide into small foraging groups that fuse together again. Terborgh (1986) cites chimpanzees in Africa, orangutans in Asia, and spider, woolly spider and woolly monkeys in South America. I would add that this is relevant to terrestrial species as well: chacma baboons, the largest ground-dwelling monkeys, chronically appear to exhibit fission and fusion throughout the winter and under various other conditions (Anderson, 1981). Mandrills, too, appear to regularly subdivide and reaggregate on a seasonal basis (Hoshino et al., 1984). If body size is one of the primary movers of social evolution (Terborgh, 1983), the New World species are vital because they enable us to carry a regression of body size on behavior out much further than we could with only Old World data points to enter, since the difference between the largest and the smallest species is much less extreme in the Old World (Fleagle, 1988).

In several New World species, it is females who emigrate from their natal group, in addition to or instead of males (Kinzey, 1986; see also Strier, this volume, on woolly spider monkeys). It was when I began reporting female transfer among Suikerbosrand baboons that I was first told that "Baboons don't do that." Female transfer is now recognized for many Old World species as well (see reviews by Moore, 1984, and Wrangham, 1987). Where males transfer, there is greater recognition that females may also transfer, either typically or occasionally, as in baboons (Anderson, 1987; Nash, 1976; Rasmussen, 1981; Stoltz, 1977). This intraspecific variation is one of the ways in which New World species like the howler monkey (Crockett, 1987) have been relevant to understanding female transfer in baboons, apes, modern hunter–gatherers, and probably australopithecines.

Infanticide is another behavior that 10 years ago was unknown in baboons, but by now at least some Old World workers have shown that it does occasionally occur even among baboons (Busse and Hamilton, 1981). In the New World, infanticide appears to be confined to howlers (Izawa and Lozano, 1989; Struhsaker and Leland, 1987). Such a primate taxon in which infanticide is present but rare seems to be a more appropriate model for baboons or for humans than is a taxon where infanticide apparently occurs much more frequently, as in many colobines.

Relevance to Humans

Although Robinson and Janson (1987) stated that large aggregates of several troops like those observed in squirrel and woolly monkey populations do not occur in cercopithecines or colobines, I believe that the largest aggregates of hamadryas and gelada baboons are analogous to these and have parallels among historically known hunter–gatherers. The more genera in which this pattern occurs, the more likely we are to be able to understand its causes and to identify its correlates in early humans, especially if fusion occurs both in arboreal forest-dwellers and in terrestrial steppe- and plateau-dwellers (e.g., Dunbar, 1984; Kummer, 1968). One of the New World species, the bearded saki, may also be relevant to humans, as it appears to be the only other primate species besides our own in which large groups form that may be composed of monogamous pairs (Robinson et al., 1987).

New World observations have influenced the extent to which paternal care, or, more generally, male investment in kin or in unrelated immatures, is believed to occur among baboons or among early or extant humans. Paternal care, and/or fraternal care, has been reported so frequently in the New World (e.g., Goldizen and Terborgh, 1986; Snowdon and Suomi, 1982) that all of us Old World primate researchers have finally begun to accept it as a normal part of the primate repertoire, given certain conditions. We now observe pair-bonding and male investment practically everywhere, even in animals as *macho* as baboons are thought to be.

Male investment occurs for a number of different reasons in New World primates, providing us with several relevant lines of explanation for possible application to early humans. It may be explained in terms of simple Darwinian fitness when the male involved is almost certainly the father of the infant, as in most monogamous, territorial species (e.g., Robinson et al., 1987). In other cases, paternity is uncertain, but related males, mainly possible fathers and older siblings, must help the mother carry large infants, or successful reproduction may be impossible (e.g., Goldizen and Terborgh, 1986; Kleiman, 1977, 1985; Wright, 1984). Among Old World baboons, recent work has emphasized the benefits males may receive from association with infants even if they are unrelated, including protection from attack (e.g., Anderson, 1992; Smith and Whitten, 1988; Smuts, 1985; Strum, 1984). The second and third of these points seem particularly applicable to male investment in humans.

Female choice does not seem to be fully accepted at this point, however. Although frequently observed among New World species, especially capuchins and spider monkeys (Robinson and Janson, 1987), active female choice is sometimes denied for Old World species, including very dimorphic species like the baboons (e.g., Busse and Hamilton, 1981; Smith, 1986), especially by male observers. However, it has been claimed for some populations studied by women (e.g., Anderson, 1992; Smuts, 1985; Strum, 1987).

All known anthropoid social organizations ever described can be found in one or more New World species (Kinzey, 1986; Terborgh, 1986). Some of these appear to be unique among nonhuman primates, and it is these in which anthropologists should be particularly interested because human social organization also includes many rare types, such as polyandry. Among primates, polyandrous breeding systems are found only in humans and in New World species (Sussman and Garber, 1987). The type of functional polyandry that results from the tremendous investment required for each pregnancy in New World tamarins and marmosets is analogous to multiple-male investment networks found among the poor in modern nations (compare Lancaster, 1991, and Stack, 1974, with Sussman and Garber, 1987).

It is noteworthy that no New World species *typically* has one-male groups of one male and several females, though such groups are occasionally observed in almost all primate species (Kinzey, 1986). One-male groups have probably been the most frequently hypothesized organization for extinct hominids, from *Australopithecus* to Neandertals. This is especially true among writers of popular books on human evolution and the presumed human biological heritage (e.g., Barash, 1977; Symons, 1979), even though one-male groups are not typical of either New World monkeys or any Old World ape either. This assumption clearly needs reexamination.

New World primates do show an adaptive complex of features that characterizes both the ethnographically known hunter–gatherers and our closest relatives, chimpanzees and bonobos. These include fission–fusion grouping, with several adult males per group (e.g., spider monkeys, McFarland Symington, 1987; muriquis, Strier, this volume), male cooperation in opposition to other groups (Robinson and Janson, 1987), less intensive and wider-ranging land use by males than by females (e.g., Wrangham, 1979b), reduced sexual dimorphism (Kinzey, 1986), tool use (Beck, 1980; McGrew, 1989), and long interbirth intervals (e.g., Goodall, 1986). It seems probable that all these features are adaptations to the combination of large size, dispersed food, and energetically expensive feeding and locomotion.

If the New World data show anything, they demonstrate that a species is most unlikely to exhibit only one mating or rearing system that remains unresponsive to local circumstances. I feel quite secure in suggesting that australopithecines, like baboons, were organized both in monogamous pairs and also in one-male groups, as well as in yet other forms of mating system. I had been trying to determine what conditions might produce monogamous baboons for several years, as a way to determine whether australopithecines could have been typically monogamous or not, when a number of New World researchers finally worked it out for me. In night monkeys and titis, small feeding patches and male help in rearing young are the main correlates of monogamy (Wright, 1986). However, this is not a new finding: small food-patch size had been offered 25 years ago to explain monogamy in gibbons (Ellefson, 1968). The difference now is that 20 years ago gibbons were thought to be aberrant and irrelevant, partly because of their monogamy, and their small size, so their monogamy was not usually referred to food patch size but was instead seen as just another gibbon peculiarity. But now we know monogamy is not aberrant in primates, and there are so many monogamous species recognized, especially in the New World, that it is much easier to separate the significant causal factors from the background noise. Goldizen and others (Goldizen and Terborgh, 1986) are now specifying exactly the factors leading to monogamy for particular species.

There are other ways in which New World primates are behaviorally more similar to humans than are the Old World monkeys or apes. For one, group sizes of 10 or below are very rare in the Old World (Terborgh, 1983), but have been observed among modern hunter–gatherers for part of the year in all habitats (Bicchieri, 1988). If early hominid group size was as small as this, we would have to look to the New World for sufficient samples to test models predicting other variables related to small group size.

There are more dietary similarities between humans and the New World primates as a group than between us and the Old World primates as a group. In both New World monkeys and early hominids, diets are very diverse (Sussman, 1987), and New World species consume more animal protein than do most Old World anthropoid species (Kinzey,

1986). Sussman (1987; see also Anderson, 1991) has recently provided data to show that early hominids must have had to rely on a certain amount of animal protein. Although most of this protein is provided by insects for New World species, many vertebrates are eaten as well, and the capuchins (genus *Cebus*) are the main predator on small arboreal animals in New World forests (Kinzey, 1986).

Finally, much of our best information on communication among non-human primates in the wild has come from not just one, but several South American species (Kinzey, 1986; Snowdon, this volume). Since most of these species are small enough and arboreal enough to be rather easily studied in captivity, there is a great potential for in-depth under-standing of the language capabilities of nonhuman and early human primates. Surely, whatever marmosets can communicate must also have been within the capabilities of australopithecines! Once again, discover-ing just how similar the New World species are to ourselves gives us valuable insight into our own distant ancestors, whose abilities often must be revised upward to take account of the demonstrated level of performance of our much more distant relatives in the New World.

In summary, I would like to suggest that what was once a weakness of the New World species when it came to attracting anthropological researchers may have turned out to be their greatest strength. Because the New World primates were not thought to be particularly relevant to human evolution, it was easier for those who studied them to observe things that the rest of us had assumed were not there. I believe that a consensus on the form of early human society and on the subsequent course of human evolution affected the observations made by anthro-pologists into the early 1970s. We went to the Old World expecting to find male dominance, linear hierarchies, polygyny, and lack of paternal investment, among other things, and we did. Those who went to the New World were just trying to describe and explain whatever patterns they found, and they were guided more by the excellent work that had been done previously on birds and other well-studied arboreal creatures (e.g., Fitzpatrick and Woolfenden, 1988; Wilson, 1975) than by assump-tions about human ancestors. They were therefore freer to deviate from the conventional wisdom to demonstrate that pair-bonding and male investment, for example, were not rare, and that polyandry and female choice not only occur but do so often enough to be worth a comparative study. New World studies are important to Old World research for two reasons: they confirm some of our hypotheses, and they contradict oth-ers. We have to take New World primates seriously now, even when they teach us something new.

8

Behavioral and Ecological Comparisons of Neotropical and Malagasy Primates

PATRICIA C. WRIGHT

Introduction

Strepsirhine primates on the island of Madagascar and platyrrhine primates in South and Central America have been isolated geographically since the Oligocene, and have taken quite different evolutionary pathways (Fleagle and Kay, this volume). At first glance the platyrrhines and strepsirhines share traits including arboreal lifestyle, small body size, and a wide variety of social structures. In this chapter various aspects of the radiation in each geographic region will be compared. Differences and similarities in ecology, diet, anatomical constraints, locomotor behavior, social structure, and reproductive patterns of these two distantly related primate groups will be examined.

Broad Comparisons

Geography and History

Both the platyrrhines and strepsirhines occur in forested areas of the southern hemisphere, but almost a quarter of the platyrrhines also occur in the forested areas of the northern hemisphere. In contrast with the African monkeys, no living primate species in Madagascar or the Neotropics is specialized to live in savannah.

Strepsirhines, the ancestral primates, have a long history that stretches back 50 million years to the Eocene, when they were abundant in forested areas of North America and Europe (Gingerich, 1984; Schwartz and Tattersall, 1985). But today the only assemblage of primates composed entirely of strepsirhines occurs in Madagascar, where there exist at least 30 species, nearly half of which are nocturnal (Mittermeier et al., 1995).

Small bodied nocturnal communities of strepsirhines do exist in Africa and Asia, but the daytime forms appear to have been outcompeted by their monkey and ape relatives (Charles-Dominique, 1977).

Understanding the history of these primate assemblages on Madagascar and in the Neotropics is important to our insight into the living groups. Although the diversity of Madagascan primate subfossils is great, our knowledge of the chronology is relatively poor. Recent studies of continental drift suggest that this large island more than a thousand miles long broke away from the Kenyan coast more than 120,000,000 years ago, but how long it took to drift the 350 miles from the mainland is unknown (Rabinowitz et al., 1983; Krause, in press). It is also unknown when the first primates arrived on the island, or who they were (Cartmill, 1972, 1975; Dewar, 1984). The limestone cave deposits in the north, marshes in the central highlands, and coastal marshes in the south and western margins of the west in Madagascar have left an incomplete but brutal history (Simons et al., 1992; MacPhee and Raholimavo, 1988). At least 16 species of lemurs have gone extinct within the last millennium. These lemurs lived on the now barren central high plateau and ranged in size from the lemur-sized *Mesopropithecus* (Simons et al., 1992, 1995; Jungers, 1980; Godfrey, 1988) to the chimpanzee-sized *Megaladapis* (Jungers, 1977; Simons et al., 1990). Because these now missing genera were all large, slow-moving, and diurnal, it is probable that they fell prey to the recent arrival of human hunters in Madagascar (Tattersall, 1982; MacPhee and Raholimavo, 1988). In contrast to the Neotropical primate communities, which have remained intact since the Pleistocene (Kinzey, 1982; Rose and Fleagle, 1981), some of the Malagasy primate communities may have lost a third of their species (Mittermeier et al., 1995; Tattersall, 1982), largely as a result of deforestation (Green and Sussman, 1990). It is possible that the primate community found at Ampasambazimba in the forests of the central plateau about a thousand years ago contained 18 species of primates, while the richest primate community today contains 17 species in west and central Africa (Simons et al., 1995; Charles-Dominique, 1977; MacPhee et al., 1985).

The primate fossil record in South America gives us a longer historical progression of evolution than does the Malagasy record. There is fossil evidence from the last 30 million years, but the number of specimens is so small that all of them can be "put in a shoebox" (Fleagle, 1988). The oldest known South American monkey is approximately 25 million years old and recently found in Bolivia (Kay and Williams, 1995). Unlike the Malagasy species, the 14 species of Neotropical fossils are almost all small-bodied, 3 kg or less, except for the howler monkey-like *Stirtonia* (about 10 kg), and a large 20 kg monkey from the pleistocene of Brasil

(Hartwig and Cortelle, 1995; Kay et al., 1987). There is no indication that any of the Neotropical fossil primates were terrestrial or lived in savannah (Rose and Fleagle, 1981). An additional recent fossil find is a giant tamarin from the Miocene of Columbia (Kay, 1994). Perhaps the most striking feature of the fossil platyrrhines is the overall similarity of extinct species to modern lineages (Setoguchi and Rosenberger, 1987; Kay, 1987; Fleagle and Bown, 1983; Fleagle, 1988). There is no evidence that many lineages of extant platyrrhines have been distinct since the Miocene. Extinction probably did not occur because of a major climate change, or to selection pressures such as human hunting; rather, lineages slowly evolved into more modern forms.

But the origins of these Neotropical primates is as enigmatic as the origins of the Malagasy primates. At present there is no convincing explanation of the origin of South American monkeys, but dispersal across the Atlantic Ocean from Africa in the early Oligocene seems to be most likely (Ciochon and Chiarelli, 1980; Rose and Fleagle, 1981; Fleagle and Kay, this volume).

It should be noted that each of these two groups has evolved in isolation from pressures from other groups of primates. There is no indication that there have been monkeys or apes in Madagascar (Tattersall, 1982; Dewar, 1984), or that there have been strepsirhines or apes, or old world monkeys in the Neotropics (Rose and Fleagle, 1981). This is an important distinction from the primate communities of Asia and mainland Africa.

In addition, both the Neotropical and Malagasy primates share the occurrence of human arrival only late in their evolution. Humans arrived in Madagascar about 1500 years ago (Tattersall, 1982; Dewar, 1984) and in the Neotropics from 11,500 years ago (Pielou, 1991). In contrast, the course of the evolution of the Asian and mainland African primate communities includes hominids, probably as predators, for over a million years (Isaac, 1978).

Morphological Comparison of Platyrrhines and Malagasy Strepsirhines

There are distinguishing characters that separate strepsirhines from all monkeys, such as the toothcomb, lack of postorbital closure, unfused mandibular symphysis, grooming claws, acute olfactive abilities, wet rhinarium, tapetum lucidum in the eye, and small cranial capacity (Fleagle, 1988). These traits give strepsirhines not only behavioral constraints relative to platyrrhines, but also some comparative advantages. For example, the fact that all the Malagasy lemurs have a complete or partial *tapetum lucidum*, itself probably a retention of an ancient mammalian inheritance, makes it possible for even the most diurnal to see well in the dark if necessary (Sussman and Tattersall, 1976; Tattersall, 1988; Overdorff, 1988; Overdorff and Rasmussen, 1995). New World monkeys do not have this capacity and, therefore, except for the owl monkey *Aotus* (with its expanded orbits), do not have nocturnal activity as an option (Wright, 1989, 1994a,b).

The difference in cranial capacity seems to be reflected in the relatively limited recall abilities in lemurs. For instance, lemurs are slow to explore novel objects that they cannot see readily, while New World monkeys exhibit great curiosity and ability to remember for long periods of time (Kappeler, 1987; Jolly, 1985; P.C. Wright, unpublished data). Jolly (1966b) pointed out that this limited primate intelligence does not restrict the complexity of social interactions.

In some cases different anatomical features are seen to perform the same function in Malagasy lemurs and New World monkeys. Malagasy primates groom with toothcomb and grooming claws, while the more dexterous monkeys groom efficiently with their hands.

Differences in Digestion and Diet

Composition of the diet had been correlated with morphology of the gastrointestinal tract in monkeys and other mammals (Chivers and Hladik, 1980; Chivers and Langer, 1994; Eisenberg, 1981; Milton, 1980). In general, mammals subsisting on animal matter have evolved a simple stomach and a colon and a long small intestine, folivorous species have either a complex stomach or an enlarged caecum and colon, and frugivorous species have an intermediate morphology. Folivorous species need more gut capacity to provide fermentation chambers to extract protein from vegetation (Bauchop, 1978).

Although all the lemurs have not been adequately studied, it is known that many of the Malagasy primates have very specialized digestive systems that allow them (as a group) to take advantage of a wide range of food items (Ganzhorn, 1988, 1989). For example, one family, the Indriidae, are specialized midgut fermenters with extensive small intestines and capacious caecums reminiscent of equids (Hill, 1953). The diet of species of indriids is more than 50% vegetation (Pollock, 1975, 1977; Richard, 1978; Ganzhorn et al., 1985; Wright, 1987). Even more specialized diets are found in several of the lemur species, such as the golden bamboo lemur (*Hapalemur aureus*), which eats parts of some species of bamboo containing high doses of cyanide (Wright, 1987, 1989; Glander et al., 1989). Another example of specialization in diet is a pair of nocturnal folivorous species, yet one (*Lepilemur mustilinus*) prefers leaves high in alkaloids, while the other species (*Avahi laniger*) prefers leaves high in tannins (Ganzhorn et al., 1985; Ganzhorn, 1989). Even two closely related sympatric lemurs (*Eulemur fulvus* and *Eulemur rubriventer*) may partition their feeding by the ability to digest different secondary compounds (Overdorff, 1990, 1991, 1993).

New World monkeys all conform either to an intermediate frugivorous gut morphology with a tendency to simplify and shorten the gut in the more insectivorous species such as *Callithricidae* and *Saimiri* (Chivers and Hladik, 1980; Chivers and Langer, 1994). Even the most folivorous New World monkey, *Alouatta*, the howler monkey, has a generalist frugivore digestive system (Milton, 1978, 1980, 1981b).

In contrast, many strepsirhines have an enlarged large intestine or caecum to provide chambers for the bacterial fermentation of cellulose and the absorption of volatile fatty acids and other metabolites (Chivers and Hladik, 1980). However, one of the most folivorous of the strepsirhines, *Lepilemur*, has the shortest small intestine of all primates. This genus has evolved a system of caecotrophy that allows a diet very high in fiber content (Charles-Dominique and Hladik, 1971; Charles-Dominique et al., 1980).

Locomotor Behavior

The locomotor specializations of these two groups allow a different access to the same types of resources such as leaves, fruits, flowers, and insects. The primates of Madagascar are quadrupedal with specializations for leaping among most groups (Walker, 1972; Dagosto, 1995). The New World monkeys are also quadrupedal, but the smaller primates, callithricids, have claws that facilitate gripping broad trunks of trees, and the larger primates have prehensile tails that facilitate the ability to reach out and pick fruits from more distant branches (Garber, 1980). Anatomically the intermembral index shows this relationship. Most of the extant strepsirhines have longer legs than arms, which facilitates a locomotor behavior described as vertical clinging and leaping (Napier and Walker, 1971; Glander et al., 1992; Dagosto, 1995). This allows even the largest species to jump from trunk to trunk a meter or two from the ground. There are New World primates that exhibit this behavior, but these include only a few species of the small bodied callithricids (Kinzey et al., 1975; Garber, 1980) and their intermembral index does not reflect this behavioral difference.

When the complete assemblage of both extant and extinct Malagasy primates is considered, the locomotor diversity is much greater than seen in any other group of primates (Gebo, 1987; Fleagle, 1988). The locomotor abilities include prodigious leapers, arboreal and terrestrial quadrupeds, long-armed suspensory species, and some such as the koala-like *Megaladapis* and the ground sloth-like *Archeoindris*, which have no living analogous counterparts among primates (Jungers, 1977, 1980; Fleagle, 1988; Godfrey, 1988; Jungers et al., 1991, 1995). Locomotor flexibility within the group has allowed for exploitation of a wide diversity of habitats and microhabitats (Dagosto, 1995; Demes et al., 1995).

Reproduction

There are obvious contrasts in reproductive characteristics between the extant strepsirhine primates on Madagascar and the New World monkeys. Females of both groups have estrous cycles in contrast to most Old World monkeys and apes, which have menstrual cycles (Asdell, 1964). However, the estrous cycles of New World monkeys occur twice as frequently as those of strepsirhines, averaging 16 days (Wright and Bush, 1977; Robinson et al., 1987; Robinson and Janson, 1987). The cycles in Malagasy strepsirhines are seasonal and correlated with day length (Van Horn, 1975, 1980; Van Horn and Eaton, 1979; Rasmussen, 1985). This photoperiodicity is a constraint that contrasts markedly with other primates. Both males and females cycle, with testicles beginning to increase in size 3 months before the breeding season (Foerg, 1982; Kappeler, 1987). Females of each species begin to cycle at approximately the same month each year, and if they do not conceive during the first cycle, are capable of cycling a second, and sometimes a third time. But neither males nor females of Malagasy species cycle throughout the year.

New World monkeys often have birth seasons during which the majority of infants are born (Terborgh, 1983; Goldizen, 1987b; Goldizen et al., 1988; Glander, 1980). This seasonality, except perhaps in the case of squirrel monkeys, seems to be correlated with the onset of the wet season (Baldwin and Baldwin, 1972; Coe and Rosenblum, 1978; Boinski, 1987; Mitchell, 1989). The same species and individuals are capable of cycling and mating all year round in captivity. Males remain potent, and healthy females continue to cycle throughout the annual cycle (Wolf et al., 1975). This results in differences in behavior that should not be underestimated. Male–male competition occurs throughout the annual cycle in all of the New World monkeys except squirrel monkeys, while in the Malagasy primates all male–male competition is concentrated into a single month, or even a few weeks corresponding with the period of estrous. Often severe wounding, or frequent male emigration from group to group (Jolly, 1966a, Richard, 1978, 1985, 1992; Sauther, 1991; Overdorff, 1991; Richard et al., 1991), occurs in the breeding season.

There are physiological contrasts in placentation between the two groups of primates. The strepsirhine primates have epitheliochorial placentation, in which several layers of tissue separate the uterine wall of the mother from the diffuse membrane of the fetus (Luckett, 1975). In all higher primates, the developing embryo forms a placental disk that invades the lining of the uterine wall (hemochorial placentation), providing a more intimate interchange between fetal and maternal circulations (Luckett, 1975). The complexity of the placentation system is related to the transfer of antibodies from mother to infant. All eutherians with epitheliochorial placenta transfer antibodies and passive immunity largely through the milk, while in the taxa with hemochorial placentas they pass the entire maternal antibody complement from the maternal circulation to the fetal circulation and less is transmitted in the milk (Buss, 1971; Eisenberg, 1981).

Gestation length is shorter in both Malagasy and Neotropical primates than in Old World monkeys and apes (Ardito, 1975). In general, the Malagasy primates have the shortest gestation lengths among primates, ranging from 60 to 179 days (Tattersall, 1982; Bourliere et al., 1961; Leutenegger, 1973; Haring et al., 1987; Glander, 1994; Wright, 1995). The New World monkeys gestate from 133 days (Hershkovitz, 1977; Hunter et al., 1979; Kinzey, 1981; Crockett and Sekaulic, 1972) to 225 days (Eisenberg, 1981). Although shorter gestation suggests a less costly pregnancy for the strepsirhine mother, the basal metabolic rates of strepsirhines are lower than those of most New World monkeys (McNab and Wright, 1987; Daniels, 1984; Young et al., 1990; Le Maho et al., 1981), which could have an effect on gestation cost. It has been suggested that low metabolic rate may increase the cost of maternal investment, and could be one of the factors leading to female dominance in the Malagasy primates (Young et al., 1990). Comparisons of the energetic differences between epitheliochorial placentation and hemochorial placentation do not exist, and so the effects on behavior remain unclear. One possible advantage to having the prosimian placental system is that toxins from the mother's food will not pass so readily from mother to infant during gestation (Tilden and Oftedal, 1995).

Litter sizes and interbirth intervals in these two primate groups are similar. In general, litter sizes in small strepsirhines range from 1 to 3 and the litter sizes of small New World monkeys also range from 1 to 3 (Hershkovitz, 1977; Rasmussen, 1985). Larger-bodied Malagasy and Neotropical primates give birth to one offspring. Interbirth interval in the small primates of both groups is 1 year or less, while the interbirth interval of the larger primates ranges from 1 to 2 years (Pollock, 1975; Symington, 1987; Wright, 1995).

The period of lactation for the Malagasy strepsirhines varies from 45 days for *Microcebus murinus* (Bouliere et al., 1961) to 6 months for

Propithecus verrauxi and *Propithecus diadema* (Richard, 1978; Meyers and Wright, 1993). In New World monkeys the shortest lactation period documented is for the callithricids, 90 days (Hershkovitz, 1977; Goldizen et al., 1988), and the longest is 330 days for the large spider monkeys *Ateles* (Eisenberg, 1981). Malagasy strepsirhine females nurse infants for fewer days than same sized New World primates (Wright and White, 1990). Since lactation is generally more energetically expensive than gestation, it might be assumed that the reproductive burden for female monkeys is greater than for strepsirhine females (Millar, 1977; Randolph et al., 1977). Marmoset milk and squirrel monkey milk have higher levels of protein, carbohydrate, and lipids than human or macaque milk (Turton et al., 1978; Buss and Cooper, 1972). Studies by Tilden and Oftedal (1995) suggest that lemur milk is more dilute than either New World monkey or galago milk. It seems intuitively puzzling that infants with a faster rate of weight gain and development would be drinking more dilute milk than slower growing species (Wright and White, 1990). Unfortunately, little is known about milk yield in these species. Lemurs may compensate for dilute milk by producing large volumes (Tilden and Oftedal, 1995). Other factors may also play an important role. For example, it could be that the larger brain of the New World monkeys requires more energy for development than a brain with less capacity (Martin, 1990).

In general, the basal metabolic rate of strepsirhine primates is below the Kleiber line, while the New World primates, except for *Aotus*, fall above the line (Young et al., 1990; Daniels, 1984; Milon et al., 1979; Le Maho et al., 1981). While a lower metabolic rate requires less energy, there may be other costs, perhaps in reproduction (Young et al., 1990). Although primatologists have based many theories on the reproductive costs of the mother, the truth is that our knowledge regarding the relationships among lactation and nutritional content of milk, infant growth patterns, and basal metabolic rates is limited.

The longevity of New World monkeys is about one-third longer than for comparable sized strepsirhines. For example, in captivity *Cheirogaleus medius* has lived 12 years and *Callithrix* jacchus 17 years (Eisenberg, 1981; DUPC records). *Lemur fulvus* has lived 30 years (DUPC records), while a *Cebus* monkey has lived 50 years and spider monkeys over 24 years. These captive longevity records reflect potential only, but very old animals are observed in the wild, (Wright, 1995) and, as our long-term data are collected, we will be able to focus on demography and factors such as lifetime reproductive success. At present, it appears that platyrrhines have a greater reproductive potential than their shorter lived strepsirhines of the same body size.

Behavioral Ecology

Keeping in mind the anatomical constraints and advantages that influence behavioral capabilities, the behavior of each of these two primate groups will be examined in an ecological context. Neotropical communities of primates have been studied in several sites (Mittermeier and van Roosmalen, 1981; Fleagle et al., 1981; Terborgh, 1983; Robinson and Janson, 1987; Peres, 1993). Malagasy primate communities have been examined at several field stations also (Ganzhorn, 1988, 1989; Pollock, 1989; Charles-Dominique et al., 1980; Sauther, 1991; Overdorff, 1993; Wright, 1995; Wright and Martin, 1995). In a general way there are two types of forest in each geographic location. Dry, deciduous forest is a contrasting environment to the forests of the humid tropics in both the Neotropics and Madagascar.

One case study, a comparison of two rainforest communities (one in Madagascar, one in the Amazon region of South America), may help us understand the relationship between ecology and behavior for these two primate radiations. To understand differences seen between these two distantly related primate radiations in the Neotropics and in Madagascar, it is useful to compare full communities that have been studied for 10 years in their forest contexts. New World primates have different predators, competitors, and even forest structure than those found in Madagascar, and this may affect the community structure.

Comparison between the two rainforest communities, one in the rainforest at Cocha Cashu Biological Research Station in Peru at 400 m altitude and 2000 m rainfall and Ranomafana Biological Research Station in Madagascar at 1000 m altitude and 2500 mm rainfall reveals many similarities. At the Peruvian site there are 13 species of primates, whereas at Ranomafana National Park in Madagascar there are 12 species. This is consistent with rich mammal assemblages worldwide, with maximum number of sympatric species usually 12 or 13 (Emmons, 1995). Body size ranges from 100 g to 9 kg in Peru (Terborgh, 1983) and from 30 g to 7 kg in Madagascar (Wright, 1992; Glander et al., 1992; Atsalis et al., 1993). Even though the Neotropical species, on average, are twice the body size of the Malagasy rainforest species, the body size of both is small compared to the much larger old world monkeys and apes. (It should be kept in mind that the subfossil lemurs now extinct were within the body weight range of old world monkeys and apes.) At the Neotropical site two of the species are vertical clingers and leapers, and 11 are quadrupedal runners (Kinzey et al., 1975; Garber, 1980; Fleagle, 1988), whereas in Madagascar six of the species are vertical clingers and leapers and six are quadrupedal runners (Walker, 1972; Dagosto, 1994, 1995).

Diet

Both primate communities are similar in having species that eat fruits, leaves, flowers, insects, and nuts. The Neotropical monkeys, in general, are more frugivorous (Terborgh, 1986a,b). Only *Brachyteles* and *Alouatta* are known to consume more leaves than fruits in their diet at any time of the year (Milton, 1981a,b, 1984; Strier, 1985, 1992). *Callicebus* is also known to ingest foliage (25% of its feeding minutes) (Wright, 1985; Kinzey and Gentry, 1979). About half the Malagasy species eat up to 50% foliage in their diet (Ganzhorn, 1989; Wright, 1995).

In both communities insects are eaten. All species of callithricid, *Saimiri*, *Cebus*, and *Aotus*, eat many prey items daily (Terborgh, 1983; Wright, 1985; Mitchell, 1989). *Daubentonia, Microcebus, Mirza*, and *Cheirogaleus* also rely on insects for protein (Charles-Dominique et al., 1980; Mittermeier et al., 1989; Harcourt, 1987).

In both communities nectar feeding and sap feeding are important to the small-bodied primates (Terborgh, 1983; Charles-Dominique et al., 1980; Overdorff, 1990). There are also species in each group that eat nuts (Iwano and Iwakawa, 1988; Terborgh, 1983; Sterling, 1993).

Social Structure

Almost all types of social structure described for mammals are found in both these primate communities (Clutton-Brock and Harvey, 1977). In Madagascar some of the nocturnal species have a solitary, but social structure that may be most like that of galagos (Hladik, 1980; Bearder, 1987; Bearder and Martin, 1979). In contrast, there are no monkey species that normally forage alone, not even the nocturnal monkey *Aotus* (Wright, 1989, 1994a,b). However, a social structure that is unique in primates is polyandry, found only in the Neotropics (Terborgh and Goldizen, 1985; Sussman and Kinzey, 1984). The small-bodied callithricids have a cooperative breeding system with one breeding female, but possibly two or more breeding males that assist in infant care (Garber, 1994; Goldizen, 1987a,b, 1990; Goldizen et al., 1988; Sussman and Kinzey, 1984).

In both primate groups there are strictly monogamous, as well as multimale, multifemale polygamous groups (Robinson et al., 1987; Kinzey, 1987; Robinson and Janson, 1987; Morland, 1991; White et al., 1995; Strier, 1992). In Madagascar and in the Neotropics one-male–multiple female groups, as seen in many Old World monkey species, are absent. However, strictly monogamous and multimale–multifemale groups occur in Madagascar and the New World. None of the Malagasy prosimians is sexually dimorphic in either body or canine length (Albrecht et al., 1990; Kappeler, 1990; Glander et al., 1992; Richard, 1992), while about half of the New World monkeys have males with larger body and canine size (Kay et al., 1987; Lemos de Sa, and Glander, 1993). In the New World monkeys tooth dimorphism is correlated with social structure and degree of male–male competition (Strier, 1991), but in Malagasy lemurs there is no correlation (Kappeler, 1993). This lack of canine dimorphism in the lemurs may be related to the high levels of female–female competition in Malagasy primates, and indirectly related to differences in predation pressure in the two environments.

Feeding group size varies dramatically among species in the New World from monogamous pairs with group size from two to six to large polygynous groups with more than 35 individuals (Wright, 1989; Terborgh, 1983; Mitchell, 1989). In both radiations there are some species that have a fission–fusion type of social structure, where parties forage in subgroups (White et al., 1995; Morland, 1991; Ahumedo, 1989; McFarland, 1986). In the Madagascar rainforest feeding group size varies between one individual in some nocturnal species to more than 18 individuals (Wright et al., 1987; Overdorff, 1991). Estimated biomass (kg/km^2) of primates at Cocha Cashu is 651 while the estimated biomass of primates at Ranomafana is 355 (Janson, 1975; Terborgh, 1983; Wright, 1992) (Figure 8.1). Neotropical rainforest primates have, on average, twice the body size, group size, and biomass than the Malagasy primates. However, there are a nearly equal number of species.

The data suggest that there may be a difference in forest structure and forest productivity. Phenology and rainfall records have been kept at Cocha Cashu and Ranomafana (Figure 8.1). The data suggest that there is a difference in amount of fruit production between the sites (Wright and Martin, 1995). The Peruvian rainforest produces more than twice the kilograms per square kilometer of fruit for the months January through June. However, in July and August there is low production in both forest and the November through December months when the rains begin are nearly identical.

The infant lemurs are being weaned during peak fruit season for nearly all species of lemurs. Timing of weaning, which may be the most difficult period in infant development, is less critical in a forest that has abundant food in all but 2 months. The data from Cocha Cashu show that birth seasons for all species are variable and occur within a 6-month period (Terborgh, 1983; Goldizen, 1987a,b; Wright, 1985). The Malagasy primates have more limited windows. Each species at Ranomafana gives birth at a different time and appears to adjust this birthing and lactation so that the infant is weaned before the fruit productivity dips beginning in February. Even though infant diademed sifakas are born in June and infant golden bamboo lemurs are born in early December, all infants are weaned by March (Wright, personal observation). During the less fruit-

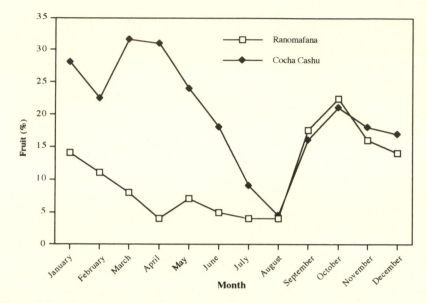

Figure 8.1. Fruiting seasonality in Ranomafana National Park (Madagascar) and
Cocha Cashu (Peru). Percentages of traps containing fruit each month. Data
from Wright (1985) and Overdorff (1991).

ful months of March through June only the sifaka are gestating. Timing
of gestation and lactation to coordinate with food abundance has been
documented for New World monkeys: tamarins at Cocha Cashu (Goldizen et al., 1988) and howler monkeys in the llanos of Venezuela (Crockett and Eisenberg, 1987). This tight association has been documented in
Old World monkeys as well (van Schaik and van Noordjwik, 1985).

This strict seasonality of breeding and birthing in the Malagasy primates has been associated with a behavioral phenomenon called female
priority (Jolly, 1966a, 1984; Pollock, 1977; Richard, 1978). Most Malagasy
primates have female leadership, feeding priority, and female dominance. In New World monkeys there are also many species where females lead travel, and enter food trees first (Boinski, 1987; Wright, 1984;
Milton, 1984), but there are also species such as *Ateles* where males
harass females (Symington, 1987). Feeding priority and seasonal breeding are not necessarily linked in the New World primates, and female
leadership may be related to energetic demands of reproduction (Wright,
1984). Female mate choice is also observed in some species and may be
influenced by factors such as protection from predators and paternal care-
giving advantages (Janson, 1984). Sperm competition may also be a factor

in the larger species of New World primate and in some prosimians (Milton, 1985; Kappeler, 1993).

A comparison of parental care patterns for the New World primates and Malagasy lemurs shows a variety of forms. In the largest species of lemurs and monkeys females are the primary caretakers while many of the small or medium sized lemurs park their infant in nests or alone, while the mother forages (Pereira et al., 1987; Gould, 1992; Wright, 1990; Morland, 1990; Wright and Martin, 1995). Many of the small and medium sized New World monkeys have extensive paternal care and sibling helpers (Goldizen, 1987a,b; Wright, 1984, 1990). The relationship between infant maternal/weight ratio and male care is clear in New World monkeys (Leutenegger, 1973; Wright, 1984, 1990), but weight alone does not explain why the small strepsirhines park rather than have paternal assistance (Gursky, 1994). Predation pressure could have an effect (Sauther, 1989; van Schaik and van Noordwijk, 1985; Terborgh, 1983; Wright, 1990).

In general, the behaviors that have been observed in Old World monkeys and apes are also seen in New World primates and in Malagasy lemurs. For example, classic cases of infanticide by immigrant males have been seen in New World howler monkeys, and in diademed sifaka (Crockett and Rudran, 1987; Wright, 1988a,b, 1995). Primates in both forests occasionally form mixed species foraging units (Garber, 1986; Terborgh, 1983; Overdorff, 1991). Testicular size is associated with social structure in both New World monkeys and Malagasy lemurs, with testicles being much smaller in monogamous species (Glander et al., 1992; Goldizen et al., 1988). Sex ratio of offspring may vary in both primate groups, depending on emigration pattern and fruit abundance (Symington, 1987; Glander, 1980; Richard et al., 1991).

Primates are key players in the ecosystem of both the Neotropics and the Malagasy forests. Primates are known to pollinate important vines and tree species (Sussman and Raven, 1978; Overdorff, 1990, 1991; Janson et al., 1981; Wright and Martin, 1995). Primates are seed dispersers in both rainforests (Janson, 1983; Garber, 1986; Terborgh, 1986a,b; Wright and White, 1990; White et al., 1995), and are important food items for carnivores and raptors (Rettig, 1978; Goodman et al., 1991; Albignac, 1972; Emmons, 1987; Wright, 1995; Wright and Martin, 1995). Leaf-eating primates in both radiations have to cope with toxins and secondary compounds used as protection against predators by the plants (Coley et al., 1985; Bauchop, 1978; Glander, 1978; Ganzhorn, 1988, 1989; Harcourt, 1991; Milton, 1981a,b).

However, there is one major difference that might affect the impact of primates on the ecosystem. Although the biomass of primates in dry forests is high in both geographic areas, the primate biomass in the rainforests of Madagascar is much lower than that of South America (Wright, 1992; Terborgh and van Schaik, 1987). This difference may be attributed to historical reasons, i.e., the absence of certain taxa that would be competitors for food resources. However, the data suggest otherwise. There are fewer species of birds (Langrand, 1990) and mammals to compete for resources in Madagascar, and primates have radiated to fill niches left available by the absence of ungulates, squirrels, rabbits, and woodpeckers (Cartmill, 1974; Charles-Dominique and Hladik, 1971; Wright, in press).

A second and more plausible reason for lower primate biomass in the rainforests of Madagascar is lower fruit production (Terborgh and van Schaik, 1987). A comparison of fruit trap data suggests this could be the case. The fact that there are more folivorous primate species in Madagascar than in South America may be related to the fact that there are new leaves available year-round (Terborgh, 1986b; Overdorff, 1991; Meyers and Wright, 1993). But the question then arises as to why the Malagasy rainforest is so low in fruit production. Analysis of soil samples and rainfall in each area does not reveal major differences. Perhaps the unpredictability of fruits in Madagascar is limiting to the lemurs.

By comparisons with the radiation of primates in Madagascar, we can better understand some aspects of the ecology, evolution, and behavior of Neotropical primates. The Neotropical group is more frugivorous, less nocturnal, and less seasonal breeders, and has less locomotor diversity. Offspring are raised with a variety of parental care systems, and grow up slower in the Neotropics. Although social structures and social interactions are complex in both radiations, curiosity and the ability to investigate new situations are limited in the Madagascar primates. Investigations into differences of the abilities of the Neotropical and Malagasy primates in cognition, deception, long-term memory, etc. may be the subject of future research.

Conservation Status

In this chapter, I have discussed the history of these two distantly related primate radiations, and I would like to end with a discussion of their future. In the Neotropics 16 genera and 65 species are currently recognized and two genera, 12 species, and a total of 28 taxa are already endangered (Marsh and Mittermeier, 1987). Although healthy populations of many of these monkeys exist, the taxa that live in the southeast forest of Brazil are nearly extinct, especially the two highly endangered genera *Brachyteles* and *Leontopithecus* (Mittermeier et al., 1995; Strier, 1991). In 1994, a new species of callithricid was discovered in Brazil and in 1979, the yellow tailed woolly monkey *Lagothrix flavocauda* was discovered. In Madagascar, there are 13 genera and 30 species (Richard and Sussman, 1988). Within the past 5 years new species have been discovered, or rediscovered in Madagascar (Meier et al., 1987; Wright, 1987; Simons, 1988; Meier and Albignac, 1991). The Primate Specialist Group of IUCN considers 10 species of Malagasy primate highly endangered, and another 15 species as vulnerable, a figure unmatched by any country (Harcourt and Thornback, 1990; Mittermeier et al., 1995).

The endangered groups in both the Neotropics and Madagascar are threatened by a combination of habitat destruction and hunting humans. Conservationists have revised their approach to this impending extinction. Action plans for conservation of primates now include sustained development programs for the humans who are destroying the habitat and species (Pollock, 1986; Mittermeier et al., 1989; Wright, 1992, in press). In order to produce adequate management plans for preserving these endangered species, we must know about their needs in the wild. The majority of the most endangered primates in Madagascar and the Neotropics have not been studied long term in their natural habitats. We have been denied the privilege of knowing about the diet, reproductive patterns, or social structure of *Megaladapis*, *Paleopropithecus*, and *Archeoindris*, taxa that existed in recent times, but are now extinct. This chapter would have been more definitive if I could have included these sixteen species. The priority of primatologists in the next decade should include not only conservation of habitat and species, but also intense research in the wild of the most endangered taxa and their relationship to their environment.

Acknowledgments

My appreciation to the Department of Agriculture of Peru and the Department of Water and Forests and ANGAP of Madagascar for permission to work in the national parks and reserves of each country. The University of Antananarivo and Madame Berthe Rakotosamimanana are also gratefully acknowledged for assistance in our primate research in Madagascar. Raymond Rakotonindrina, Director of ANGAP, Joelina Ratsirarison of Water and Forests, Benjamin Andriamihaja, National Coordinator of the Ranomafana National Project, and Henri Finoana, Director of Water and Forests are acknowledged for their support and assistance. Special thanks to Amanda Wright, Jukka Jernvall, John Terborgh, John Oates, Charlie Janson, Deborah Overdorff, and John Fleagle. Jackie Stephens is especially appreciated for her assistance with this manuscript. The support and inspiration of the late Dr. Warren Kinzey guided this research at all of its stages.

9

The Human Niche in Amazonia: Explorations in Ethnoprimatology

LESLIE E. SPONSEL

Introduction

More than two and a half centuries after Carl Linnaeus classified *Homo sapiens* within the Order *Primates* (Greene, 1959), science has yet to embrace fully the idea that our species is a primate. For example, a systematic search of the literature in primatology does not reveal a single article based on fieldwork that considers *Homo sapiens* among the species of primates as an integral part of the faunal community of an ecosystem.[1] As another example, in a superb textbook on primate behavior and ecology, Richard (1985a) includes two chapters (10–11) on symbioses between primate species, and also between primates and other species; however, *Homo sapiens* is not considered.[2,3] In this and other omissions both biology and anthropology are each, in its own way, anthropocentric, ambivalent, and even hypocritical.

Paradoxically, biologists tend to consider *Homo sapiens* as part of nature in an evolutionary sense, yet apart from nature in an ecological sense. Biological ecologists usually ignore humans as if they were unnatural and readily excluded from the natural environment for analytical purposes. Apparently the presence of humans in an ecosystem somehow contaminates pristine nature. Although coevolution is a popular concept in contemporary biology, it has yet to be applied to human predator–prey dynamics, even though such a relationship may have persisted for thousands or millions of years in many regions (e.g., Marks, 1976, 1977). Conservation is the main context in which biologists may consider humans; however humans are usually treated simply as agents of environmental degradation, resource depletion, habitat destruction, and species extinction.

In recent decades the division of labor in anthropology between the

143

biological and cultural subdisciplines has become increasingly rigid to the point that the relationship between them is seldom probed despite occasional appeals to holism. Cultural anthropology focuses on the uniqueness of *Homo sapiens* by virtue of mind, language, and culture, while biological anthropology focuses on human biological evolution, variation, and adaptation. These two subdisciplines tend to be mutually exclusive. Yet human phenomena are not so neatly compartmentalized in reality. *Homo sapiens* is simultaneously an organic, mental, linguistic, and cultural being, and all of these four are important components in adaptation to the natural environment (Sponsel, 1981: Chs. 2–4, 1987).

Ecological approaches have generally followed the academic compartmentalization of disciplines and subdisciplines. In anthropology the main ecological approaches according to sudiscipline include paleoecology in archaeology, primate ecology and human adaptability in biological anthropology, cultural ecology in cultural anthropology, and ethnoecology in linguistic anthropology and, to some extent, in cultural anthropology (Sponsel, 1987). There has been little if any exploration of the interfaces between the ecological approaches of these subdisciplines or those of anthropology and other disciplines like biology and geography, although such efforts might generate useful insights. To illustrate this potential, this chapter explores the mutual relevance of primate and human ecology, and, more broadly, of the disciplines of biology and anthropology, and of the subdisciplines of biological and cultural anthropology.

In particular, six lines for inquiry at the interface between human and primate ecology are provisionally outlined: (1) *comparative ecology*—similarities and differences between the ecology of human and other primate species as integral components of the faunal community of a particular ecosystem, regardless of whether or not these species interact with one another; (2) *predation ecology*—subsistence hunting of nonhuman primates by traditional indigenous societies; (3) *symbiotic ecology*—additional types of relationships (direct and indirect) between human and other primate species in the same ecosystem; (4) *cultural ecology*—the relevance of nonhuman primates at the cultural level of interactions between a human population and its ecosystem; (5) *ethnoecology*—indigenous knowledge, beliefs, and values regarding environmental domains such as nonhuman primates; and (6) *conservation ecology*—human use and management of nonhuman primate populations as a renewable natural resource including the positive as well as negative impact of humans on the populations and their habitats. These six heuristic categories emphasize ecological aspects (see May and Seger, 1986). They are distinguished for purposes of analysis, but of course they overlap as components of a broader field of inquiry, which might be called *ethnoprimatology* for lack of a better designation.

Beyond the above delimitation, to restrict further the scope and aims of this study to manageable boundaries, the focus is on the subsistence economy of traditional indigenous societies in interfluvial Amazonia. Traditional indigenous human populations are the original peoples of the forest, the most intimately involved with nature, and usually live in relative harmony with their environment, although they are increasingly endangered by external forces that are alien to the Amazon and its indigenous societies (Sponsel, 1986a, 1995). In Amazonia, interfluvial refers to the interior forest ecosystem beyond the floodplains, which is also called *terra firme*.[4] The first part of this chapter develops several ideas about the human adaptive complex in the interfluves.

Amazon Ecology[5]

Demineralization

From a broad geological perspective, the interior Amazon region is one of the most ancient and stable terrestrial environments on the planet. The soils formed from ancient Tertiary parent material that is about 65 million years old. During this long time they have been subjected to high temperature, humidity, and rainfall, which provide optimum conditions for chemical weathering and leaching of nutrients. The result is that most of the soils are very low in nutrients. This process has been called demineralization (Schwabe, 1968:121, 127).

On whitewater river floodplains or várzea, soils are annually rejuvenated by fresh nutrients deposited as sediments during the wet season. Rivers that flow from the geologically young Andes are fairly rich in nutrients, whereas those from the ancient Brazilian and Guiana Highlands are poor. However, the interfluvial zone is not seasonally rejuvenated by flooding, hence its name, *terra firme*. Also in the interfluves, since over great periods of geological time the existing soils and sediments have been deeply weathered and leached, fresh parent material is not likely to become available to provide new nutrients. The soils provide little more than physical support for plant growth. Such considerations led to the characterization of Amazonia as a counterfeit paradise (Meggers, 1975; cf. Moran, 1993).

Biomass

Nearly 50% of the terrestrial biomass (organic material) of the Earth is concentrated in about 6% of its land area in the form of the tropical rainforest (Myers, 1992). In this region most climatic conditions are optimum for plant growth with high temperature, humidity, and rainfall, albeit there are zones with frequent cloud cover that reduces solar energy for photosynthesis. While rainfall is abundant throughout the year, there are usually marked wet and dry seasons.

Although the tropical rainforest ecosystem has high productivity (biomass produced/unit area/unit time), by far most of the nutrients are not in the soils but in the biomass (including litter on the ground). Yet the plant biomass is about 98% wood (Fittkau and Klinge, 1973:11), so most of the nutrients are inaccessible for food to animals other than a few specialized consumers, such as termites. Nor is the 2% of the plant biomass, which is mostly leaves, readily available to animals for food, since many leaves contain toxic substances as defenses against foragers (Richard, 1985a:132).

Biodiversity

The great antiquity of the tropical rainforest ecosystem, intense competition for solar energy and nutrients, and other factors such as perturbations (lightening stikes, wind storms, floods, etc.) have contributed to the highest biodiversity of any terrestrial environment on the planet by measures such as the number of species per unit area (Collinaux, 1989; Diamond, 1988; Myers, 1992; Wilson, 1988, 1989). While species are abundant, individuals are not. Population density is low and individuals are distributed unevenly. This can be illustrated by fruit, which is seldom available in large amounts and is sporadically distributed in time and space. Thus, for example, spider monkeys (*Ateles*), which emphasize ripe fruit in their diet, forage over an area 10 times as great as howlers (*Alouatta*), which are mainly leaf eaters (Milton, 1981a).

Limiting Factors

A hierarchy of several factors limits populations in tropical forests. The particular combination and ranking of these factors vary in time and space with the specifics of the population and ecosystem (Sponsel, 1986a:74). For the human species, attacks from predators would seem to be a minor problem at best, given the combination of large body size, large group size, projectile technology, fire, and, in many societies, dogs. Disease, although present in some forms, was probably not an important limiting factor prior to European contact because of the relatively small, dispersed, and mobile populations (Warren and Adel, 1984;

cf. Coimbra, 1995). Thus food, especially proteins (Gross, 1975, 1982) and carbohydrates (Beckerman, 1979; Milton, 1984d), is likely to be the more important limiting factor on human populations.

Animal resources are not readily available to provide quality protein with the full complement of the amino acids required for growth, development, and maintenance of the human organism. Animal species are abundant, but individuals within a species are not. Moreover, their behavioral and ecological attributes may render them less available. Likewise, edible plants are not abundant, but are scattered in time and space at low density and in small amounts. Thus both wild animal and plant resources are expensive for humans in terms of the time and energy required to locate, acquire, distribute, and process them in the interior forest ecosystem. These and other considerations have led some anthropologists to speculate about whether humans can readily adapt to tropical rainforests by foraging, although these speculations tend to ignore variations within and between the different biogeographic regions of the tropical rainforest biome (Bailey, 1989; Headland, 1987; Headland and Reid, 1989; Headland and Bailey, 1991; Hutterer, 1988; Sponsel, 1989).

In the fluvial zone of Amazonia, most human populations concentrate on getting their food from sources other than the forest; that is, carbohydrates from swidden gardens and protein from fish in the rivers. In the interfluvial zone, most human populations derive most carbohydrates from gardens and most protein from hunting. In Amazonia the interplay of the properties of the interfluvial ecosystem and the bio-behavioral attributes of *Homo sapiens* tend to limit the possibilities for adaptation in this region to either (1) a mixed subsistence economy (including foraging and farming) or (2) foraging complemented by trade with riverine farmers. Foraging and farming may be necessarily complementary components of the human adaptive strategy in interfluvial Amazonia. At the minimum, foraging and farming facilitated the dispersal of humans into the interfluvial forests. At the maximum, foraging and farming may have been prerequisites for this dispersal (Sponsel, 1989). The limitations and possibilities for human adaptation in the interior Amazon are further elucidated through comparing human and monkey species.

Comparative Ecology

Continuities

Homo sapiens is clearly a primate by any objective scrutiny. Many of the megatrends or basic patterns in primate evolution during the last 55 million years continue in the human line, and some are even accentuated such as increased mobility of the digits, prehension (grasping function of the hand), elaboration of the visual apparatus, and reduction of the olfactory apparatus (Richard, 1985a:30).

Diurnality. At the behavioral level, diurnal habit is the most important attribute common to humans and monkeys, except for the unique night or owl monkey (*Aotus*). Moreover, diurnality distinguishes *Homo sapiens* from the other top carnivores in the interfluves, thereby reducing competition for prey (Sponsel, 1981:190–193).

Fission–Fusion. The fission–fusion dynamic, the process whereby a group splits into subgroups for foraging and later reunites into the larger group, is a response to fluctuations in resource abundance in some monkey (Kinzey, 1986a:132) as well as human populations (e.g., Good, 1986, 1995).

Discontinuities

While humans are primates, that does not mean they are nothing but primates; the human mind, language, and culture are unique (Adler, 1967; Stevenson, 1987; Wilson, 1983). Also there are biobehavioral attributes that distinguish *Homo sapiens* from other species of primates. These include terrestriality, large group size, large body size, and greater importance of hunting. Storing, sharing, and cooking food are also important human distinctions.

Terrestriality. Humans are terrestrial, in contrast to Amazonian monkeys, which are among the most arboreal of primates. Large body, group sizes, and sexual dimorphism may have originally evolved as part of the protohominid antipredator strategy, judging from a comparison of terrestrial primates in the forests and savannas of Africa (Clutton-Brock and Harvey, 1977; Richard, 1985a:343, 354, 459).

The contrast between the terrestrial emphasis of the human primate compared to its arboreal relatives in the Amazon has at least two consequences: (1) Direct competition for food resources between humans and other primates is greatly reduced, although not eliminated. (2) Arboreal primates have greater access to most food resources than humans, because the bulk of productivity (plant and animal) is concentrated in the upper strata of the forest. However, to some degree technology (bow and arrow, blowpipe, ax, etc.) allows humans to reach game and other food as high as the canopy.

Group Size. Among primates, group size varies as much within as between species (Richard, 1985a:335–336). At the interspecific level group size varies from 2 to 6 individuals in the titi (*Callicebus*) to 100 or more in the squirrel monkeys (*Saimiri*) (Kinzey, 1986a:134). The larger species of monkeys like the spider (*Ateles*) and howler (*Alouatta*) seldom have a group size larger than two to three dozen (Kavanagh, 1983; Moynihan, 1976b). In contrast, the group size of humans may be ten times greater. For example, in general, Yanomami villages range from 35 to 250 individuals with a mode of 70 (Sponsel, 1981:89); however, the group size of foraging units that use the village as a home base may be much smaller.

Body Size.   Body size is a significant factor in the niche differentiation of nonhuman primates as well as in many behavioral differences (Gaulin, 1979; Jungers, 1985; Kinzey, 1986a:129,136). Neotropical (New World) primates range from the 100 to 120 g pygmy marmoset (*Cebuella*) to the 15 kg woolly spider monkey (*Brachyteles*) (Kinzey, 1986a:121). Although surprisingly variable, the body size of humans in Amazonia is at least three to four times greater than that of the largest known monkey species.

Density.   In Amazonia the highest population density as well as the highest species diversity for monkeys have been reported from Cocha Cashu in Peru. Thirteen species of nonhuman primates are found at Cocha Cashu, and their population densities range from 5 to 60/km² while their group sizes range from 1 to 40 individuals (Terborg, 1985:286). Elsewhere Robinson and Janson (1987:71) report that densities for the larger monkey species range from 5 to over 100 individuals, but the mode is less than a dozen. The density for *Homo sapiens* tends to be much lower than that of other primate species in interfluves, something to be expected given the much larger body and group sizes of the former. Density for precontact human populations has been estimated at 0.2/km² in most of the interior forests of lowland Amazon, 0.5 or less in the upland forests of the Guiana and Brazilian Highlands, and 14.6 to 28.0 in floodplain zones (Denevan, 1992:205–234). For instance, among the Yanomami density is estimated at 0.19–0.5 (Sponsel, 1981:89).

Sensory Perception.   Most species of primates emphasize visual over olfactory and auditory senses. However, for the human predator in the tropical rainforest, sound assumes primary importance over vision because foliage is usually too dense to allow scanning for prey over any distance. [For the same reason, sound may be more important for the arboreal neotropical monkeys. See Snowdon, this volume.] Prey are usually detected first by sound, then by sight, although in some situations smell may also be important. The hunter often imitates prey calls (territorial, mating, alarm, etc.) to focus on the location of prey and to attract them to within firing range (Sponsel, 1981:259–260, 1986b:17).

Competitors. Another complication is the fact that humans have no monopoly on the prey species, but compete with large cats, snakes, and eagles, as well as other top carnivores (cf. Ross, 1976). For example, humans and jaguars (*Panthera onca*) prey on exactly the same mammalian species, but their respective diurnal and nocturnal habits affect niche differentiation (Sponsel, 1981:190–193).

Cooking. The use of fire to cook food is an obvious yet neglected advantage that humans have over other primate species. Cooking may not only make foods easier to chew, but also render some plants edible that might not otherwise be. In addition, cooking may increase the quantity and quality of nutrients that can be assimilated in the human digestive tract. Thus cooking is an important factor in increasing dietary breadth for the human primate. (An interesting research project would be to compare the nutritional value of cooked and uncooked food items, especially where the latter are the same as those consumed by monkey species.)

Farming. The domestication of manioc (*Manihot esculenta*) surely facilitated human dispersal and adaptation to the interior Amazon with its impoverished soils. Manioc provides a reliable and adequate supply of carbohyrates throughout the year, circumventing the seasonal fluctuations in wild plants that may limit other primate populations, especially during the dry seasons. Home ranges of many monkey groups are much larger in the dry season, reflecting the greater scarcity of food resources. During the dry season monkeys rely on a diverse but very small number of plant species (Kinzey, 1986a:130,135). In contrast, for human populations the reliability of manioc and the need for animal protein throughout the year shift the emphasis in seasonality to fluctuations in animal resource availability. For human populations in the fluvial zone, the wet season is the time of scarcity as fish and game resources are more widely dispersed with the flooding of large sections of the forest along major rivers (Sponsel, 1986a:69; Sponsel and Loya, 1993). During the wet season hunting activity and productivity decline in both the interfluves and the várzea, because terrestrial prey are less active during rains and the noise of the rainfall also reduces auditory detection of the prey by the human predator (Sponsel, 1986b:9).

The advantages of manioc are balanced by various disadvantages, including the fact that it is notoriously low in quantity and quality of protein. Moreover, the residual toxins that remain after processing require complementary protein from other sources (Spath, 1981; cf. Dufour, 1995). Consequently, when manioc is used as the staple cultigen, it magnifies the importance of hunting wild animals as the principal source of quality protein in interfluvial Amazonia. In the case of the Yanomami of Venezuela, who inhabit mainly the interfluves and have a mixed subsistence economy including hunting, fishing, and gathering as well as farming (many with manioc), hunting accounts by weight for about 11% of their food, 14.8% of their calories, and 68.8% of their protein. About 22% of the working time of males is spent hunting (Lizot, 1977).

Predation Ecology

Nonhuman Predation on Primates

The influence of nonhuman predators on nonhuman primates is receiving increasing attention, although there is much theoretical controversy over the subject (e.g., Cheney and Wrangham, 1987:227). The importance of predation on nonhuman primates has probably been grossly underestimated, since most evidence for predation on nonhuman primates is anecdotal or indirect in the form of alarm calls and other supposed antipredator tactics. Predation is rarely observed directly by fieldworkers, perhaps because most predators may be nocturnal (Cheney and Wrangham, 1987:231; Jolly, 1985:72; McKenna, 1982:72).

As McKenna (1982:73) concludes in an earlier review of the literature about nonhuman predation on primates: "Predation must have been, and still must be, a selective pressure to which groups must adapt." For instance, one study of a harpy eagle (*Harpia harpia*) nest in Guyana revealed 47 cebid monkeys out of 141 kills. This is about 30 times what might be expected from relative prey abundance (Retting, 1978). Beyond the harpy eagle and other raptors, alligators (*Caiman crocodilus*), boas (*Boa constrictor*), tayras (*Tayra barbara*), and felids such as the jaguar also prey on monkeys.

Cheney and Wrangham (1987:231–232) suggest four general trends in nonhuman predation on primates: (1) carnivores and raptors are the most common predators, (2) smaller species are more likely to be captured by raptors than larger ones, (3) larger primates are the prey for a greater variety of carnivore species than are smaller ones (cf. Redford, 1989), and (4) arboreal primates are more vulnerable to raptors, whereas terrestrial ones are more vulnerable to carnivores. Amazonia is an exception to this fourth trend. There, *Homo sapiens* is the only species of terrestrial primate, and it preys on arboreal primates through projectile technology (blowpipes and/or bow and arrow).

Various studies have suggested that nonhuman predation on primates may influence arboreality, diurnality, geographic distribution, body size, sexual dimorphism, group size, social organization (monogamy, multimale group structure), intragroup spacing and progression order, parental care, foraging, sleeping style and sites, birth timing and location, polyspecific associations, and so on (Cheney and Wrangham, 1987; Jolly, 1985:38, 119–120, 138; Richard, 1985a:60, 276, 343, 354, 459; van Schaik, et al., 1983).

Either the absence of predators, or their presence and level of pressure, may influence the behavior, ecology, and demography of nonhuman primate populations in various ways. Predators may be a limiting factor on primate populations (Cant, 1980). Richard (1985a:274) notes that "Because most primates live at low densities in relatively small social groups, three or four successful kills in a year could cause major changes in the composition of a group or even in the structure of the whole population." Where predation is not the primary limiting factor on primate populations, then food may be instead (Cheney and Wrangham, 1987; Kinzey 1986a). Two or more species may coexist under predation pressure, whereas without it, competitive exclusion would occur (Jolly, 1985:36). Also human hunters may reduce predator populations, thus relieving some of their pressure on monkey populations (Cheney and Wrangham, 1987). However, humans are also an important predator on monkey populations.

Human Predation

Human predation on nonhuman primates has been grossly neglected, as if *Homo sapiens* were not a natural part of the faunal community of forest ecosystems. For example, the three reviews of predation on primates (Anderson, 1986; Cheney and Wrangham, 1987; McKenna, 1982) almost completely ignore the human predator. In part this neglect may result from the fact that many primate studies have been conducted in game reserves and parks where hunting is illegal. Also the usual approach to primate behavior and ecology, naturalistic observation of habituated groups, is impossible where the population is subjected to human predation (Terborgh, 1985:284). Nevertheless Cheney and Wrangham, (1987) note that of all the predators on primates, humans probably account for the greatest number of deaths.

In the Old World, human predation on nonhuman primates has an enormous antiquity extending back to the australopithecines in Africa and the pithecanthropines in Asia. The faunal remains recovered from many of the archaeological sites of fossil hominids contain one or more nonhuman primate species that may well have been hunted (Day, 1986;

Shipman et al., 1981). Human predation was also an important influence on primates in the prehistory and history of Madagascar (Walker, 1967).

In the Amazon, the antiquity of human occupation appears to extend back only to about 6,000 to 8,000 years ago (Meggers, 1985:310; Roosevelt, 1989). However, this is based on very limited archaeological evidence and human antiquity may be extended substantially by further research following the trend in other regions. In any case, it is a sufficient time depth to suggest the coevolution of predator–prey dynamics between human and other primate species in Amazonia. The potential influence of *Homo sapiens* as an integral part of the faunal community of ecosystems in Amazonia can no longer be ignored by field researchers.

Beckerman and Sussenbach (1983) suggest several factors that influence human predation in Amazonia: seasonality, residence longevity and location (mainstream tributary, headwater, etc.), degree of agriculture, and food taboos. Extraneous factors include the presence of shotguns, flashlights, and outboard motors; nucleation of settlements around a mission station or trading post; commercial hunting by indigenes; and hunting in area by nonindigenes (Beckerman and Sussenbach, 1983:343).[6]

Human groups require much larger quantities of protein than monkeys, given the larger body and group sizes of the former. Human predators are distinctive among primates in capturing prey that are larger in body size than themselves by using projectile technology. Coordinated group hunting may also be important in hunting some species such as the tapir (*Tapirus terrestris*) and peccary (collared species = *Tayassu tajacu* and white-lipped species = *T. pecari*).

While the diet of most monkey species focuses on fruits and/or leaves, some quality protein is acquired from the capture of small game like insects, birds, bats, frogs, and lizards. For example, squirrel monkeys (*Saimiri oerstedi*) in Costa Rica prey on tent-making bats (Boinski and Timm, 1985). Capuchin monkeys (*Cebus apella*) prey on small species of amphibians, reptiles, birds, bird's eggs, and insects (Izawa, 1975, 1978a). Unlike humans, monkeys are restricted to prey that are much smaller in body size than themselves.

Mammals comprise the overwhelming majority of prey taken by indigenous hunters in interfluvial Amazonia (Table 9.1). An inventory of the behavioral and ecological attributes of the 41 mammalian species that are most commonly taken as human prey reveals that about 39% are small in body size (5 kg or less), 54% solitary, 44% arboreal, and 73% nocturnal. Also litter size and frequency are very low for most of these mammalian prey species (Sponsel, 1981:171–189).

There are only two outstanding exceptions to these generalizations— two species of peccary and the larger monkeys, which include the cap-

Table 9.1.   Classes in Predation Records*

Class	Number (%)	Weight (%)
Mammals	64–74	91–92
Birds	21–33	4–6
Reptiles	3–5	3–4

*a*Data from Vickers (1984:366, 369–372); comparison of eight indigenous and four mestizo communities in Amazonia.

uchin (*Cebus*), howler (*Alouatta*), woolly monkey (*Lagothrix*), spider monkey (*Ateles*), and woolly spider monkey (*Brachyteles*). Both pigs and these monkeys are diurnal, weigh well above 5 kg, and live in relatively large groups. The most prized prey is the wild pig, which weighs up to 30 kg and may have a group size up to 100 or more (Nowak and Paradizo, 1991).

Whereas pigs are terrestrial, monkeys are arboreal. The disadvantage of the arboreal attribute of monkeys is circumvented by the human predator through the technology of projectile weapons—the blowpipe and the bow and arrow, and the coating of the points with the drug curare. In addition, some societies like the Yanomami notch the arrow points so that they break when a monkey grasps it to remove it, thus leaving part of the point embedded in the wound where the drug continues to diffuse into the blood stream.

From the perspective of the above analysis, it is no surprise that pigs and monkeys, respectively, rank first and second by number, and first and third by weight in the predation record of eight indigenous and four mestizo communities compared by Vickers (1984). From the records of these 12 communities, among the mammalian prey, the most frequent were pigs in seven communities, monkeys in three, and rodents in two (Vickers, 1984:371). Monkeys were the most frequent prey and ranked second in weight (after pigs) for the Ache, Bari, and Waorani societies (Vickers, 1984:366, 369–372). In another comparative study Beckerman and Sussenbach (1983:334) found that 6 of 10 cases (or 5 out of 8 cultures) focused on the white-lipped peccary.

A survey of the ethnographic literature reveals that the species of primates that are most commonly subject to human predation are the howler, woolly, spider, capuchin, and owl or night monkeys (Sponsel, 1981:81; also see Wolfheim, 1983). The larger monkey species may weigh up to 10 kg and have a group size up to 30 or more (Kavanagh, 1983; Moynihan, 1976; Napier and Napier, 1967; Terborgh, 1985:286). In another comparative study, Beckerman and Sussenbach (1983:338–341)

Table 9.2. Waorani Predation on Primates[a]

Name	Rank (number)	Number (%)		Rank (weight)	Weight (%) (kg)	
Woolly monkey (*Lagothrix lagotricha*)	1	562	(17.7)	2	3,873.5	(21)
Howler monkey (*Alouatta seniculus*)	3	246	(7.7)	4	2,197.4	(12)
Capuchin monkey (*Cebus albifrons*)	11	116	(3.6)	8	400.8	(2)
Red-mantled tamarin (*Saguinus fascicollis*)	13	67	(2.1)	29	18.7	(0.1)
Saki monkey (*Pithecia monachus*)	15	51	(1.6)	17	100.0	(0.5)
Squirrel monkey (*Saimiri sciureus*)	18	31	(0.9)	24	23.7	(0.1)
Titi monkey (*Callicebus moloch*)	19	28	(0.8)	32	17.2	(0.1)
Owl monkey (*Aotus tirvirgatus*)	21	25	(0.7)	31	17.6	(0.1)
Spider monkey (*Ateles belzebuth*)	28	13	(0.4)	16	110.1	(0.6)
Total monkey prey		1,072			6,759	
Total prey		3,165			18,781.1	
Monkey prey/total prey (%)		33.8			35.9	

[a]Data from Yost and Kelley (1983:208, 210–211) based on 867 hunts by 13 hunters. (The two species of peccary accounted for 9% by number and 41% by weight of the total prey.)

found that woolly monkeys yield 1% of the meat by weight in the diet of Siona-Secoya, monkeys provide 5% for Yanomami and 5% for Shipibo, and spider monkeys 10% for Bari, 10% for mainstream Makuna, and 19% for headwater Makuna.

The Waorani of Ecuador take an unusually high percentage of monkeys as prey. In 867 hunts 13 hunters killed nine different species of monkeys out of a total of 37 prey species. Monkeys comprise 33.8% by number and 35.9% by weight of their total prey (Yost and Kelley, 1983:208, 210–211) (see Table 9.2). Even in indigenous groups that do not focus on monkeys, hunters still occasionally kill them for food and other purposes (Steward, 1944).

Thus monkeys can be an important source of quality protein for indigenous human populations in Amazonia, even though they usually comprise a small proportion of the mammalian biomass in the forest ecosystem. The faunal composition of areas in Surinam, Barro Colorado Island, and the Central Amazon are estimated to range from 10 to 35% for rodents, 18 to 50% for edentates (anteaters, sloths, armadillos), and 5 to 10% for primates (Beckerman and Sussenbach, 1983:344; see also Bourliere, 1985; Jolly, 1985:31; Richard, 1985a:431–432). Monkeys are also preferred for taste, especially some like the spider monkey, which is usually the first species to be depleted as a new area is colonized (Wolfheim, 1983:249).

The subsistence hunting of primates appears to be more intensive in Amazonia than other tropical rainforest areas. About 80% of the callitrichid (marmoset and tamarin) species and 100% of the cebid species are subject to human predation in the Neotropics (Wolfheim, 1983:739). This compares to an overall average of 34% (51 species) of nonhuman primates, which are hunted in more than half the countries of their ranges in the Old and New Worlds (Wolfheim, 1983:739).

The higher percentage of primates hunted in Amazonia may reflect the relatively low availability of alternative prey as suggested by the analysis of the behavioral and ecological attributes of potential prey previously presented (see above). Furthermore, fewer ungulate species are available to hunt in Amazonia compared to the tropical rainforests of the Old World. For example, in the Amazon there are only nine species of ungulates, whereas in the African rainforest there are 27 (Bourliere, 1973:281). Also the Neotropics lack megafauna, unlike the African and Asian forests (Beckerman and Sussenbach, 1983:337; Hershkovitz, 1972:372).

Even the primate resources of the Amazon are poorer than those of the Old World. New World monkeys have a smaller body size (Kinzey, 1986a:128), smaller group size (Jolly, 1985:117–119), and lower population densities (Jolly, 1985:30–31) than their Old World counterparts, something that may be related to the distinctive phenomena of demineralization (cf. Terborgh, 1983:205–208; Jolly, 1985:18, 21, 39).

Human predation on Amazonian primates may be an important selective pressure determining body size. The larger the body size of the monkey the more vulnerable it is because it is easier to detect and hit as a target, and because the cost/benefit ratio is more favorable for the hunter in terms of the food returned for his investment of time and energy. The largest of the cebids, the muriqui or woolly spider monkey (*Brachyteles arachnoides*), is near extinction. The smallest of the cebids, the squirrel monkey and night or owl monkey, are the least threatened. The largest of the marmosets, the lion tamarin (*Leontopithecus rosalia*), is the most endangered (Wolfheim, 1983:733). Human predators prefer larger monkeys, whereas nonhuman predators more frequently prey on smaller ones (Cheney and Wrangham, 1987:234; Terborgh, 1983:194–197). For example, the Waorani hunt larger monkeys in preference to smaller ones; however, sometimes they also take smaller species like the squirrel and night monkeys (Yost and Kelley, 1983:208, 210–211) (see Table 9.2). Nevertheless, in areas where larger species have been depleted, hunters turn to the smaller ones such as the squirrel monkey (Soini, 1972).

The large spider monkeys, which appear relatively invulnerable to most predators (Cheney and Wrangham, 1987:238; Terborg, 1983), are prized by the human hunter. Also many communities consider the meat of the spider monkey to be particularly delicious. This species is often the first to be depleted as a new area is settled (Freese, 1975; Wolfheim, 1983:249). It is especially vulnerable to local extinction because of its slow reproductive rate. Females do not breed until their fourth year and they give birth only once every 2 years. Consequently population recruitment is fairly slow in this genus (Klein and Klein, 1976).

Also human predation may reduce the number of polyspecific associations in an area since they are more likely to be detected (Jolly, 1985:172). On the other hand, humans may reduce some of the limiting factors on monkey populations through hunting their predators and competitors (Richard, 1985a:80–81).

Symbiotic Ecology

Beyond predation, other kinds of symbioses may exist between human and nonhuman primate populations in Amazonia (cf. May and Seger, 1986).

Succession

Fluvial dynamics such as the meandering of rivers and oxbow lakes create a mosaic of plant communities in different stages of succession in floodplain zones (Colinvaux et al., 1985; Remsen and Parker, 1983; Salo, 1986).[7] Other natural perturbations such as lightening strikes, wind storms, and floods also trigger succession and generate biodiversity in Amazonia (Collinaux, 1989; Denslow, 1987). The influence that the mosaic of successional communities has on monkeys is being explored at Cocha Cashu, Peru (Terborgh, 1983).

Swiddening (slash-and-burn horticulture) also creates a mosaic of plant communities in different stages of succession, and this, in turn, surely influences nonhuman primate populations in various ways. Swiddening may extend back 6,000 years or more in Amazonia (Meggers, 1985:310; Sanford et al., 1985; Roosevelt, 1989:82). Also it probably had a significant environmental impact since, prior to depopulation through epidemics of Western diseases and other factors, the indigenous population of the Amazon was substantial and concentrated in riverine zones (Roosevelt, 1989:76). The precontact population of Amazonia is estimated at 6 million (Denevan, 1992). The human impact on tropical forests in aboriginal times has probably been grossly underestimated, even if it was largely benign, especially compared to the direct and indirect impact of Western civilization (Clay, 1988; Hallsworth, 1982). Balée (1989) estimates that about 12% of the Amazon is anthropogenic.

The role of swiddening in habitat modification and diversification may be especially important for some monkey species. Howlers are among the species that are attracted to secondary growth (Kinzey, 1986a:136). In some cases, indigenous societies like the Kayapó even plant fruit and other trees to attract game to gardens as the frequency and intensity of harvesting decline over several decades. This practice is correlated with forest regeneration through succession, a process often encouraged by indigenous agroforestry (Posey, 1982; Linares, 1976).

Competitors

The competition between humans and monkeys is probably reduced by the terrestrial habitat of the former. Nevertheless, populations of humans and monkeys may be competitors in some situations. Both use palms as major food resources, for example (Kinzey, 1986a:136). Also monkeys may be crop pests in some areas (Wolfheim, 1983:743).

Disease

Another symbiotic relationship between human and monkey populations in Amazonia may be disease. Monkey populations may act as reservoirs for human diseases such as malaria (Warren and Adel, 1984; Coimbra, 1995).

Symbiotic interactions among species of plants and animals are an important frontier for research in tropical ecology (Janzen, 1986; Richard, 1985a: Chs. 10–11). Clearly symbioses between humans and monkeys also merit much more systematic investigation.

Cultural Ecology

Some Possibilities

Unfortunately information on the role of monkeys in the cultural ecology, ethnoecology, and ethnography of Amazonian indigenous societies is anecdotal and scattered in an occasional sentence or paragraph in the literature. Perusal of the *Handbook of South American Indians* (Steward, 1944) through references to monkeys in the index reveals that, at the minimum, monkeys are hunted for food and sometimes proscribed by taboos, kept as pets, used for body ornamentation (teeth, fur, tails), used to make projectiles (bones), and figure in myths and folklore. (However, the particular species of monkey is almost never identified in the literature, a common deficiency in ethnography.) A search of the more recent ethnographic literature has not revealed any more substantial coverage of the subject. However, monkeys may also be as culturally important for some societies in Amazonia as they are elsewhere, such as in many parts of Asia (McNeely and Wachtel, 1988; Ohnuki-Tierney, 1987).

Wolfheim (1983) identifies a number of interesting uses of nonhuman primates in Amazonia: the rendered subcutaneous fat of the woolly monkey (*Lagothrix lagothricha*) is used for frying, seasoning, and medicines (p. 315); the howler monkey (*Alouatta seniculus*) is used for bait in jaguar and ocelot traps in Bolivia, Brazil, and Peru, while its hyoid bone is used as a drinking and medicinal device in Colombia, Peru, and Surinam (p. 231); the uakari monkey (*Cacajao melanocephalus*) is used for bait for fish, turtles, and cats in Brazil (p. 266); the fur of the night or owl monkey (*Aotus trivirgatus*) is used to decorate brides in Colombia (p. 239); the skulls, teeth, and paws of the squirrel monkey (*Saimiri sciureus*) are made into necklaces for sale at markets in Iquitos, Peru (p. 330); the tails of the saki monkey (*Pithecia monachus*) are used for dusters in Brazil and Peru and its teeth and fur are used for ornaments (p. 320); the bearded saki monkey's (*Chiropotes satanas*) tail is used as an ornament by some indigenes in Venezuela (p. 307); and the white-nosed saki monkey's (*Chiropotes albinasus*) tail is used for a duster (p. 304).

These instances hint at some of the cultural possibilities for monkeys in the traditional indigenous societies of Amazonia. To explore this territory further, the natural place to begin is with those indigenous societies, such as the Waorani, for which monkeys are an important prey species in the diet. In such societies monkeys may be important in the various components of the cultural system from subsistence, economy, and social organization to myth, ritual, and symbol (see Urton, 1985).

Relevance of Cultural Anthropology for Primatology

Through investigations such as those cited above, cultural anthropology can be relevant to primatology in providing information that is otherwise neglected, yet that is essential to understanding human–monkey interactions. It is noteworthy that indigenous societies do not appear to harvest the classes of mammals in direct proportion to their natural abundance (Beckerman and Sussenbach, 1983:344). Culture is an important factor to consider in explaining this disparity.

Studies of the relationship between human and monkey ecology are important even if a primatologist is interested only in monkey behavior and ecology. The primatologist cannot know what is "undisturbed" natural behavior and ecology until anthropogenic influences on the monkey species are known, if for no other reason than to factor them out or to establish their absence in areas where monkey populations are not "disturbed" (e.g., Seidensticker, 1983). It would be especially useful to make a controlled comparison by studying the same species in similar habitats in areas that are subject to human predation and in other areas that are not (see van Schaik and van Noordwijk, 1985).

Although primatologists usually avoid studying populations that are subjected to human predation, actually valuable behavioral and ecological data can still be collected. While it is rare for a fieldworker to witness nonhuman predation on monkeys, human predation on monkeys can be observed regularly and in detail, provided the primatologist and/or ethnographer is willing to reside in indigenous communities such as the Waorani and accompany people on hunts as much as possible. This situation provides a unique opportunity for the collection of data on monkeys such as age, sex, weight, stomach contents, ecto- and endoparasites, and disease. It is unique because to obtain some of this information monkeys would have to be sacrificed, a measure that few field primatologists would be willing to take in the usual naturalistic studies of the modern era, especially given the problems of primate conservation. During periods when monkeys are not taken as prey the fieldworker can still collect relevant data through intensive interviews with hunters and other indigenes.

In the context of primate conservation, Wolfheim (1983) is concerned with the lack of quantitative measures of the intensity of human predation on primates. Moreover, "Reliable estimates of predation rates can be obtained only if the predators themselves are systematically observed" (Cheney and Wrangham, 1987:231). Fieldwork based at indigenous villages would allow such observations. Inadvertently, some relevant data are already available from several cultural ecological studies, the most detailed ones being on the Siona-Secoya (Vickers, 1983, 1984, 1988) and Waorani (Yost and Kelly, 1983).

Relevance of Primatology for Cultural Studies

The framework of primatology is also relevant to cultural anthropology. It can generate new kinds of data, questions, and analyses for studies of human ecology and adaptation. For example, even in the best studies of cultural ecology, often information such as population density, group size, and home range is missing. Yet such information is elemental for an ecological understanding of the populations of any animal species including our own. The collection of such data is a prerequisite to effective and penentrating comparisons between *Homo sapiens* and other species, including monkeys. Moreover, as noted earlier, the niche of a traditional indigenous population, which is basically a feeding strategy, can be adequately defined only with attention to the ecology of the faunal community, especially in the case of societies that engage in any foraging (Sponsel, 1981:156–197). Thus, primatology can contribute to cultural anthropology relevant material that is otherwise neglected.

Ethnoecology

Successful hunting by indigenes requires detailed empirical knowledge of the behavior and ecology of the prey species. Thus, if fieldworkers can transcend the ethnocentric superiority complex of Western science, they may learn a lot about nonhuman primates through indigenous experts on ethology and ecology, albeit the empirical knowledge of the latter may be mixed with mythological, symbolic, and other cultural content. Unfortunately, most ethnoecology research has been preoccupied with ethnosemantics (folk classifications) and has not related environmental ideas to actions and the ecosystem (Fowler, 1976; Spradley, 1979). However, ethnoecology has considerable potential to contribute to a better understanding of human adaptation (Patton et al., 1982; Reichel-Dolmatoff, 1971; Taylor, 1974).

Conservation Ecology

The relationship between human and nonhuman primate ecology is not of purely academic interest, but has important implications for the conservation of primate populations and their forest habitats. From the foregoing discussion it should be clear that indigenes as predators and as agents of habitat modification and diversification are an important element to consider in primate conservation.

Likewise, nonhuman primates as a source of quality protein are an important consideration for those concerned with promoting the survival and welfare of indigenous societies. It would be a mistake to frame this relationship as a conflict of interest between monkeys and indigenes, or between primatologists and cultural anthropologists. From predator–prey studies, it is well known that moderate predation may actually benefit a prey population in the long term (Deshmuhk, 1986: 123–124). It is also well known that in the absence of predators, which help regulate a prey population, some mammalian species may actually degrade their habitat and undermine their own resource base. Wild pigs are notorious in this regard (e.g., Diong, 1973). Moreover, subsistence hunting may be accomplished on a sustained-yield basis, as a remarkable 10-year study among the Siona-Secoya demonstrates (Vickers, 1988). [Subsistence and market hunting of primates should not be confused, since the latter may rapidly deplete populations (Mittermeier 1987).]

Thus, systematic investigation of symbiotic relationships between human and nonhuman primate species, including predation, is of great importance for applied as well as basic science. It is one of the fundamental factors to consider in the conservation of monkeys and their forest ecosystems in Amazonia and elsewhere. Indeed, most of the attention to human influences on nonhuman primates has come from conservation-oriented studies, although the subject has much broader interest as indicated here in the previous discussion.

In the conservation context, human predation is often considered to be somehow unnatural (Wolfheim, 1983:738–739). Wolfheim (1983:741–742) makes an interesting distinction:

> The status of each animal population is determined by a constellation of ultimate and proximate causes: original geographic range, body size, density, and habitat are the ultimate determinants of population status, while habitat alterations and human predation are the proximate ones. Ultimate causes are inherent in the evolutionary history of the species, dependent on geological, ecological, and genetic characteristics. These traits are specific to species or subspecies and determine the degree to which proximate causes affect a species. Proximate causes are those directly responsible for the expansion, reduction, extinction, merging, or isolation of populations. In the absence of human interference, spontaneous events, such as changes in riverbeds, soil, or climate, could lead to modifications in the ranges or habitats of primate groups. But human enterprises are now overshadowing these natural processes.
>
> Thus the proximate influences considered here are those resulting from human activities: alteration of the habitat on which the species depend, and exploitation of the animals themselves.

However, this distinction may not be so clear and useful if one considers the antiquity of human predation, swiddening, and other activities in tropical forests, as discussed earlier. Traditional indigenous habitat alteration and predation on primates for subsistence purposes are no less natural ecological and evolutionary processes than the other factors mentioned by Wolfheim. Only from colonial times until the present with the development of the commercial extraction industries in Amazonia, which irreversibly deplete resources and degrade environments, is there clearly a destructive process operative that can be labeled unnatural (Bunker, 1984; Sponsel, 1992; Stone, 1986).

Unfortunately, Wolfheim does not distinguish between *subsistence* and *commercial* hunting, and between indigenous and other hunters. Wolfheim (1983:743) distinguishes three types of predation: species killed for meat, those killed as crop pests, and those trapped for export. This is an economic typology, not an ecological one. Also it applies more to acculturated indigenes and mestizos than to traditional indigenes. This is not to ignore the fact that primates were part of an extensive network of exchange in Amazonia and from the Amazon into Andean communities prior to European contact (Murra, 1944:794).

When Wolfheim and other primate conservationists neglect the realities of indigenous economy and cultural ecology, they miss a significant factor for primate ecology and conservation. As Aiken and Leigh (1985:19) observe: "Conservation policies should take into account the environmental knowledge, attitudes, and values of indigenous forest dwellers." (See also Brokensha et al., 1980; Denslow and Padoch, 1988; McNeely and Pitt, 1985; McNeely and Watchel, 1988.) To the extent that indigenous peoples are knowledgeable about monkeys and have an impact on their behavior, ecology, and demography, they cannot be ignored. For instance, in the case of endangered species and areas, one way to relieve some of the predation pressure on monkey populations is to encourage indigenous societies to shift to other protein resources, if available.

Also, one way to protect monkey habitat is to protect traditional indigenous lifestyles that usually do not irreversibly deplete resources and degrade environments (Clay, 1988). Another way to protect monkeys is to employ and train indigenes as nature reserve or park guards, foresters, etc. Much more attention needs to be given to collaboration between primate conservationists and indigenous societies, building on their common interests. Manu National Park in Peru provides an example of successful collaboration between conservationists and indigenes.

Conclusion

It is curious that in spite of Linnaeus and subsequent progress in viewing our species from a primatological perspective, field primatologists have largely ignored *Homo sapiens* while cultural anthropologists have largely ignored nonhuman primates. From this chapter it should be clear that both in the natural processes and in the corresponding research, there is significant mutual relevance in the ecology of human and monkey populations in Amazonia. Ethnoprimatology, the interface between human and primate ecology, is a new frontier with enormous potential for basic and applied research in the future.

The human niche as a feeding strategy can be defined only in relation to other species in the faunal community. The special character of the human primate and the interfluvial forest ecosystem renders monkey ecology especially important for understanding the adaptation of many indigenous societies in Amazonia.

Such a view of humanity need not be reductionist. The comparative ecology of primates including *Homo sapiens* in Amazonia not only reveals continuities, but highlights discontinuities as well. Indeed, the present exploration emphasizes the unique attributes of *Homo sapiens* among the species in the primate community, and it points to the adaptive importance of technology and other aspects of culture. It also reveals the neglect of research on the cultural context and meaning of monkeys in traditional indigenous societies of Amazonia.

The survival and welfare of both indigenous humans and monkeys in Amazonia are closely interrelated. Primate conservation ecology must consider human predation pressure and environmental impact. Advocacy anthropology must consider monkeys among the renewable natural resources for indigenous populations. The two concerns may be quite compatible rather than in conflict.

Finally, to consider indigenous human populations as an integral part of nature, a component of the biotic community of the ecosystem, is to point to their adaptive wisdom compared to the superiority complex of Western civilization, which is at the heart of the spiraling ecological disequilibrium facing modern humanity on a global scale (Sponsel, 1995).

Acknowledgments

Although the author assumes full responsibility for this chapter including any deficiencies, special thanks are given to Thomas Hogue and Kent Redford who provided very helpful comments on the manuscript.

Notes

1.   In 1989 the author searched the complete series of the following journals: *American Journal of Primatology, Folia Primatologica, International Journal of Primatology,* and *Primates.* A bibliographic search was also made by the Primate Information Center of the Regional Primate Research Center, University of Washington, Seattle, Washington. These systematic and exhaustive efforts did not reveal a

single field study comparing the ecology of human and nonhuman primate species in the same ecosystem anywhere in the world.

2. Symbioses are any kind of interactions between populations of different species or between populations within a species (Odum, 1971:211). The interactions may be beneficial, harmful, or neutral to one or both parties involved. They include cooperation (+/+), mutualism (+/+), commensalism (+/0), amensalism (−/0), competition (−/−), parasitism (+/−), and predation (+/−). Mutualism is distinguished from cooperation in that the interaction is necessary for the survival and maintenance of each species.

3. Richard (1985a) aims to examine the Order *Primates* as an end in itself, rather than the usual route, as a means to the end of understanding human evolution and behavior (Jolly, 1985). Nevertheless, this zoological attempt, while admirable in many ways, remains anthropocentric by arbitarily excluding *Homo sapiens* from its survey of the Order *Primates* and from the realm of natural ecological phenomena.

4. The author is aware of the problems inherent in comparing data on populations of species derived from different regions, given the extent of intraspecific and environmental variation in Amazonia. Seasonal and habitat differences in diet and behavior may be greater at the intraspecific than the interspecific level in monkeys (Kinzey, 1986a:138). Nevertheless, in this chapter, in general, variations within this ecosystem will be largely ignored given the reliance on a literature review, since there are no field studies that encompass the local human population as an integral part of the primate community of an ecosystem.

5. Prance and Lovejoy (1985) provide general background for understanding Amazon ecology. For recent surveys of human ecology see Moran (1993) and Sponsel (1995).

6. See Hames (1979b), Saffirio and Hames (1983), Saffirio and Scaglion (1982), and Yost and Kelley (1983) for the impact of Western technology on indigenous hunting.

7. Succession refers to a dynamic ecological process wherein successive stages of plant and associated animal communities develop in an area until a climax stage is reached, which is stable baring new perturbations. The climax stage represents a dynamic and temporary equilibrium between abiotic and biotic factors in the ecosystem (Odum, 1969).

II

SYNOPSIS OF NEW WORLD PRIMATES (16 GENERA)

WARREN G. KINZEY

10

Classification of Living New World Monkeys

Classification of Living New World Monkeys[a]

	Common name
Infraorder	
Superfamily	
Family	
Subfamily	
Tribe	
Subtribe	
Genus	
species[b]	*Common name*
Platyrrhini	
Ateloidea	
Atelidae	
Atelinae	
Atelini	
Atelina	
Ateles	Spider monkeys
A. paniscus	Black spider monkey
A. belzebuth	Long-haired spider monkey
A. fusciceps	Brown-headed spider monkey
A. geoffroyi	Black-handed spider monkey
Brachyteles	
B. arachnoides	The muriqui
Lagotrichina	
Lagothrix	Woolly monkeys
L. lagotricha	Common wooly monkey
L. flavicauda	Yellow-tailed woolly monkey
Alouattini	
Alouatta	Howler monkeys
Alouatta seniculus group	
A. belzebul	Black and red howler

[a]Rosenberger et al. (1990). See also Rosenberger (in press).
[b]Species according to Mittermeier et al. (1988).

169

Classification of Living New World Monkeys[a]

Infraorder	
Superfamily	
Family	
Subfamily	
Tribe	
Subtribe	
Genus	
species[b]	Common name

A. fusca	Brown howler
A. seniculus	Red howler
Alouatta palliata group	
A. coibensis	Coiba Island howler
A. palliata	Mantled howler
A. pigra	Guatemalan howler
Alouatta caraya group	
A. caraya	Black howler
Pitheciinae	
Pitheciini	
Pithecia	Saki monkeys
Pithecia pithecia group	
P. pithecia	Guianan saki
Pithecia monachus group	
P. aequatorialis	Red-bearded saki
P. albicans	Buffy saki
P. irrorata	
P. monachus	Black-bearded saki
Chiropotes	Bearded saki monkeys
C. albinasus	White-nosed saki
C. satanas	Black-bearded saki
Cacajao	Uakaris
C. calvus	White uakari and Red uakari
C. melanocephalus	Black-headed uakari
Homunculini	
Aotus	Night monkeys
Gray-neck group	
A. brumbacki	
A. lemurinus	
A. trivirgatus	
A. vociferans	

(continued)

Classification of Living New World Monkeys[a]

	Common name
Infraorder	
Superfamily	
Family	
Subfamily	
Tribe	
Subtribe	
Genus	
species[b]	
Red-neck group	
A. azarae	
A. infulatus	
A. miconax	
A. nancymai	
A. nigriceps	
Callicebus	Titi monkeys
C. cupreus	
C. torquatus	
C. oenanthe	
C. caligatus	
C. moloch	
C. brunneus	
C. modestus	
C. olallae	
C. donacophilus	
C. personatus	
Cebidae	
Cebinae	
Cebus	Capuchin monkeys
Tufted group	
C. apella	
Untufted group	
C. albifrons	
C. capucinus	
C. olivaceus	
(=*C. nigrivittatus*)	
Saimiri	Squirrel monkeys
Roman type group	
S. boliviensis	
Gothis type group	
S. sciureus	
S. oerstedi	
S. ustus	

(continued)

Classification of Living New World Monkeys[a]

	Common name
Infraorder	
Superfamily	
Family	
Subfamily	
Tribe	
Subtribe	
Genus	
species[b]	*Common name*

	Common name
Callitrichinae	
Callitrichini	
Callitrichina	
Callithrix	Marmosets
Callithrix argentata group	
C. argentata	
C. emiliae	
C. humeralifer	
Callithrix jacchus group	
C. aurita	
C. flaviceps	
C. geoffroyi	
C. kuhli	
C. jacchus	
C. penicillata	
Cebuella	Pygmy marmoset
C. pygmaea	
Leontopithecus	Lion tamarins [?]
L. chrysomelas	
L. chrysopygus	
L. rosalia	
Leontocebina	
Saguinus	Tamarins
Hairy-face tamarins	
Saguinus nigricollis group	
S. fuscicollis	
S. nigricollis	
S. tripartitus	
Saguinus mystax group	
S. imperator	
S. libiatus	
S. mystax	
Saguinus midas group	
S. midas	

(continued)

Classification of Living New World Monkeys[a]

Infraorder	
Superfamily	
Family	
Subfamily	
Tribe	
Subtribe	
Genus	
species[b]	*Common name*
Mottled-face tamarins	
S. inustus	
Bare-face tamarins	
S. bicolor	
S. geoffroyi	
S. leucopus	
S. oedipus	
Callimiconini	
Callimico	Goeldi's monkey
C. goeldii	

11

Alouatta

Introduction

Alouatta (the howler, or howling monkey) is one of the largest (average weight = 6.4 kg) and the most sexually dimorphic monkey in the Neotropics. Body weight dimorphism $(100[\male - \female]/\female)$ varies from 29% in *A. seniculus* to 76% in *A. pigra* (Ford and Davis, 1992). Canine dimorphism is greatest in *A. fusca* and *A. belzebul* (Kay et al., 1988). There is also sexual dimorphism in adult coat color in *A. caraya* ($\male\male$ are black; $\female\female$ are buff) and in *A. fusca clamitans* ($\male\male$ are rufous; $\female\female$ are brown). The most significant morphological feature is the enlarged hyoid bone, which acts as a resonator for howling. The hyoid is much larger in the male than in the female, and largest hyoid is found in male *A. seniculus* (Crockett and Eisenberg, 1987). Howlers are frequently referred to as "folivores," but the term "folivore-frugivore" is more appropriate since they eat approximately equal portions of fruit and leaves. Howlers are polygynous, but vary in male composition between and within species; both males and females emigrate (Crockett and Eisenberg, 1987).

Since Carpenter's field observation of *Alouatta palliata* on Barro Colorado Island (BCI) from 1931 to 1932 (Carpenter, 1934), which was the first field study of a nonhuman primate, *Alouatta* has been the most studied genus in the Neotropics, especially from the perspective of foraging ecology and demography (Kinzey, 1986; Southwick and Smith, 1986). In the past 5 years, as many as 800 articles and abstracts have been published on this genus.

Taxonomy and Distribution

Alouatta is the most widespread neotropical primate genus. Its range extends from southern Mexico (Maps 15 and 17) (although it does not extend as far north as *Ateles*) to southeastern Brazil and Argentina (Map 1). The genus is generally thought to contain six species (Wolfheim,

174

1983; Crockett and Eisenberg, 1987; Neville et al., 1988). Based on dermatoglyphics and zoogeographic considerations (Froehlich and Froehlich, 1986, 1987), there may be a seventh species, *A. coibensis*, in Panama (Mittermeier et al., 1988b). The species are essentially allopatric, but areas of sympatry have been reported to occur in southern Mexico between *A. palliata* and *A. pigra* (Smith, 1970), in northwestern Colombia between *A. palliata* and *A. seniculus* (Hernández-Camacho and Cooper, 1976)(see Map 1), and in northeastern Argentina between *A. caraya* and *A. fusca* (Crespo, 1982). The southern limit of the distribution of the Amazonian populations of *A. belzebul* remains uncertain (Langguth et al., 1987; Bonvicino et al., 1989).

The only endangered species of *Alouatta* is *A. fusca* (Mittermeier and Cheney, 1987; Mittermeier et al., 1989). *A. palliata* and *A. pigra* are seriously threatened (Crockett and Eisenberg, 1987), and based on Landsat image modeling their habitat will be lost by the year 2025 (Cuaron, 1992).

Evolutionary History

The tribe Alouattini is represented in the fossil record as early as 15–16 mya (MacFadden, 1990) by two species of *Stirtonia* in the Colombian Miocene (Setoguchi et al., 1981; Kay et al., 1987), each showing strong folivorous adaptations (Rosenberger and Strier, 1989). Thus, the ancestor of *Alouatta* was already separated by that time from the atelin clade. It is somewhat surprising that no relatives of *Alouatta* have yet been found in the Patagonian fossil deposits (Rosenberger and Strier, 1989).

With the closing of the Central American land bridge in the Pliocene, howler monkeys spread to Central America, probably in two separate waves (Smith, 1970). The first invasion presumably involved island hopping leading to the differentiation of *A. pigra*, and the later invasion involved *A. palliata*.

Previous Studies

Captive Studies

There have not been many captive studies of howler monkeys, since they do not socialize, survive, or reproduce well under captive conditions. Among platyrrhine primates howler monkeys are the least adaptable to captivity (Schapiro and Mitchell, 1986). Over a 5-year period, 28 red howler monkeys were introduced at Monkey Jungle, a zoological park in Miami, Florida, but only five survived. In France, the Verlhiac Primate Center had a colony of *A. seniculus* and *A. caraya*. At the River-

banks Zoological Park, Columbia, South Carolina, a breeding colony of
A. seniculus and *A. caraya* has been the subject of physiological as well as
behavioral studies (Shoemaker, 1979; Jones, 1983a; Neville et al., 1988).
The Japan Monkey Center, Inuyama, Japan, houses a group of *A. palliata*, and some of the results have been published in Japanese. Kohda
observed allomothering behavior in *A. palliata* (Masataka and Kohda,
1988). The Argentinean Primate Center (CAPRIM) in Corrientes Province has a breeding colony of *A. caraya* (Colillas and Coppo, 1978).

Field Studies

The most extensively studied species is *A. palliata*. Since 1931 they
have been studied frequently on BCI (Barro Colorado Island), Panama,
where they are sympatric with *Cebus capucinus, Ateles geoffroyi, Alouatta
palliata, Saguinus oedipus,* and *Aotus lemurinus*. Studies conducted on BCI
were summarized in Carpenter (1965), Kinzey (1986), and Neville et al.
(1988). To these should be added Dudley and Milton (1990). Mantled
howlers have also been studied at five other sites in Central America.

1. Hacienda La Pacifica, Guanacaste, Costa Rica; lowland tropical
dry forest; initiated by Glander in 1972 where he, Clarke, and others
have accumulated demographic data that allow analysis of population
trends (Clarke, 1990; Clarke and Glander, 1984; Clarke et al., 1986, 1991;
Glander, 1975a,b, 1977, 1978a,b, 1979, 1980, 1981; Milton, 1978; Rockwood and Glander, 1979); other studies are on social relations (Jones,
1978, 1979, 1980a,b, 1982, 1983b; Clarke, 1981, 1982, 1983), vocal behavior (Whitehead, 1986, 1987, 1989), births (Moreno et al., 1991), parasites
(Stuart et al., 1990), and feeding (Bilgener, 1988; Teaford and Glander,
1991). There are no other nonhuman primates at the site.
2. Santa Rosa National Park, Guanacaste, Costa Rica; tropical dry
forest with sympatric *Ateles geoffroyi* and *Cebus capucinus* (Freese, 1976);
howlers are restricted to evergreen forest; Fedigan and others have conducted research since 1983 (Chapman, 1987, 1989a, 1990; Chapman and
Chapman, 1986, 1990; Fedigan, 1986).
3. Palo Verde National Park, Guanacaste, Costa Rica; lowland tropical dry forest with sympatric *Ateles geoffroyi* and *Cebus capucinus*; brief
report of howlers drinking from ground water (Gilbert and Stouffer,
1989).
4. Estación Biológica la Selva, Heredia, Costa Rica; tropical dry forest with sympatric *Ateles geoffroyi* and *Cebus capucinus*; brief surveys only
(Fishkind and Sussman 1987; Wilson, 1991), and a study of *Alouatta* by
Stoner (1991).
5. Los Tuxtlas, Veracruz, Mexico; sympatric *Ateles geoffroyi*; Estrada,

Coates-Estrada, and others have conducted research since 1977 (Costello, 1991; Estrada, 1982, 1986; Estrada and Coates-Estrada, 1984a, 1985, 1986a,b; Estrada and Trejo, 1978; Estrada et al., 1984).

A. pigra groups have been studied by Bolin (1981), Cant (1977), Coelho et al. (1976a,b, 1977), Schlichte (1978), and others at Tikal, Guatemala where *Ateles geoffroyi* is sympatric, and by Horwich (1983a,b) and Horwich and Gebhard (1983) in Belize.

A. seniculus has been most extensively investigated at Hato Masaguaral, Guárico, Venezuela since 1969 by Sekulic, Crockett, and others. *Cebus olivaceus* is the only sympatric primate. Studies at this cattle ranch in the llanos have provided important information on demography, emigration, and immigration, reproduction, male takeover, and infanticide (Crockett, 1984, 1985, 1987, 1991; Crockett and Eisenberg, 1987; Crockett and Pope, 1988; Crockett and Rudran, 1987a,b; Crockett and Sekulic, 1984; Sekulic, 1982a,b,c,d, 1983a,b,c; Sekulic and Chivers, 1986; Sekulic and Eisenberg, 1983). Recent studies have involved paternity exclusion (Pope, 1989, 1990) and the relationship of diet to dental microwear (Ungar, 1990). The red howler has also been studied at nine other sites in South America:

1.   At Hato "El Frio," Apure, Venezuela, Braza and his co-workers have studied reproduction, social behavior and feeding behavior of red howlers (Braza, 1978: Braza et al., 1981, 1983). No other nonhuman primates are found at this site.
2.   At Guri Lake, Bolívar, Venezuela (Kinzey et al., 1988), Peetz (1990) studied a group of howlers that was eventually decimated by jaguar predation (Peetz et al., 1992). *Cebus olivaceus, Pithecia pithecia,* and *Chiropotes satanas* are sympatric.
3.   At La Macarena National Park, Meta, Colombia, beginning in 1967 the Kleins studied *A. seniculus, Cebus,* and *Saimiri* on the Río Guayabero while concentrating on *Ateles* (Klein and Klein, 1975, 1976). More recently Japanese and Colombian primatologists have organized a cooperative site on the Río Duda, a tributary of the Río Guayabero, where they are studying six sympatric species: *A. seniculus, Cebus apella, Ateles belzebuth, Callicebus moloch, Lagothrix,* and *Saimiri* (Ceballos, 1989; Figueroa, 1989; Hirabuki and Izawa, 1990; Izawa, 1989; Izawa and Lozano, 1989, 1990a,b; Izawa and Nishimura, 1988; Nishimura, 1988; Yoneda, 1988, 1990).
4.   In the area of the Río Peneya (tributary of the Río Caquetá) and Río Putumayo, Colombia, Japanese researchers have conducted studies on primate density and ecology (Izawa, 1975, 1976; Nishimura, 1988; To-

kuda, 1968). In addition to *A. seniculus*, there are eight other species in the area: *Lagothrix lagotricha, Ateles belzebuth, Cebus apella, C. albifrons, Pithecia monachus, Saimiri sciureus, Aotus vociferans,* and *Saguinus nigricollis.*

5. In the Andean cloud forest at Finca Merenberg, Reserva La Plata, southwestern Colombia, the Gaulins (Gaulin, 1977; Gaulin and Gaulin, 1982) have studied activity budgets, diet, and ranging behavior. *Cebus apella* is the only other primate at the site.

6. In El Tuparro National Park, eastern Colombia, Defler (1981) obtained density and home range data on *A. seniculus* while studying *Cebus apella, C. albifrons,* and *Callicebus torquatus.*

7. In Manu National Park (Madre de Dios) Peru, there are at least 11 species of nonhuman primates (Terborgh, 1983). *A. seniculus* was studied by Milton (1980), and predation on howlers was noted by Eason (1989) and by Sherman (1991).

8. Raleighvallen-Voltzberg Nature Reserve, Surinam was a study site from 1974 intermittently until 1987 when the civil war terminated its use. The site has been reopened recently to continue research activities (Mittermeier, personal communication). Eight species, including *A. seniculus,* were part of a synecological study (Fleagle and Mittermeier, 1980; Mittermeier, 1977; Mittermeier and van Roosmalen, 1981).

9. Neville (1972a,b, 1976a,b) studied *A. seniculus* in Bush Bush Forest, Trinidad where *Cebus albifrons* is also found.

A. caraya was studied on islands in the Río Paraná (Pope, 1966, 1968), and more recently as part of the Argentinean Primate Center studies (Colillas and Coppo, 1978; Milton, 1980; Rumiz et al., 1986; Rumiz, 1990; Thorington et al., 1984; Zunino, 1989).

A. fusca has been studied by Kuhlman (1975), Da Silva (1981), and Mendes (1985).

A. belzebul is the least investigated species, but has been studied around Tucurui Lake, which resulted from flooding of the Rio Tocantins above a hydroelectric power dam (Bonvicino, 1989; Bonvicino et al., 1989; Johns, 1986b; Schneider et al., 1991).

Not all species of *Alouatta* have been equally well studied. In fact, *Alouatta* contains two of the least studied platyrrhine species, *A. fusca* and *A. belzebul.* Thus, many findings may not represent the genus as a whole (Crockett, 1987). Most of the above sites are described in more detail in Neville et al. (1988).

Habitat and Ecology

Alouatta have the widest range of habitats of any neotropical primate (Crockett and Eisenberg, 1987), including swamp forest, seasonally flooded forest, gallery forest, wet evergreen to seasonal semideciduous habitats, and dry deciduous thorn forest. Because they derive most of their necessary moisture from leaves, they are not bound to areas near permanent water (Eisenberg, 1991). They are known to cross large open spaces of a kilometer or more between forest patches (Neville et al., 1988). They are "pioneer" species (Eisenberg, 1979) capable of colonizing rapidly in a wide variety of habitats, possibly through interdemic selection (Pope, 1989).

Alouatta occur from sea level to over 3,200 m (Hernandez-Comacho and Cooper, 1976). In most areas, howler monkeys are sympatric with other primates, but they are one of few primates that successfully exist as the only nonhuman primate species in an area (Crockett and Eisenberg, 1987).

Although howlers spend most of their time in the middle to upper strata, they do come down to the forest floor to drink water (*A. fusca*: Galetti et al., 1987; *A. palliata*: Gilbert and Stouffer, 1989) and to ingest termite soil and clay that are high in mineral content (Hirabuki and Izawa, 1990; Izawa and Lozano 1990a).

Diet

Howler monkeys are either folivore-frugivores or frugivore-folivores, depending on the population and the season. They are extremely selective feeders and rarely eat the most common food (Milton, 1987). At the same time, they are flexible over time and space in their diets. Chapman's 4-year study of feeding behavior of three species of Costa Rican primates showed that in most months, red howlers ate a mixture of fruits, flowers, and leaves. However, in one month, they fed almost exclusively on flowers, and in two months, almost exclusively on leaves. Their choice of food was not always consistent with food availability. Furthermore, their diet was not consistent from one year to the next (Chapman, 1987). The most important factor influencing food choice is apparently protein content, and howlers require, on average 15.6% protein (by dry weight) in their diet (Bilgener, 1988).

Howler monkeys prefer young leaves to mature leaves. Mantled howlers at BCI spent 62% of the feeding time on leaves. They may spend as much as 65–80% of the feeding time eating leaves. However, in other habitats they were reported to spend as much as 80% of time eating fruit. They ate leaves significantly less (35%) and ate flowers and flower buds more in the dry season. There was no outstanding difference in the amount of time they spent eating fruits between the dry and wet season (Milton, 1978, 1987c).

Howlers were reported to feed on more immature fruits than other sympatric primates, but quantitative data are unavailable (Terborgh, 1983; Crockett and Eisenberg, 1987). Howler monkeys are important seed dispersal agents for at least 15 plant species (Estrada and Coates-Estrada, 1986a). A slow food passage rate ensures that ingested seeds are deposited away from the parental plants (Estrada et al., 1984). Arthropods are only inadvertently ingested along with plant food. Some reports, however, suggest that howlers eat birds' eggs (Neville et al., 1988).

Although howlers are well known to obtain water from various arboreal reservoirs, such as bromeliads (Neville et al., 1988), they recently have also been observed to come to the ground to drink (Gilbert and Stouffer, 1989).

Howler monkeys are behavioral-physiological folivores. They lack complex gastrointestinal tracts, but some hindgut enlargement is present. The food passage rate is extremely slow (Milton, 1984e). As much as 31% of their required energy may be derived from fermentation end products. They possess large salivary glands that may solve the problem of leaf tannins by binding many such polymers before the food bolus reaches the gut (Milton, 1987).

Recently, field studies have been conducted of the microwear of howler molars (Runestad and Teaford, 1990; Teaford and Glander, 1990) and incisors (Ungar, 1990). Howler microwear data have confirmed the infrequent use of incisors and established the rate at which wear occurs. Such data allow taxonomic differences in wear patterns to be correlated with actual field observations of feeding and foraging (Walker and Teaford, 1989).

Interspecific dietary differences among howlers seem to be smaller than population differences due to habitat and season. In wetter habitats, both red and mantled howlers spend similar amounts of time feeding on leaves and fruit. In contrast, the mantled howlers in a highly seasonal semideciduous habitat spend comparatively more time feeding on leaves and flowers and less time on fruits (Crockett and Eisenberg, 1987). Details of dietary characteristics of each howler species are provided in Neville et al. (1988).

Predation

The harpy eagle, *Harpia harpyja*, the largest raptor in the world, has been reported to attack and kill the adult male of a group of *A. seniculus* (Peres, 1990b). Other incidences of attack on red howlers by harpy eagles were reported at Cocha Cashu (Eason, 1989; Sherman, 1991), and in Guyana (Rettig, 1978). Peetz et al. (1992) reported a jaguar killing howler monkeys.

Behavior

Social Behavior

Group size is interspecifically and intraspecifically variable, but generally larger where the population density is higher (*A. palliata*: 6–23, Crockett and Eisenberg, 1987; *A. pigra*: 4–6, Horwich and Johnson, 1986; *A. seniculus*: 3–16, Crockett and Eisenberg, 1987; *A. caraya*: 2–19, Rumiz, 1990). The composition and size of mantled, red, and black howler groups are significantly different, even when population densities are similar. Mantled howler groups usually contain more adult females per male than red and black howler groups (Crockett and Eisenberg, 1987; Rumiz, 1990). Rumiz's (1990) study of *A. caraya* shows that solitary individuals and small associations of adult and subadult individuals were found in inferior quality habitats. Surprisingly, multimale groups are less frequently seen in the more dimorphic species (*A. pigra*, *A. seniculus*, and *A. caraya*) than in the least dimorphic species, *A. palliata*. In *A. palliata* a single studied troop was reported to be led most often by subordinate females, but it was not thought that they directed or initiated the progression (Costello, 1991). Howler groups may not be as stable as once thought. *A. seniculus* groups studied by Chapman (1990) divided into subgroups during foraging, showing characteristic fusion–fission organization.

A. palliata, *A. pigra*, and *A. seniculus* rarely groom, and it has generally been thought to be typical of all species of howlers. Most grooming bouts are short and are performed by females, especially adults. The principal groomees are infants or adult males. Male members rarely groom other group members (Crockett and Eisenberg, 1987; Goosen, 1987). In contrast, *A. fusca* (Mendes, 1985), and *A. caraya* in both the field (Thorington et al., 1984) and captivity (Neville and Gunter, 1979) are frequent groomers.

Both sexes leave the natal group before sexual maturity; thus, most adult group members are probably unrelated (Glander, 1980; Clarke, 1990). Females emigrate and may form an extra group association or join an already established group; males invade other groups by forceful takeovers (*A. seniculus*: Rudran, 1979; Crockett, 1984, 1991; *A. palliata*: Glander, 1980; *A. caraya*: Rumiz, 1990). The rate of male emigration increases with increase in population density (Eisenberg, 1991).

Male red, mantled, and black howlers have been reported to commit infanticide during or after a group takeover. This occurs both in one-male and multimale groups (Collias and Southwick, 1952; Clarke, 1983; Crockett and Sekulic, 1984; Rumiz, 1990). In red howlers, infanticide is a major cause of infant mortality (Crockett and Rudran, 1987b). Izawa

(1989) reported adoption of a newborn by the dominant female after loss of her own infant by an infanticidal male. The rate of troop takeover, displacement of the resident male, and subsequent infanticide is density dependent (Eisenberg, 1991).

Male–male agonistic interactions within a group are rare in *A. palliata* (Clarke, 1986). However, intraspecific physical aggression seems common in *A. seniculus*. The high incidence of injuries in subadults is consistent with the individuals most likely to be emigrating or immigrating. Nevertheless, many of them recover from injuries and successfully reproduce (Crockett and Pope, 1988). In multimale *A. palliata* groups, male tenure in the alpha position averages 12.5 (range 3–36) months (Glander, 1980).

Ranging Behavior

Average day range varies among species and within species living in different habitats, but usually it is less than 600 m (Crockett and Eisenberg, 1987). Within-site daily variability is more significant than between-site variability. There is considerable seasonal difference in day range length between drier habitats and seasonal forests; longer in the wet season in the drier habitats and longer in the dry season in seasonal forests with higher rainfall (Crockett and Eisenberg, 1987; Neville et al., 1988).

There is considerable interspecific as well as intraspecific variation in size of home range (*A. palliata*: 3–60 ha; *A. pigra*: 11–125 ha; *A. seniculus*: 4–25 ha; *A. fusca*: 5.8 ha) (Crockett and Eisenberg, 1987; Neville et al., 1988). As is the case with day range, interspecific differences are smaller than intraspecific differences. This suggests that environmental differences have a major influence on home range size. The degree of home range overlap seems to be smaller in red howler groups than in mantled howler groups (Crockett and Eisenberg, 1987). Howler monkeys do not defend territories.

Activity Pattern

Traditionally howler monkeys are considered to be diurnal primates; however, a recent study provides evidence that a population of *A. pigra* observed in Belize was cathemeral (Dahl and Hemingway, 1988; Tattersall, 1987).

Mating and Reproduction

Female *A. palliata* and *A. seniculus* produce subtle cyclic swellings and color changes, whereas female *A. caraya* do not show any signs of reproductive receptivity (Glander, 1980; Crockett and Eisenberg, 1987; Crockett and Sekulic, 1982).

Howler monkeys possess odor-producing scent glands. Although pheromones have not been identified, olfaction seems to be important in determining reproductive states of individuals. Males were reported to rub the throat to substrates and muzzle and lick female genitalia, and *A. palliata* males were reported to taste female urine. Female howler monkeys also wash their hands in urine. Both sexes solicit mating partners, for example, by tongue flicking (Crockett and Eisenberg, 1987; Crockett and Sekulic, 1982; Glander, 1980; Sekulic and Eisenberg, 1983).

Most females give birth to their first offspring at around 5 years of age, but as early as 4 years. The youngest known father was a captive *A. caraya* who sired an infant at 37 months (Crockett and Eisenberg, 1987; Shoemaker, 1979). Chapman and Chapman (1986) report the birth and development of twin infants in *A. palliata* in Costa Rica.

The gestation length ranges from 180 to 194 days (average equals 6.3 months) in *Alouatta*. Females become receptive for 2 to 4 days every 16–20 days (Crockett and Eisenberg, 1987; Crockett and Rudran, 1987a; Izawa and Lozano, 1989; Shoemaker, 1979). Postconception copulation was observed to occur in mantled howler females (Glander, 1980).

The interbirth interval (IBI) for females with a previous surviving infant is variable in *Alouatta*. Mean IBI in *A. seniculus* was 17 months (Crockett and Rudran, 1987a). In Rumiz's (1990) study, the mean IBI of *A. caraya* in northern Argentina was 15.93 months (the shortest IBI was 12 months). In *A. palliata*, the mean IBI was 22.5 months, longer than either *A. seniculus* or *A. caraya* (Glander, 1980). (IBI with a previous dead infant is as short as 7 months in captive *A. caraya*; Neville et al., 1988.) The variation in IBI seems to be due primarily to habitat and seasonal differences (Crockett and Rudran, 1987a; Crockett and Sekulic, 1984; Glander, 1980; Milton, 1982).

Moreno et al. (1991) observed a breech delivery (legs first) in *A. palliata*. The experienced mother was able to pull on the tail of the fetus during birth and successfully retrieved the newborn. She ate the placenta and chewed the umbilical cord. The delivery took 5 minutes, longer than that of a normal delivery of 1–2 minutes.

In the population of mantled howlers at BCI, births occur throughout the year; however, in red howlers at Hato Masaguaral, Venezuela and in black howlers in northern Argentina, a birth peak occurred in the dry season. The neonatal sex ratio of infants was found to be 1:1 (*A. seniculus*: Crockett and Rudran, 1987a; *A. caraya*: Rumiz 1990).

Growth and Development

Information regarding ontogenetic development of free-ranging howler monkeys is rare. Clarke (1990) reports growth and development of red howler (*A. seniculus*) infants in Guanacaste, Costa Rica. At birth, the natal pelage is silver and starts changing to gold within a few days. At 8 weeks of age, an infant weighs about 500 g. Female infants begin feeding independently at 8–10 weeks, but male infants do not begin feeding independently until 24 weeks. During this period, an infant starts traveling independently. Weaning also begins, but suckling may not cease until 72 weeks (15 months). Between the age of 52 and 86 weeks (12–20 months), males emigrate from the natal group and may become solitary. Females start emigrating between 90 and 98 weeks (22–24 months). Sexual maturity takes place between 26 and 34 months in females and 36 and 40 months in males. Similar data are described for *A. palliata* (Crockett and Eisenberg, 1987; Glander, 1980; Neville et al., 1988), and for *A. caraya* (Crockett and Eisenberg, 1987; Neville et al., 1988; Rumiz, 1990).

In *A. palliata*, the testes do not descend until after puberty, whereas in other species, they descend before or during infancy (Crockett and Eisenberg, 1987).

Communication

Howler monkeys have an impressive vocal communication system in addition to the olfactory communication system mentioned above (Snowdon, 1989). Both males and females produce long calls (also called dawn calls, roaring, howls, etc.). The enlarged hyoid bone acts as a resonator and amplifies the long calls. The howling behavior has been extensively studied in *A. palliata* and *A. seniculus*. Loud calls, which are most frequently produced at dawn, are directed toward solitary individuals and other group members. Probable functions of long calls are for announcing the group location, spacing the groups' distance, and advertising group composition. Duets between mates strengthen their pair bonds. Howling may occur during intratroop interactions, which may incite competition among males (Crockett and Eisenberg, 1987; Snowdon, 1989). The calls of *A. palliata* often exceed 90 dB. Frequencies are below 1 kHz (Whitehead, 1987). The rate of calling is higher at night than during the day, at least for red howlers in French Guyana (Vercauteren and Gautier, 1992). Extensive details on howler vocalizations are available in Neville et al. (1988).

Play

The most extensive report on play behavior is Baldwin and Baldwin (1978). Summaries of play behavior are available in Neville et al. (1988). Juveniles have special play calls (Masataka and Kohda, 1988).

Posture and Locomotion

Howler monkeys use the prehensile tail primarily to balance their posture for terminal branch hang-feeding; it is also used for travel during above branch pronograde quadrupedalism. They are slow cautious arboreal quadrupeds (Cant, 1986; Kinzey, 1986; Rosenberger and Strier, 1989). In red and mantled howler monkeys, the tail is also used for crossing gaps and "bridging" gaps to enhance safe crossing for immatures. They rarely suspend by the forelimbs or leap (Crockett and Eisenberg, 1987; Rosenberger and Strier, 1989). Both red and mantled howler are also excellent swimmers (Froehlich and Thorington, 1982; Neville et al., 1988; Izawa and Lozano, 1990b).

12

Aotus

Introduction

Aotus (the night monkey, owl monkey, or douroucouli) is the only nocturnal primate in the neotropics and the world's only nocturnal monkey. Night monkeys weigh about 1 kg and there is no difference in average weight between male and female (Ford and Davis, 1992). *Aotus* is easily identified by its large eyes and black and white facial markings. Above the eyes are two white or buffy semilunes outlined by three black stripes that converge on the crown; the exact pattern varies individually and taxonomically (Wright, 1981). Male canine teeth were reported to be significantly larger than those of females (Orlosky, 1973; Swindler, 1976). More recent data do not appear to support that conclusion and the low level of canine dimorphism may correlate with the low level of competition in this genus (Kay et al., 1988); however, patterns of dimorphic canine variability do differ among *Aotus* populations (Plavcan and Kay, 1988).

Night monkeys are completely arboreal and occur in a wide range of forested habitats. They are the only cebids susceptible to infection by malarial parasites (Hershkovitz, 1983), and thus are used extensively as a research model for human malaria. Night monkeys have a monogamous social organization.

Taxonomy and Distribution

Until 1983 *Aotus* was generally thought to contain a single, widespread polytypic species, *A. trivirgatus* (e.g., Napier and Napier, 1967; Hershkovitz, 1972; Moynihan, 1976b; Thorington and Vorek, 1976). As such, it had the widest distribution of any "species" of neotropical primate, from about 81°W in Panama to the Gran Chaco in Argentina (Wright, 1981). Based on differences in pelage pattern and karyotype, Hershkovitz (1983) divided *Aotus* into two natural groups, comprising nine allopatric species. A "gray-neck" species group (4 species) is found north of the Amazon (Rios Amazonas, Solimões, and Marañón) (Map 2), extending into Panama (Map 15), and a "red-neck" species group (5 species) is found south of the same three rivers (Map 2). Additional karyotypes have been found subsequently (Ma et al., 1985). Chromosomal variation, however, does not always correlate with taxonomy; Schneider et al. (1989) found a consistent karyological difference between animals from opposite sides of the Rio Tocantins, despite the fact that all were classified as *A. infulatus*.

Ramirez-Cerquera (1983) briefly described four specimens of a fifth gray-neck form, *Aotus hershkovitzi*, collected at 1750 m elevation in Boyacá Department, Colombia. The chromosome number is $2N = 58$, a diploid number not previously reported for *Aotus*. A complete description of the taxon is not yet available. Obviously, the taxonomic tale of *Aotus* has not yet been completed.

Evolutionary History

The presence of many characteristics of a diurnal visual system in *Aotus* (e.g., a retinal fovea, a macula lutea, lack of a *tapetum lucidum*, cones in the retina) implies that night monkeys evolved from a diurnal ancestor (Polyak, 1957; Ogden, 1975; Noback, 1975; Jacobs et al., 1993a). The sister group of *Aotus* is probably *Callicebus* (Rosenberger, 1992; Ford and Davis, 1992); however, see Kay (1990) and Tyler (1991) for alternate views.

The presence of *Aotus dindensis* (Setoguchi and Rosenberger, 1987) [including foot bones (Gebo et al., 1990)] in the Colombian Miocene and the "*Aotus*-like" *Tremecebus harringtoni* (Rusconi, 1933; Hershkovitz, 1974; Fleagle and Bown, 1983) with enlarged orbits in the Late Oligocene implies a long separation of the *Aotus* lineage. An early separation is also suggested by biochemical data (Sarich and Cronin, 1980).

Ma (1981a,b) suggested that extant variability in *Aotus* might be explained as the result of isolation of populations in Pleistocene forest refugia. An evolutionary scheme of relationships among living taxa, based on karyotypes, is provided by Galbreath (1983).

Previous Studies

Captive Studies

Dukelow and Dukelow (1989) suggested that night monkeys may have an elevated reaction to stress in captivity, since they have relatively high levels of abortion and stillbirth (Johnson et al., 1986). Although it is known that small neotropical primates tend to have high levels of circulating cortisol, apparently the level in *Aotus* has not been reported (Coe et al., 1992). A more likely explanation for high stillbirth rates is mating of animals with different karyotypes (Ma, 1981a,b).

Most studies of captive animals have either involved their use in the study of malaria (e.g., Schmidt 1973), or concerned neuroanatomy primarily related to their unusual nocturnal visual system (e.g., Jacobs, 1977a,b, 1993; Jacobs et al., 1993a; Weller, 1988; Allman and McGuinness, 1988). Other studies have been conducted on vocalizations (Moynihan, 1964), circadian rhythm (Erkert, 1991), reproduction (Gozalo and Montoya, 1990), and infant development (Wright, 1984, 1990).

Field Studies

The first field study in the wild was a 9-week study conducted in the Department of Pasco, Peru by Wright in 1976 (Wright, 1978, 1979; Sivertsen et al., 1982). Other studies were subsequently conducted at Cocha Cashu, Peru (Wright, 1984, 1985, 1986, 1989, 1990, 1992a; Sivertsen et al., 1982), northeastern Peru (Aquino and Encarnacion, 1986a,b, 1988; Aquino et al., 1990), Paraguay (Wright, 1985, 1989), and Bolivia (Garcia, 1988; Garcia and Braza, 1987, 1989). Thorington et al. (1976) detailed the home range and activity rhythms of an *Aotus trivirgatus* male, which was reared in captivity and released into its natural Panamanian habitat. Earlier anecdotal reports were summarized by Wright (1981).

Habitat and Ecology

Night monkeys are found throughout most forested areas of the Neotropics, except the Guyanas and the southeast coast of Brazil. *Aotus* is a versatile genus and occurs in primary, secondary, and seasonally deciduous scrub forest (Heltne, 1977; Green, 1978), subtropical dry forest (Wright, 1985), and gallery forest (Rathbun and Gache, 1980), from sea level up to 3000 m altitude (Hernandez-Camacho and Cooper, 1976), at all canopy levels (Wright, 1981), and even at temperatures as low as — 5°C (Krieg, 1930). Rathbun and Gache (1980) suggest that *Aotus* tend to inhabit areas of high plant species diversity, and a relatively dense canopy, irrespective of whether the canopy is low or high.

Diet

All reports indicate that *Aotus* is primarily frugivorous, preferring small ripe fruit whenever available; the diet is supplemented with flowers, insects, nectar, and leaves (Hladik and Hladik, 1969; Hladik et al., 1971; Wright, 1985). Seasonal differences in diet have been reported (Wright, 1989). If seeds are eaten they usually pass through the digestive tract and are dispersed, however, seeds of *Brosimum alicastrum* (Moraceae) were chewed by *Aotus azarae* in Peru (Wright, 1985).

Aotus feeds mostly in large-crown trees [in contrast with sympatric *Callicebus*, which fed mostly in tree crowns of 10 m diameter or less (Wright, 1989)]. *Aotus* forage for insects primarily at dawn, dusk, and on moonlit nights (Wright, 1989). They appear to have an advantage in feeding at night—little interference from other frugivores, and the availability of large nocturnal insects (Wright, 1989).

Predation

Wright (1985, 1989) documented the availability of both nocturnal and diurnal predators. She found that by being nocturnal *Aotus* avoids a large array of diurnal predatory raptors; furthermore, nocturnal predators rarely eat monkeys.

In subtropical dry forest of Paraguay, where large diurnal predators were rare or absent, but where the nocturnal great horned owl was a formidable predator on small monkeys, *Aotus* changed its behavior, and spent up to 3 hours foraging during daylight hours (Wright, 1985, 1989). Garcia and Braza (1987) suggested that the diurnal activity might be the result of cold nights forcing the animals to feed during the day.

Aotus should choose safe sleeping sites to avoid diurnal predators (Wright, 1989). Aquino and Encarnacion (1986b) documented sleeping sites used by *Aotus vociferans* in Peru, and found that protection from predators, easy access, shelter from adverse weather, and sufficient space were the primary factors affecting sleeping site selection.

Behavior

Social behavior, including intragroup calling, playing, and intergroup fighting, occur most often when the moon is bright (Wright, 1989).

Ranging Behavior

Ranging patterns of *Aotus* are probably limited by luminosity, for they are more active when the moon is bright (Wright, 1984). They repeatedly travel the same routes, suggesting that memorization of routes may be important, especially on dark, moonless nights (Wright, 1989). Erkert (1989) demonstrated in the laboratory that maximum nocturnal activity occurred in *Aotus* when illumination approximated that of the full moon and fell off at both higher and lower light intensities. Details of adaptation to a nocturnal niche are provided in Wright (1989).

Home range and path length average about the same as for sympatric *Callicebus* in Peru and are much smaller than for all other species of monkey (except *Cebuella*) in Manu National Park, Peru (Wright, 1989). Average territory size was 9.2 ha ($N = 9$); average nightly path length was 708 m ($N = 60$). In a 0.33-ha island forest in Bolivia path length was only 337 m ($N = 10$) (Garcia and Braza, 1987). Territories are used exclusively by one family group and defended from conspecifics (Wright, 1986).

Reproduction and Mating

Aotus has a monogamous mating system, with the adult pair remaining together for a number of years, probably for life. As in the case of *Callicebus*, monogamy in *Aotus* "is correlated with (1) distribution of food into small, predictable, uniform patches, and (2) the need for male assistance to raise offspring successfully" (Wright, 1986:166).

Gozalo and Montoya (1990) summarized 10 years of data from the breeding colony of *Aotus nancymai* in Iquitos, Peru. Mean age at first birth was about 3½ years (40.6 ± 7.8 months for females, and 42.2 ± 10.7 months for males at the time their mates produced offspring). The earliest fertile copulation in a female occurred at 20.5 months. Mean interbirth interval was 12.7 ± 5.7 months. Their results agree essentially with those of Hunter et al. (1979) and Dixson (1983).

Gestation is 133 days (Hunter et al., 1979). One infant is born each year; there is no strict breeding season, but there is a birth peak at the beginning of the rainy season (Wright, 1986). Subadults leave their natal group during the third year of age (Wright, 1986), without any evidence of agonistic behavior at or prior to departure (Robinson et al., 1987). The process of mate selection is not known (Robinson et al., 1987).

Growth and Development

The male assumes most of the parental care: carrying, playing, instructing in feeding, and protecting from danger from the first week after birth throughout the first year of the infant's life (Dixson and Fleming, 1981; Wright, 1984). He is the primary infant carrier from the first day of birth (Wright, 1984), and stops carrying the infant by the fifth month of age. The mother nurses the infant until it is 7 or 8 months old. Each infant remains with the family group until it is 2½–3½ years of age; both males and females disperse. Further details are provided in Wright (1984).

Communication

Aotus pairs do not give dawn call duets like *Callicebus* or gibbons. Individual animals give the long low (200–400 Hz) loud "hoot" call that can be heard by humans for 500 m (Wright, 1985). It has little harmonic structure, and consists of a short sequence of pure tones repeated monotonously for over an hour (Wright, 1989). It functions as sexual advertisement and is given by either subadult male or female when looking for a mate. It is also given by a single adult animal if the mate is lost, or by a juvenile if alone. The call of the male may differ from that of the female (Wright, 1985:192).

The "resonant whoop" or "resonant grunt" is a series of 10–17 low notes building to a climax, and accompanied by arch-postures and stiff-legged jumping. It is used to defend fruit trees (not to defend a territory per se). Both *Aotus* adults participate equally in intergroup confrontations while giving resonant whoops (Wright, 1984).

Grooming is rare in *Aotus*, except in a sexual context and associated with copulation (Moynihan, 1964; Wright, 1979, 1984). Additional vocal, visual, olfactory, and tactile behavior patterns are summarized in Wright (1981).

Play

Juvenile play behavior is described in Wright (1981). Males play with their infant and juvenile offspring even after the infant is independently locomoting, 4–5 times per day in bouts of 5–20 minutes (Wright, 1984). Mothers rarely, if ever, play with infants.

Posture and Locomotion

Aotus are predominantly quadrupedal, but are skillful leapers (Fleagle, 1988). Details of posture and locomotion are provided by Wright (1981).

13

Ateles

Introduction

Spider monkeys (genus *Ateles*) are among the largest New World monkeys (up to 11 kg, adult body weight; Ford and Davis, 1992). They are completely arboreal, and long-limbed with a long prehensile tail and a vestigial or absent thumb. Their color ranges from beige to black and there is virtually no sexual dimorphism in color or size (Fleagle, 1988), although in the species *A. paniscus* females are slightly larger than males (Ford and Davis, 1992). Spider monkeys are primarily frugivorous and specialize on ripe fruit. The social organization of *Ateles*, together with other members of the Atelini, may be unique among New World monkeys in having a fission–fusion organization similar to that of chimpanzees (McFarland, 1986; Symington, 1990). They live in large multimale, multi-female communities and forage in smaller subgroups of varying size and composition. Males are philopatric and females emigrate when they are sexually mature. Because their habitat is threatened, and they are among the most frequently hunted monkeys, spider monkeys are among the most endangered neotropical primates (Wolfheim, 1983). The species, *A. fusciceps*, is one of only two species of South America monkey that have not yet been studied in a natural habitat.

Taxonomy and Distribution

Spider monkeys are geographically widespread throughout Central and South America, from southern Mexico to Central Bolivia and Brazil. They occur farther north than any other neotropical primate, in Escandon (23°N, 99°W) in southern Tamaulipas, Mexico and the southernmost population is in Buena Vista, Bolivia (17°28'S, 63°37'W) (Hershkovitz, 1977). See Map 16 for the distribution of *A. geoffroyi* in Central America, and Map 3 for the distribution of the other three species in South America.

The genus *Ateles*, following the classification of Kellogg and Goldman (1944), is generally thought to contain four allopatric species: *A. geoffroyi* in Central America, *A. fusciceps* in Colombia, Ecuador, and Panama, and *A. belzebuth* and *A. paniscus*, which are found in several, possibly disjunct regions in South America (Konstant et al., 1985; Mittermeier et al., 1988b; van Roosmalen and Klein, 1988). Species and subspecies designations are based almost entirely on pelage characteristics. Some (e.g., Hershkovitz, 1977; Wolfheim, 1983) regard all spider monkeys as belonging to a single wide-ranging species, *A. paniscus*. Recently, Froehlich et al. (1991), based on a multivariate morphometric analysis, suggested that *Ateles* may contain only three species, one for Central America and Colombia, a ring species for Amazonia, and the most differentiated form in the Guianas. The integrity of many captive spider monkey subspecies is threatened due to the large amount of hybridization in captive populations. Obviously, this genus needs further investigation and taxonomic revision.

Evolutionary History

Ateles is most closely related to *Brachyteles*, and these two genera probably separated from the *Lagothrix* lineage in South American lowland rainforest, to evolve their unique locomotor system (Rosenberger and Strier, 1989). Although there are no fossils on continental South America to support this evolutionary scenario, there is evidence that spider monkeys may have inhabited in Cuba in Recent times (Ford, 1990b).

Previous Studies

Captive Studies

There are relatively few captive studies of *Ateles*, especially compared with the number of field studies. Social behavior has been studied in zoo colonies of *A. geoffroyi* (Klein and Klein, 1971; Rondinelli and Klein, 1976; Dare, 1974, 1975). Eisenberg (1976) studied communication of a captive population of *A. fusciceps*. More recently, Masataka and Kohda (1988) studied the play vocalizations of *A. geoffroyi*. Kaemmerer et al. (1987) studied parturition of *A. geoffroyi* and *A. fusciceps* at the National Zoo and the Louisiana Purchase Gardens and Zoo. Chevalier-Skolnikoff (1989) studied tool use of *A. belzebuth*, *A. fusciceps*, and *A. geoffroyi* at the San Diego Wild Animal Park. *A. fusciceps* has been studied primarily in captive situations. Unfortunately for conservation of spider monkeys, only *A. geoffroyi* is breeding in captivity at a sustainable rate (Konstant, 1986).

Field Studies

Field studies have concentrated mainly on *A. geoffroyi* in Central America and *A. paniscus* at two sites in South America. Many aspects of the ecology and social behavior of *A. geoffroyi* have been studied at Tikal National Park in Guatemala by Cant (1977, 1978, 1986, 1990), Bramblett et al. (1980), and Fedigan and Baxter (1984). Chapman (1987, 1988a,b, 1989a,b, 1990) has been studying the ecology, feeding behavior, social organization, reproduction, and communication of *A. geoffroyi* in the Santa Rosa National Park, Costa Rica since 1983. Milton (1981c) studied the reproductive parameters of free-ranging *A. geoffroyi* on Barro Colorado Island. van Roosmalen (1985) studied behavior and ecology, and Mittermeier and van Roosmalen (1981) studied habitat utilization of *A. paniscus* in Surinam. Many aspects of the behavior and ecology of *A. p. chamek* have been studied at the Manu National Park, Peru (Terborgh, 1983; McFarland, 1986; Symington, 1987a,b,c, 1988a,b, 1990). *A. belzebuth* was studied in the La Macarena National Park, Colombia by Klein and Klein (1977) and more recently by Ahumada (1989).

Habitat and Ecology

Ateles are found most frequently in evergreen tropical rain forest and are more abundant in wet than in dry forests (Hershkovitz, 1977). They are thought to prefer primary, lowland, humid evergreen rainforests, but also occupy old secondary, highland rainforests in Surinam (Kinzey and Norconk, 1990), or dry, or deciduous, and even swamp forests (Hershkovitz, 1977; van Roosmalen and Klein, 1988). *Ateles* are found at a wide range of elevations, from sea level to 2500 m (Hershkovitz, 1977). Rylands and Keuroghlian (1988) surveyed *Ateles* in four central Amazonian forest reserves and found that *Ateles* were found only in continuous and not in isolated forests. Spider monkeys utilize the upper canopy of the forest most frequently (van Roosmalen and Klein, 1988; Yoneda, 1990).

Predation

The most dangerous predators of *Ateles* are probably humans (Mexico: Garcia et al., 1987; Cuaron, 1987; Estrada and Coates-Estrada, 1984; Brazil: Johns, 1986a; Ecuador: Madden and Albuja, 1987). Raptors are generally thought to be an inconsequential threat, because of the large body size of *Ateles* (Robinson and Janson, 1987).

Diet

Spider monkeys are primarily frugivores, specializing on ripe fruit. Fruit comprise 75–93% of the diet (Klein and Klein, 1977; Chapman,

1987; Nunes, 1987; van Roosmalen 1985; Symington, 1988a; Kinzey and Norconk, 1990). Immature fruit are eaten only when mature fruit are unavailable. They feed primarily in large trees [i.e., those with large diameter at breast height (DBH)] in the upper part of the crown (Chapman, 1988b; Yoneda, 1990). Although spider monkeys feed on a large variety of plant species, only a small number of species represents the majority of the diet (van Roosmalen and Klein, 1988). When fruit are less abundant, spider monkeys resort to flowers and/or young leaves. Chapman and Chapman (1991) found that *A. geoffroyi* leaf-eating was highest in the late morning and late in the day, immediately prior to resting times. See van Roosmalen (1985) and van Roosmalen and Klein (1988) for extensive lists of the plant and animal species eaten by *Ateles*, and detailed discussion of seasonal variation in feeding in *A. paniscus*.

Ateles probably play an important role in the forest as seed dispersers. Small fruit are usually swallowed whole without mastication, or the exocarp is removed with the teeth and the seed is swallowed with the aril; even relatively large seeds are usually swallowed whole and eliminated with the feces (Kinzey and Norconk, 1990; van Roosmalen, 1987a). Several studies have investigated seed dispersal in *Ateles* (Boucher, 1981; Muskin and Fischgrund, 1981; Howe, 1983; Fleming et al., 1985; Chapman, 1989a).

Abundance and distribution of food resources can greatly affect social organization. The food resources of *Ateles* are widely distributed, patchy, ephemeral, and occur at low densities (Chapman, 1987; Symington, 1988a; Silva-Lopes et al., 1987). For *A. geoffroyi* in Costa Rica, 53% of the plant species (constituting more than 1% of feeding time) occurred at densities of less than 1 individual tree per hectare. Food patch size and density determine the number of individuals able to forage together in one patch. Thus, foraging subgroup size is positively correlated with resource patch size for *A. geoffroyi* (Chapman, 1990) and for *A. paniscus* (Symington, 1988a). The flexibility of fission–fusion social organization of *Ateles* solves the problem of a large-sized primate feeding predominantly on ripe fruits that are generally small, widely dispersed, and occur in low densities (van Roosmalen, 1985, 1987a; Symington, 1988a; Chapman, 1990). Feeding on mature fruits also affects other behavioral traits (see below).

Behavior

Ranging Behavior

Ateles have large home ranges, up to 250 ha with 10–15% overlap between neighboring groups for *A. paniscus* in Peru (Symington, 1988b).

Home ranges of males are substantially greater than those of females. *A. paniscus* females ranged over core areas of 34–66 ha (representing 20–33% of total home range) while male ranges were twice that and evenly distributed over the entire area (Symington, 1988b). The average range size was 37 ha for *A. geoffroyi* females and 98 ha for males (Fedigan et al., 1988). Daily path length is highly variable, sometimes exceeding 4000 m, and depends on subgroup size and composition, weather, season, and availability and distribution of food resources (van Roosmalen and Klein, 1988; Chapman, 1988b; Symington, 1988b).

Ateles forage in small subgroups that are highly variable in size and composition. Subgroups of *A. belzebuth* contained 1–6 individuals 90% of the time with 2 being the most frequent number (Ahumada, 1989). Subgroups were composed of isolates (20% of the time), all males (13%), all females, with or without infants (32%), and mixed sexes (35%). Compositions of subgroups of other species, reviewed by van Roosmalen and Klein (1988), are very similar to that of *A. belzebuth*. van Roosmalen (1987) characterized the foraging behavior of *A. paniscus* as determined by a few dominant females that monitor food sources, plan economic foraging routes, including up to 30 food-plant species, and take into account maturation rates and spatial distribution of fruiting plants.

Choice of a sleeping site is another variable that affects ranging pattern. Chapman et al. (1989a) studied the selection of sleeping sites of *A. geoffroyi*. Sites were generally located in the center of the range, most often (82%) used repeatedly, and used once only 18% of the time. They generally slept at the site that was closest to their current feeding area. Based on these data and the fact that ranging tended to be in a circular pattern, often returning to the same sleeping site as the night before, they concluded that this population was best described as "multiple central place foragers."

Growth and Development

There are very limited data on growth and development of *Ateles*. Chapman and Chapman (1987) describe the activities of a juvenile male that had lost three-fourths of its tail. They found that development of the injured infant was significantly impaired due to the injury. van Roosmalen and Klein (1988) provide a list of six developmental stages from various studies of all four species. Both males and females become sexually mature between 4.5 and 5.5 years of age (Robinson and Janson, 1987). The rate of reproduction is low, since infant dependency lasts 3 years (Robinson and Janson, 1987), the longest known of any cebid. Delayed maturation in spider monkeys in many ways parallels the evolutionary trend in the hominoids (Kinzey, 1971).

Reproduction

The multimale, multifemale social organization implies that *Ateles* are polygamous. Aggression between males is infrequent, estrous females actively choose their mating partners, and both promiscuous matings and consortships have been observed (Robinson and Janson, 1987). Although *Ateles* have the expectedly low levels of testosterone and progesterone for their large body size, they have the highest level (relative to body size) of circulating cortisol of any primate (Coe et al., 1992). Unlike most other primates, male *Ateles* lack a baculum (Dixson, 1987).

Recently the significance of the sex ratio bias in adults and at birth in *Ateles* has received more attention (Hrdy, 1988). All *Ateles* adult sex ratios favor females for all four species (Symington, 1987a; Chapman et al., 1989b, 1989c).

Data on sex ratios at birth, however, are limited and conflicting. In *A. paniscus* (Symington, 1987a) there was a significant difference between sex ratios at birth for high and low ranking females, with low-ranking females only producing females, and the highest ranking females (top 4) having bias toward producing males. Symington suggested that longer interbirth intervals after the birth of male infants was due to more intensive preweaning investment.

Chapman et al. (1989c) predicted that if females are limiting the number of male infants produced because they present a greater competition threat as adults than female infants, then sex biases will vary with food abundance in different habitats, measured by rainfall. They found a negative relationship between sex ratio and rainfall for the five major study sites, thus supporting the local resource competition hypothesis to explain the skewed sex ratios (Chapman et al., 1989b).

Communication

Recent data focus on vocal communication, but *Ateles* are also known to communicate using olfactory, tactile, and visual cues (van Roosmalen and Klein, 1988). Some olfactory behavioral patterns have been observed in a sexual context. Klein and Klein (1971), for example, suggested that the enlarged clitoris of *Ateles* is adapted to collect drops of urine as scent marks.

Eisenberg (1976) provided a detailed vocal analysis of *A. fusciceps*. More recent data on other species, however, indicate that there are major structural differences in vocalizations among species (van Roosmalen and Klein, 1988).

"Long calls," which are only made by males, can be heard up to 500 m away (Symington, 1987c). There is significant interindividual variation in long-range vocalizations such that individuals could be recognized by human observers 50% of the time (Chapman and Weary, 1990). Long calls are made in various contexts, including spacing between subgroups, when a male loses contact with his subgroup, as alarm calls (van Roosmalen, 1985), and as food calls to attract individuals to food patches (Chapman and Lefebvre, 1990). Chapman and Lefebvre (1990) suggested that *A. geoffroyi* are able to manipulate foraging subgroup size through long-distance food calls. They observed a positive correlation between the frequency of calls and density of food resource, size of tree entered, and rank of the most dominant individual in the group. They suggested that because 50% of the time calling was withheld and calls often produced no joining, individuals were both deceptive and knowledgeable about resource abundance.

Symington (1987b) investigated the effect of alarm call playbacks on the decision of *A. paniscus* subgroups to stay or leave an area. Neither number of females with young nor total party size affected the decision; only the number of males increased the probability of staying. Males were more vigilant than females and more likely to mob potential predators cooperatively. Chapman et al. (1990) investigated the response to alarm call playbacks in *A. geoffroyi*. The intensity and duration of alarm calls depended on social context; call duration was positively correlated with the number of kin present, possibly benefiting kin preferentially.

Play

Play behavior in the wild has been described in detail for *A. paniscus* in Surinam (van Roosmalen and Klein, 1988). Captive *A. geoffroyi* infants do not utter play vocalizations (Masataka and Kohda, 1988).

Posture and Locomotion

The main locomotor modes of *Ateles* are arboreal quadrupedalism and suspension, including brachiation, but they also exhibit climbing, leaping, and bipedalism (Fleagle, 1988). The ancestral atelines are thought to have engaged in forelimb-dominated climbing (Rosenberger and Strier, 1989).

Cant (1986) comprehensively described the locomotor and feeding postures of *A. geoffroyi* in Tikal National Park, Guatemala. Three feeding postures were observed: (1) suspensory postures: 52% (suspension from tail only, 23% of time; tail used as an anchor, 34% of time); (2) sitting: 45%; (3) standing: 3% of the time. *A. geoffroyi* were observed to use small substrates 50% of the time, and large substrates only 8% of the time. When they were traveling quadrupedally, they moved slowly the majority of the time, on medium to large supports that were level or had an upward incline. When using tail–arm suspension (25% of the time), travel speed was slow on medium and small supports that were level or had a downward incline. "Fast" progression was utilized only 6% of the time. These results are similar to those of van Roosmalen and Klein (1988) for other species. Lemelin (1989) found muscular features in the tail of *Ateles* associated with hyperflexion of the tail during brachiation.

The suspensory postures of spider monkey are most likely an adaptation to allow the large bodied *Ateles* to exploit a widely dispersed food resource that occurs at the end of thin branches (Grand, 1972). Below branch suspension increases their maneuverability on thin supports.

14

Brachyteles

Introduction

Brachyteles arachnoides (the muriqui, or woolly spider monkey) is the largest neotropical primate, and the largest endemic mammal in Brazil (Milton, 1986). According to Aguirre (1971) adult males weigh up to 15 kg, although this may be an exaggeration; weights of recently captured wild specimens average 8.375 kg (4♀♀) and 9.613 kg (4♂♂) (Lemos de Sá and Glander, 1995). *Brachyteles* are generally thought to be sexually monomorphic in body size (Strier, 1990a), canine size (Zingeser, 1973; Lucas et al., 1986), and pelage color (Milton 1985a, Strier 1990a), but Lemos de Sá and Glander (1993) note that male canines are slightly larger in the southern subspecies. The more important point (Rosenberger and Strier, 1989) is that both male and female canines are rather small and essentially nonprojecting! The postcanine teeth have an essentially folivorous adaptation with a lingual shearing mechanism (Zingeser, 1973; Rosenberger and Strier, 1989). The pelage of *Brachyteles* is buff-gold in color with black extremities. The face is black, pink, or mottled and individuals in the northern populations can be identified by their unique facial markings (Strier 1991a, 1992a,b). Adult males are noted for their large testes, while the female has a pendulous clitoris (Hill, 1962; Nishimura et al., 1988). Muriquis have a fully prehensile tail and either lack a thumb or have a rudimentary one. They are folivorous/frugivorous with slightly more than 50% leaves in the diet on an annual basis but ripe fruit are preferred (Strier, 1991b). They live in groups of 15–49 and while some groups are cohesive (Strier, 1990a, 1991a, 1992b), others have a fission–fusion social organization. Although superficially similar to the fission–fusion organization of *Ateles* and chimpanzees (Robinson and Janson, 1987), a *Brachyteles* group usually divides into large heterosexual associations unlike those of chimpanzees and spider monkeys, which frequently are small unisexual groups (Strier, 1990a). There is remarkably little aggression among individuals in a group (Strier, 1992a). Muriquis, together with *Leontopithecus*, are the most critically endangered New World primates (Mittermeier et al., 1986, 1989).

Taxonomy and Distribution

Brachyteles is a monotypic genus, with *B. arachnoides* the only species. The muriqui appears to have a monophyletic origin with *Ateles*, both having a chromosome number of $2N = 34$, which is unique among New World monkeys (Rosenberger and Strier, 1989). Contrary to Vieira (1955) and Kinzey (1982), there now appears to be adequate evidence for two subspecies (see below). Other views of taxonomic relationships of *Brachyteles* are discussed in Rosenberger and Strier (1989).

Muriquis are endemic to the Atlantic Coastal Forests of eastern Brazil (see Map 4), disjunctly isolated from the other atelins (*Ateles* and *Lagothrix*), but they are sympatric with *Alouatta fusca* (Rosenberger and Strier, 1989). This forested region was once continuous along the coast of Brazil from Rio Grande do Norte to Rio Grande do Sul. Within this area, *Brachyteles* once was abundant from the southern part of Bahia all the way to São Paulo (Mittermeier et al., 1982, 1987). Today this forested region has dwindled to 1–5% of its original condition and, as a consequence, *Brachyteles* has suffered extensive decline in numbers and geographic range. Aguirre's (1971) survey indicated that only 3000 woolly spider monkeys survived out of an estimated 400,000 in 1500. Based on recent surveys *Brachyteles* has been located at 11 different sites in Minas Gerais, São Paulo and Espirito Santo, with an estimated total population of 386 individuals (Strier, 1992b).

Evolutionary History

The best known fossils of *Brachyteles* are some limb bones (originally referred to as *Protopithecus brasiliensis*) from a Pleistocene cave deposit in Lagoa Santa (MG), Brazil (Lund, 1841). According to Winge (1895), however, they are indistinguishable from the living species. Recently Castor (1993) described two Pleistocene skeletons of *Brachyteles*, slightly larger than the living species, from Boa Vista cave in northern Bahia.

The extant distribution of the species has been explained on the basis of the Pleistocene refuge model (Kinzey, 1982), with the muriqui found only in the Bahia and Paulista Centers of endemism (Map 4). The Lagoa Santa cave represents roughly a 150 km western extension of the more northerly Bahia Center. The Boa Vista cave extends the distribution of the muriqui further north than that of any existing Atlantic forest primate. It appears now (Strier, 1992b; Lemos se Sá et al., 1995) that there are two valid subspecies, *B. arachnoides hypoxanthus* (Vieira, 1944), which differentiated in the Bahia refuge, and *B. a. arachnoides*, which differentiated in the Paulista refuge. The northern form has a mottled face and genitalia in the adult, a vestigal thumb, no sexual dimorphism in canine size, and social grouping that is more cohesive. The southern subspecies has a black face and genitalia, completely lacks a thumb, is sexually dimorphic in canine size, and may be sexually dimorphic in body size; social grouping is more flexible.

Previous Studies

Captive Studies

Individual *Brachyteles* have been kept in captivity in Brazil (Nishimura, 1979; Milton, 1986e; Nishimura et al., 1988), and Milton (1984) studied food passage rate in a female. *Brachyteles* has only just recently bred in captivity at the Rio de Janeiro Primate Center (Strier, 1992b), as part of a plan to establish a breeding program (Jones, 1986; Santos et al., 1987). The woolly spider monkey's rarity, its highly folivorous and very eclectic diet, and the fact that learning to forage is likely to be an important component in the life of the young all constitute obstacles to successful captive breeding, especially if the ultimate goal is reintroduction to the natural habitat (Milton, 1986).

Field Studies

Aguirre (1971) conducted the first major survey of *Brachyteles* focusing on its present and past distribution. Of the 11 forests known to support populations of muriquis (Alves, 1986; Mittermeier et al., 1987; Strier 1992b) the following are and/or have been under study: (1) Fazenda Montes Claros, MG; ecological, behavioral and demographic data have been compiled continuously since 1982 in an 800-ha privately owned forest, always with a view to the conservation of the species (Fonseca, 1985, 1986; Strier, 1986a,b, 1987a–d, 1988, 1989, 1990a–d, 1991a,b, 1992a,b, 1993a,b; Strier and Stuart, 1992; Cristoffer, 1987; Mendes, 1990). [Some preliminary studies were also done prior to 1982 (Nishimura, 1979; Valle et al., 1984).] The research station at Montes Claros ensures continuous future study of this population. (2) Fazenda Barreiro Rico, SP [FBR]; ecology and mating behavior of a single muriqui group in one of five forest patches (Torres de Assumpção, 1981, 1983b; Milton, 1984a,b, 1985a,b, 1986, 1987a,b; Strier and Stuart, 1992). (3) Fazenda Esmeralda, MG; ecology of a single group in a 44-ha secondary forest patch (Fon-

seca, 1985; Lemos de Sá, 1987; Strier and Stuart, 1992). (4) Parque Estadual Carlos Botelho, SP (PECB); a 37,644-ha park with at least 132
muriquis (Strier, 1992a), censused during 8 months by Paccagnella (1986)
(Mittermeier et al., 1987; Strier and Stuart, 1992). (5) Reserva Biologica
Augusto Ruschi, ES (formerly, Reserva Florestal Nova Lombardia); a
4450-ha reserve with five species of primates, censused in 1989 by Pinto
et al. (1991) (Mittermeier et al., 1987; Strier, 1992b).

Habitat and Ecology

Strier (1987d), Milton (1984a), Cristoffer (1987), and Lemos de Sá and
Strier (1992) documented the muriqui's preference for primary or, at
least, tall secondary forest. The species shows both flexibility and adaptive capabilities, however, and is able to survive even in secondary remnant forests (Fonseca, 1985; Lemos de Sá, 1987, Strier, 1987d, 1992c).

Diet

The muriqui is strongly folivorous with leaves accounting for an average of 67% of feeding time in all months, increasing to 80–90% at the
end of the dry season at FBR (Milton, 1984a,b). At this site, leaves
chosen most frequently were from relatively few and rare plant species
(Milton, 1986). At Montes Claros leaves accounted for 51% of annual
feeding time, with a minimum of 28% in December (peak of rainy season) (Strier, 1986a). Leaves formed a substantial part of the diet even
when other preferred foods were abundant (Strier, 1991b), and leaf-
eating was not correlated with availability, suggesting that leaves were
eaten for bulk and protein and not due to their relative abundance. At
both sites immature leaves were preferred to mature ones. The muriqui
is probably the most folivorous neotropical primate, except for sympatric *Alouatta fusca* (Strier, 1992d).

At Montes Claros 32% of feeding time was spent on fruit and seeds,
11% on flowers and flower products, and 6% on bark, bamboo, buds,
and ferns (Strier, 1986a, 1991b). Strier (1989, 1991b) attributes the relatively high degree of frugivory at Montes Claros to the relatively greater
availability of large fruit patches. Although fruit was relatively scarce at
FBR, it represented over 50% of total feeding time when available. Ripe
fruit was preferred, but immature specimens from rare tree and vine
species were also consumed (Milton, 1984a). Flower eating was highly
seasonal, ranging from less than 4% of total feeding time to 30% during
the period of increased rainfall (Milton, 1984b).

Although Milton (1984a) suggested that the muriqui's highly folivorous diet conforms to predictions based on body size energetics (Mil-

ton, 1984a), this species is twice as large as *Alouatta*, and, therefore, should be even more folivorous (Strier, 1987d, 1992d). A more likely explanation is that folivory in the muriqui is imposed on an atelin strategy as a secondary adaptation for survival during seasonal fruit shortage (Strier, 1992d). During periods of fruit shortage muriquis also resorted to feeding on nectar (Torres de Assumpção, 1981; Ferrari and Strier, 1992). Muriquis apparently can tolerate relatively high levels of tannins, presumably because of a relatively rapid passage of food through the digestive tract (Strier, 1993a).

Predation

Except for human hunting, predation on muriquis has not been observed. Habitat destruction by humans, however, has been the primary cause of its decline in numbers (Milton, 1986; Mittermeier et al., 1989).

Behavior

Brachyteles live in multimale–multifemale groups (Milton, 1984a; Fonseca, 1985; Strier, 1986a, 1991a), with male philopatry and subadult female dispersal at 5–6 years of age (Strier, 1991a). The group studied at Montes Claros has a cohesive social organization (Strier, 1986a, 1989a). When first observed the group had 22 individuals; it doubled within 9 years, and age–sex class composition has changed recently (Strier, 1992b). At FBR a more fluid organization existed (Milton, 1984a, 1985b). Mothers with their infants formed permanent daily associations and remained within discrete home ranges, while males were itinerant either alone or in groups. No consistent pattern in subgroup fusion and fission was observed. The above differences in social organization are attributed to the availability of large fruit patches at Montes Claros, permitting a greater number of individuals to feed simultaneously (Strier, 1989).

Social Behavior

One striking feature of *Brachyteles* behavior is the infrequency of adult aggressive interactions (Milton, 1985b; Mendes, 1987; Strier, 1990a, 1992a). Furthermore, individuals spend much of their time in proximity to one another, especially to members of their own sex (Strier, 1987c, 1990a, 1992a). Males have particularly strong affiliative relationships (Strier, this volume); in no other primate do males spend as much time in close association without agonistic interactions (Strier, 1990a). The lack of a dominance hierarchy is attributed to the monomorphism of the species (Strier, 1987b, 1990a, 1992a). Mendes (1987), however, suggests that a dominance hierarchy may exist. Some individuals were more likely to sit near or in contact with other specific individuals. Sexual monomorphism may contribute to opportunities for female choice, since males probably do not engage in overt competition (Strier, 1990a, 1992b).

As group size increased there were tendencies toward fissioning into smaller subgroups (Strier, 1990b, 1992b). Muriquis do not groom each other, but they do offer affiliative gestures through touch, frequently between related individuals (Strier, 1992b).

Ranging Behavior

Milton (1984a) calculated an average day range of 630 ± 128 m at FBR (range 350–1400 m); Strier (1987d) found day ranges to be longer for Montes Claros muriquis (1283 ± 642 m; range 141–3403 m), particularly in the wet season. Differences between the two sites probably reflect variations in group size, group composition, and distribution of food resources. Home range for the 26 member group at Montes Claros was 168 ha (Strier, 1987d); range size expanded to 184 ha when group size increased to 34 (Strier, 1991a). The group of 7 individuals at FBR utilized a home range of 70 ha (Milton, 1984a). Home ranges of *Brachyteles* groups appear to overlap in areas where important food sources are found. Aggressive intergroup encounters can occur at such locations with both males and females participating in resource defense (Strier, 1986a, 1992b). Muriquis do not defend territories (Strier, 1990a).

Activity Pattern

Activity budgets vary depending on the group under study. Milton (1984a) found that *Brachyteles* spent 60% of the day resting, 30% feeding, 8% traveling, and less than 1% socializing. Strier (1987b) found that Montes Claros muriquis spent 49% of their time resting, 19% feeding, 29% traveling, and 2% socializing. She observed seasonal differences in diurnal activity (probably the result of climatic variation), but no change in the proportion of time spent on each activity, nor any difference between males and females. Strier (1987b) attributes intraspecific differences in activity budgets to variation in group size, and to food quality and distribution. The smaller group at FBR and the scarcity of widely dispersed food patches may explain their decreased travel time. The fact that leaves are low in readily available energy may explain why the FBR group spends more time resting.

Mating and Reproduction

Brachyteles are promiscuous with a polygamous mating system in which females mate with more than one male (Milton, 1985b; Strier,

1986b, 1991a). Based on a single female estrous cycle, Milton (1985b) reported that males are attracted to females through pheromonal cues in their urine (Milton, 1987b) and congregate, without agonistic interactions, around an estrous female. Copulation is repeated and prolonged (Milton, 1987a; Strier, 1990a, 1992b). Males lost interest in further copulation within a period of 36–48 hours despite continued female solicitation, perhaps at the point when they could detect that further copulations would not improve fertilization (Milton, 1985b, 1987b). Despite the lack of overt male aggression for estrous females, their large testes may imply the existence of sperm competition (Milton, 1986, 1987a,b).

During an 8-year study Strier (1991a) noted an increase of the study group from 22 to 43 individuals as a result of 21 births. The average interbirth interval for 6 females was 33.8 months; age at first reproduction for one female was $7^{1}/_{2}$ years. Births occur throughout the year, but most occurred during the dry season between May and September (Strier, 1990b, 1991a).

Growth and Development

Little is known about muriqui growth and development. Strier (1987a) defines the different age classes as follows: *Infants*: from 0 to 8 months, frequently nursed and always carried on the ventrum; from 8 to 24 months they are still nursing, until weaned at 14–18 months, and they are carried dorsally. *Small juveniles*: 2–$3^{1}/_{2}$ years old; not easily distinguished from infants, since both have uniformly black faces. *Large juveniles*: $3^{1}/_{2}$–6 years old. *Subadults*: distinctly enlarged scrotum in males, although not as large as those of adults; female nipples and genitalia are smaller and less pendulous than those of adults.

Communication

Nine different vocalization patterns have been described for *Brachyteles* [see Nishimura et al. (1988) for complete description]. Olfactory communication is also common. Individuals of both sexes are seen to wash their hands with their urine, which is then transferred to the substrate as they travel (Milton, 1984c). Pheromones probably convey information about sex and identity as well as reproductive status of females (Milton, 1985b, 1987b).

Play

Play behavior appears to be restricted to young and subadult animals (Nishimura et al., 1988).

Posture and Locomotion

Four categories of locomotion have been described for *Brachyteles*: quadrupedal walking and running, suspension, climbing, and leaping, with variations depending on the characteristics of the substrate (Nishimura et al., 1988). Suspensory locomotion may be an adaptation that permits the species to move rapidly, thus minimizing travel time between widely dispersed fruit sources (Strier, 1987d, 1989, 1990a). Strier (1987b) believes that the muriqui's suspensory locomotion is critical to allow them to travel rapidly between dispersed patches of high quality foods, thus exploiting a greater proportion of energy-rich foods than expected for their large body size. The distal one-third of the ventral surface of the tail is bare; the tail is fully prehensile and aids in brachiation.

Three categories of postural behavior are described by Nishimura et al. (1988). They include postures for feeding, short rests (under 30 seconds), and long rests (over 30 seconds). Suspension, utilizing three or four limbs on twigs and small branches, is the most important feeding posture while seated. Reclining positions are widely used during long rests, which take place primarily on branches or boughs.

15

Cacajao

Introduction

Cacajao (the uakari) is the largest (average weight = 3.0 kg) of the three genera of saki-uakaris (the tribe Pitheciini). The genus also has the highest degree of sexual dimorphism of the saki-uakaris (males average 23% heavier than females) (Ford and Davis, 1992; Kay et al., 1988). See Introduction to Chapter 24 on *Pithecia* for a description of this group. *Cacajao* is most notable among neotropical primates for its short tail. *C. calvus* is unique for its bright red naked face and bald head; *C. c. calvus* has white fur, and *C. c. rubicundus* has red fur. *C. melanocephalus* has brown to black fur on the head and body. *Cacajao*, like the other saki-uakaris, is a frugivorous seed predator. The genus is polygynous, with group size varying between 30 and 48 individuals. All species of *Cacajao* are endangered, but the white (bald-headed) uakari (*C. c. calvus*) is one of the most endangered of the Amazonian primates (Ayres, 1986a; Ayres and Johns, 1987).

Taxonomy and Distribution

Two allopatric species of *Cacajao* are recognized: *C. calvus* and *C. melanocephalus* (Hershkovitz, 1987a; Janson, 1984b). Many earlier taxonomists, however, regarded the white (bald-headed) uakari (*C. c. calvus*) and the red (bald-headed) uakari (*C. c. rubicundus*) as separate species (Napier and Napier, 1967; Hill, 1960).

Cacajao has a relatively limited distribution, being confined to western Amazonian flooded forests. *C. melanocephalus*, with two subspecies, is found in regions of black-water rivers, north of the Amazon, in Brazil and Venezuela; *C. calvus*, the bald-headed uakari, with four subspecies, is found in regions of whitewater rivers, south of the Amazon/Solimões, in Brazil and Peru (Ayres, 1989). *C. c. calvus*, the white uakari, is found in a limited area in Brazil between the Rios Japurá and Solimões. *C. c.*

208

rubicundus, the red uakari, is found in Brazil and Colombia, north of the Solimões and south of the Rio Apaporis. *C. c. ucayali*, the reddish-orange uakari, is found in Peru; and *C. c. novaesi*, the orange-buff uakari, is found in a limited area in Brazil, south of the Rio Juruá, between the Rios Eiru and Taraucá (see Map 5). The early taxonomic history is well described by da Cunha and Barnett (1989). It is curious that a white subspecies of *Saguinus fusicollis* is found in the same area as *C. calvus novaesi* (Hershkovitz, 1987b:429).

Evolutionary History

There are no fossils of the genus *Cacajao*. For a discussion of related fossil saki-uakaris, see Chapter 24 on *Pithecia*.

Previous Studies

Early accounts are reviewed in Fontaine (1981).

Captive Studies

The most extensive captive work was of 2000-hour study conducted between 1970 and 1975 of *C. c. rubicundus* in a 1.6-ha seminatural environment at Monkey Jungle in Florida (Fontaine and DuMont, 1977; Fontaine, 1981).

Field Studies

Only the white (bald-headed) uakari (*C. c. calvus*) has been studied in detail, at Lake Tieú, Brazil, in a 20-month study by Ayres (1986a,b, 1990b). It was also the subject of a 6-week study of locomotor behavior (Walker, 1993b). The only other subspecies of bald-headed uakari that has received brief reports is *C. c. ucayali* in Peru by Fontaine (1979), Bartecki and Heymann (1987), and Aquino (1978).

The black-headed uakari (*C. melanocephalus*) has not yet been studied extensively, but there have been three surveys. The first two, by Mittermeier and Coimbra-Filho (1977) and da Cunha and Barnett (1989; Barnett and da Cunha, 1991), was of *C. m. ouakary* in Brazil, and the third of *C. m. melanocephalus*, by Lehman and Robertson (1994) in Venezuela.

Habitat and Ecology

Uakaris are restricted to seasonally flooded forests of the Amazon Basin. The black-headed uakari is found in habitat drained by black-water rivers (igapó); *C. calvus* is restricted to habitat drained by white-water rivers (várzea) (Ayres, 1989), and only occupies limited forest areas within the várzea (Ayres, 1986a). There is some evidence of seasonal migration from flooded areas to terra firme during the dry season (da Cunha and Barnett, 1989).

Diet

Uakaris are largely frugivorous, and like other members of the saki-uakari group (*Pithecia*, *Chiropotes*, and *Cacajao*) are seed predators. In *C. c. calvus* 67% of feeding observations comprised seeds of immature fruit. Three plant families (Hippocrateaceae, Lecythidaceae, and Annonaceae) comprised half the annual fruit consumption (Ayres, 1986b). Ayres (1989) provides a list of the top 20 fruit species. During the dry season *Cacajao* forage for newly germinating seeds on the ground, made possible by the absence of large terrestrial mammals in the várzea (Ayres, 1989). Also, during times of fruit scarcity, uakaris may migrate to terra firme.

Predation

There have been no specific reports of predation on uakaris, but presumably they are prey to large raptors and felids. Along the Casiquiare Canal local hunters especially prize the black-headed uakari (Lehman and Robertson, 1994; da Cunha and Barnett, 1989), but white uakaris are not usually hunted for food (Ayres, 1986a).

Behavior

Social Behavior

Group size ranges between 20 and 50 with roughly equal numbers of males and females (Ayres, 1989). In captivity uakaris display a well-marked dominance hierarchy (Fontaine, 1981), although this has not been reported for wild animals. Allogrooming is a frequent activity (Fontaine, 1981).

Ranging Behavior

The home range of *Cacajao* is large; Ayres' main study group had a home range of 500–550 ha, and daily travel paths were as long as 5 km. During foraging, a large group may split into smaller groups that travel 1–2 km apart (Ayres, 1989).

Activity Pattern

The active day is about 12 hours, but it varies with food availability rather than with amount of daylight. The annual activity budget was 36% feeding, 35% moving, and 29% resting, with seasonal variation. For example, time spent feeding varied monthly from 28 to 44%, 21–48% moving and 18–47% resting. A troop sleeps in several trees, up to 100 m apart, with one to eight animals per tree. They sleep in the highest portion of the crowns of tall trees, such as *Piranhea trifoliata* (Ayres, 1986b).

Mating and Reproduction

In the seminatural environment of Monkey Jungle, Florida, mating was promiscuous, and births were seasonal, occurring between May and October (Fontaine and Hench, 1982). A survey of the black-headed uakari in Brazil found infants carried only during the months of March to June (da Cunha and Barnett, 1989).

Growth and Development

Stages of development in the seminatural environment are provided by Fontaine (1981). Newborn animals lack the red face of the adult, which appears gradually after the third month.

Communication

Fontaine (1981) provides an extensive list of visual expressions, vocalizations, and olfactory and tactile patterns in the red uakari. Vocalizations are frequently antiphonal. The most frequent call of the white uakari in the wild is a low pitched "Ca-Ca-Ca-Ca" given when feeding and resting; high pitched sounds are common during travel (Ayres, 1989).

Play

Infants and juveniles play several hours daily, and at least in the seminatural environment red uakaris of all age and sex categories engage in social play (Fontaine, 1981).

Posture and Locomotion

Postures and locomotion of the red uakari in the seminatural environment are described extensively by Fontaine (1981). Stern (1971) reports locomotion of uakaris as primarily quadrupedal; observations of the white uakari at Lake Teiú demonstrate that quadrupedalism comprised

34% of the samples, but that other behaviors figured importantly as well (Walker, 1993b). These include leaping (21%), pronograde clambering (21%), and dropping from higher to lower support (10%), and smaller percentages of climbing up, bridging, and hopping. *Cacajao* engages in considerably more locomotor behavior while feeding than do either of the other two pitheciins (Walker, 1993b). During resting, sitting postures occupied 59% of bouts, ventral lying 28%, and standing 3%. In contrast to *Pithecia*, vertical supports were rarely used. Pedal suspension, rarely observed in the other pitheciins, comprised 3% of feeding postures. *Cacajao*, due to its short tail and specialized hindlimb suspension, is more derived in its positional behaviors than are the other pitheciins (Walker, 1993b).

16

Callicebus

Introduction

The genus *Callicebus* (titi monkeys) are small (about 1 kg) shy monkeys, sexually monomorphic in body size (except *C. moloch*) (Ford and Davis, 1992) and sexually monomorphic in canine size (Kinzey, 1972). In parts of South America, the term "titi" may refer to other small species of Neotropical primate. Pelage color varies among the species: *C. personatus*, the largest species, has black hands; *C. torquatus*, the second largest species, is generally blackish all over, except for yellow hands in all but one subspecies; the other species are smaller and vary in color from agouti to rufous to grayish with varying facial markings (Hershkovitz, 1990). Titi monkeys are basically frugivorous, territorial, and monogamous in social structure. According to Hershkovitz (1990) *Callicebus* is the most complex and diversified genus of the platyrrhines.

Taxonomy and Distribution

Prior to 1990 only three species of titi monkey were recognized, *C. torquatus*, *C. moloch*, and *C. personatus* (Hershkovitz, 1963). In 1990 Hershkovitz revised the taxonomy to include 13 species in four species groups: *C. torquatus* (basically unchanged since 1963 except for additionally recognized subspecies), *C. moloch* (which includes *C. personatus*), *C. donacophilus*, and *C. modestus*. The *C. torquatus* group contains a lone species that differs grossly from all other cebids in its combination of external, skeletal, and cytogenetic characters (Hershkovitz, 1990). *C. torquatus* has the lowest chromosome number ($2N = 20$) of any primate (Ardito, 1979). The *C. modestus* group consists of a single species that Hershkovitz believes to be the most primitive living cebid. Recent karyological evidence (Minezawa et al., 1989) suggests that if karyological differences among species of *Aotus* and *Saimiri* provide a reliable guide; *C. donacophilus*, *C. cupreus*, and *C. brunneus* are probably valid species.

The genus *Callicebus* occurs throughout most of the tropical forests of the Amazon and Orinoco basins, the Atlantic Forest (*C. personatus*), and the Parana forests of Bolivia and Paraguay (*C. donacophilus*) (Map 6). The distribution is disrupted by the Brazilian cerrado. The geographic ranges of several species of the *C. moloch* group overlap those of *C. torquatus* (Map 6). The overlap of species within the *C. moloch* group (Hershkovitz, 1988) has been interpreted by Kinzey (1982) as secondary intergradation between subspecies.

Evolutionary History

The only fossils of *Callicebus* are some skeletal fragments from a Pleistocene limestone cave deposit in Lagoa Santa (MG) Brazil (Lund, 1839). According to Winge (1895), however, they are indistinguishable from the living *C. personatus*, despite their slightly larger size.

The extant distribution of the species has been explained on the basis of the Pleistocene refuge model (Kinzey and Gentry, 1979; Kinzey, 1982). One of Hershkovitz's new species, *C. oenanthe*, is isolated in the Huallaga Refuge of northern Peru (Lamas, 1982), together with *Lagothrix flavicauda* and *Aotus miconax*. The Lagoa Santa cave probably represents a western extension of *C. personatus melanochir* from the Bahia Center. Contrary to Kinzey (1982), Hershkovitz (1988) believes that titi monkeys spread through centripetal dispersal from highland regions. Kinzey (1992) suggested that *C. torquatus*, which is a seed predator, might serve as a model for the early evolution of the pitheciines. *Callicebus*, on the basis of both dentition (Gregory, 1955) and postcrania (Ford 1990a) is regarded (together with *Aotus*) as the most primitive living platyrrhine.

Previous Studies

Captive Studies

The colony at the Delta Primate Center (Lorenz and Mason, 1971) was moved to Davis, California in 1971 and has been the subject of many behavioral studies, especially in comparison with the captive behavior of *Saimiri* (Andrews, 1986, 1988a,b; Anzenberger et al., 1986; Cubicciotti and Mason, 1975; Fragaszy, 1978, 1986; Mason, 1971; Mendoza and Mason, 1986a,b; Menzel, 1986a,b). Anzenberger (1986, 1988) studied the pair bond of *C. moloch* in Zurich. Moynihan (1966) studied communication in captivity in Panama. Jacobs (this volume; Jacobs and Neitz, 1987b) studied color vision in *C. moloch*.

Field Studies

There has been relatively little new field work since the review of Kinzey (1981). *Callicebus cupreus ornatus* is found at the La Macarena site (Izawa and Nishimura, 1988), but has been studied only by Robinson (1977, 1979a,b, 1981a). *C. torquatus* was studied on the Rio Nanay, Peru (Kinzey et al., 1977; Kinzey, 1977a,b, 1978, 1983, 1986b; Kinzey and Redhead, 1982; Kinzey and Robinson, 1983; Kinzey and Wright, 1982; Easley, 1982; Easley and Kinzey, 1986; Snodderly, 1978; Starin, 1978), and in gallery forest in Colombia (Defler, 1983). *C. brunneus* was studied in Madre de Dios, Peru, in Manu National Park (Kinzey, 1978; Wright, 1984, 1985, 1986, 1989, 1990, 1992a) and on the Rio Tambopata (Crandlemire-Sacco, 1988). *C. personatus* was studied at Sooretama Reserve, ES, Brazil (Kinzey and Becker, 1983). Earlier studies are summarized in Kinzey (1981).

Habitat and Ecology

Callicebus is an arboreal monkey that only rarely comes to the ground. *C. torquatus* spent less than 1% of its day on the ground (Kinzey, 1977b), and most of that consisted of juveniles at play, an occasional instance of a young animal coming to the ground to obtain a large insect for food, or an occasional fall of a young animal to the ground. Other species of *Callicebus* may even be less terrestrial; Moynihan (1976b) stated that he saw "*C. moloch*" on the ground only when they fell.

The habitat clearly differs among species. At Mishana on the Rio Nanay, Peru, *C. torquatus* and *C. cupreus discolor* are habitat isolated and restricted to vegetation growing on different soil types. Differences in soil composition at this site were reported by Stark et al. (1980). Kinzey and Gentry (1979) argue that *C. torquatus* is restricted throughout its range to habitat consisting of vegetation growing primarily on white sand (podzol) soils. *C. brunneus* in Manu National Park, Peru, are often found in bamboo. *C. personatus* appear to inhabit the most diverse habitats (Kinzey, 1981).

C. brunneus, in Manu National Park, prefer lower levels of the forest canopy (Kinzey, 1981), whereas, *C. torquatus* at Mishana prefer middle to upper canopy levels (Kinzey, 1977b).

Diet

Callicebus is largely frugivorous. Studied species spend from about 50 to 80% of feeding time eating fruit (Kinzey, 1992). Generally a relatively small number of fruit species comprises the daily diet. Seed predation comprised 37% of the feeding time of *C. torquatus* (Kinzey, 1977a), and occurred in *C. brunneus* (only seeds of *Brosimum alicastrum*) as well (Wright, 1985). Species lists of fruit eaten by *C. brunneus* (Wright, 1985) and by *C. torquatus* (Kinzey, 1981; Easley, 1982) are available.

Members of a family group of *Callicebus* generally feed on the same food items at the same time (Kinzey et al., 1977). In captivity there is social facilitation of feeding; familiar cagemates have a strong tendency to feed in the same location and on the same food source (Mason, 1978; Andrews, 1986). The passive sharing of food observed in captivity (Fragaszy, 1978) has rarely been observed in the wild, except between a male and his offspring (Starin, 1978).

The most notable difference between the diets of *C. torquatus* and *C. brunneus* is in the nonfrugivorous part of the diet. In the latter much of the nonfrugivorous part of the diet consisted of young leaves, including those of two species of bamboo: 23–28% of feeding time in the wet season, increased to 40–66% in the dry season (Wright, 1985). In the dry season many other primates at Manu consumed quantities of nectar, but the titi monkeys rarely ate nectar and then only coincidentally with flowers. Because of a low capture efficiency, insects were not a significant portion of their diet (Wright 1985). *C. torquatus*, in contrast, living in a habitat where leaves are very sclerophyllus, obtained their major protein source from insects (Kinzey, 1978). Milton and Nessimian (1984), however, found relatively small quantities of insects among the stomach contents of *C. torquatus lugens* at the beginning of the dry season.

C. brunneus fed primarily (nearly 80% of feeding minutes) in trees with crown diameters of less than 10 m; although ripe fruit was preferred, some unripe fruit was eaten (Wright, 1985). *C. torquatus* at the Mishana site, in contrast, fed in large part in emergent trees with large crown diameters, such as *Brosimum* and *Clarisia* (Moraceae), perhaps because there were no other primate food competitors at the site except *Saguinus fuscicollis* (Kinzey, 1977a).

Crandlemire-Sacco (1988) studied sympatric *Saguinus fuscicollis* and *Callicebus brunneus* and noted how they minimized competition for food and space. They shared little overlap in their choices of plants or animal foods, and they used distinctive foraging styles for animal prey, exploiting different parts of the habitat.

Callicebus appears to be deficient in the red part of the visual spectrum (Jacobs and Neitz, 1987b). Snodderly (1978, 1979) has suggested that color vision in *Callicebus* may have coevolved with the green color of neotropical fruit, since *C. torquatus* makes few visual discriminations of food in the wild in the red end of the spectrum.

Predation

Diurnal raptors, cats, and *Cebus* are potential predators on *Callicebus* (Wright, 1985). Predation pressure by humans is probably less on *Callicebus* than on the larger neotropical monkeys, since they are not among the principal sources of food of rural hunters (Neville, 1974). *C. brunneus* appeared to trace paths through very dense concentrations of vegetation where they were afforded the greatest protection from arboreal predators (Crandlemire-Sacco, 1988).

Behavior

Early studies of *Callicebus* behavior are summarized in Kinzey (1981).

Social Behavior

Adult titi monkeys are pair-bonded and monogamous, presumably mating for life. Mason (1966), however, described temporary liaisons of up to several hours between a sexually receptive female and a male of a neighboring group; most often, though, such extrapair associations lasted only a few minutes. There is no evidence of dominance of one adult over the other. A number of captive studies have emphasized the strength of the heterosexual pair bond in *Callicebus* (Andrews and Rosenblum, 1988).

In the wild relationships between the two adult members of a group are virtually always amiable. Grooming is an important activity throughout the day and *C. torquatus* spent nearly 10% of their active day grooming (Kinzey and Wright, 1982).

In captivity intruder experiments were conducted by bringing unfamiliar heterosexual dyads together in the presence or absence of each member's pairmate, allowing the assessment of the motivational structure of social conflict. Individuals preferred their mates over social alternatives, and the presence of a male pairmate had a much stronger influence on all behavioral measures, compared to the presence of a female pairmate (Anzenberger, 1986, 1988).

Among neotropical primates *Callicebus* is virtually unique in its lack of association with other nonhuman primates. The only other primates they have been seen to associate with are the tamarins and marmosets, *Saguinus* and *Callithrix* (Kinzey, 1981; Kinzey and Becker, 1983; Crandlemire-Sacco, 1988).

Ranging Behavior

Small territories are used exclusively by one family group and defended from intruders of the same species (Mason, 1966; Wright, 1985).

The territory size of *C. moloch* at Cocha Cashu was 6–12 ha (Wright, 1986). The territory size of one group of *C. personatus* was 4.7 ha (Kinzey and Becker, 1983). Territory size of *C. torquatus* was 20 ha in Peru, although over a 3-year period 29 ha was utilized (Easley and Kinzey, 1986). In Colombia the territory size of five *C. torquatus* groups averaged 14.2 ha (range 9–22 ha) (Defler, 1983). Although territory size was clearly largest in *C. torquatus*, it may well be the result of their less productive habitat.

The mean daily path length of *Callicebus* moloch group 1 was 552 ± 106 m (N = 25) and that of group 2 was 670 ± 193 m (N = 64). Group members travel and feed in close proximity and group spread averaged 10 m. Path length of *C. personatus* was 695 ± 38 m (N = 6) (Kinzey and Becker, 1983). That of a *C. torquatus* group averaged 819 ± 37 m over 3 years, but increased significantly with increase in group size from three individuals (684 ± 65 m) to four (833 ± 31 m) to five individuals (1030 ± 100 m) (Kinzey, 1986b). A similar correlation between group size and path length has been found in several other primate species (Janson, 1988a). When the group lost the adult female, presumably due to predation, path length increased to 1475 m (N = 5) for the remaining adult male and his daughter as they unsuccessfully attempted to maintain their territory by frequent moving and calling (Kinzey and Redhead, 1982).

In a captive study, Menzel (1986b) gave individuals opportunities to descend from runways 2 m above the ground and travel to food placed nearby on the ground. He found that when the pathway increased substantially in length and no longer provided a direct route to food, the arboreal route was still preferred. In another captive study, foraging behavior designed to model a patchy distribution of food in *Callicebus moloch* showed reduced foraging efficiency, particularly as reflected in reduced economy of travel (Andrews, 1986). Kinzey (1987b) argued that monogamous primate species are behaviorally and ecologically less flexible than those with other forms of social organization.

Activity Pattern

In *C. moloch* (Wright, 1985) and *C. torquatus* (Easley, 1982) the amount of daily time spent feeding did not vary significantly between seasons or with size of group (Kinzey, 1986b), although the composition of the diet did vary seasonally. When resources are more plentiful travel time decreases and time spent in social interactions increases (Wright, 1985). The diurnal pattern of feeding differs: *C. moloch* and *C. personatus*, which feed heavily on young leaves, tend to have a major rest time at mid-day, whereas *C. torquatus*, which spend more time foraging for insects, tend to have many shorter rest times spread throughout the day (Kinzey, 1977a, 1981; Kinzey and Becker, 1983).

During bouts of resting and while sleeping, titi monkeys, especially the mated pair, characteristically entwine their tails [see, for example, Figure 7 in Kinzey (1977b)]. The selection of sleeping locations differs among the species. *C. brunneus* and *C. cupreus ornatus* (the entire *C. moloch* species, sensu Hershkovitz, 1963) sleep in concealment, often in a dense vine tangle (Wright, 1985; Robinson, 1977). *C. torquatus* and *C. personatus*, in contrast, sleep in the open, on a large, relatively horizontal bough (Kinzey, 1981; Kinzey and Becker, 1983).

Mating and Reproduction

All births that have been recorded in the wild for *C. torquatus* and *C. moloch* (sensu Hershkovitz, 1963) have taken place between November and March (Kinzey, 1981). In Colombia, Defler (1983) reported 10 births of *C. torquatus* all taking place in January. *C. personatus*, living in a more southerly latitude, probably have a slightly earlier birth season (Kinzey and Becker, 1983).

Although *C. moloch* (sensu Hershkovitz, 1963) reproduce well in captivity (Kinzey, 1981), *C. torquatus* and *C. personatus* have thus far failed to bear young in captivity (Jones, 1986).

In captivity male and female associations within each species typically result in reproduction. However, when a third party is introduced to an established heterosexual pair, social conflicts are inevitable (Anzenberger, 1986).

During the 5-month breeding season in captivity *C. moloch* has an estrous cycle of 17–21 days (Sassenrath et al., 1980).

Growth and Development

Group size of *Callicebus* ranges from two to five individuals and composition consists of an adult monogamous pair and one to three sequential offspring. One infant is born each year and the subadult generally emigrates by the third year. Peripheralization of a subadult from its natal group generally is without agonistic behavior, and young generally leave the group by the third year (Kinzey, 1981).

Males participate actively in the care of their offspring (Dixson and Fleming, 1981; Wright, 1984, 1990). The father carries the infant almost exclusively in the first 2 months. At 3 weeks, the infant is on the father's back 80–90% of the time, returning to the mother only for nursing bouts. As the infant gains more independence the father carries it less. By age 3 months, the infant (now over 40% of its father's weight) is carried 38% of the time and is independent for 60%. At the age of 4 months, the infant rides on the father only during long leaps between trees, or in times of danger. The lactating mother leads the group into fruit trees and eats fruit and insects at a faster rate than the father, who is burdened with an active infant (Wright, 1984, 1985). Investigation of the contribution of mother and father in captivity to parental care in C. *moloch* was conducted by Mendoza and Mason (1986a).

Communication

There have been two extensive studies of communication in *Callicebus*, one largely of captive animals (Moynihan, 1966), and one of C. *cupreus ornatus* in the natural habitat (Robinson, 1977, 1979a,b, 1981a) (see summaries in Kinzey, 1981; Robinson et al., 1987). The loudest and most remarkable call of *Callicebus* is the loud call, usually given at dawn, and often as a duet between the mated male and female. Young animals in the group frequently join in. In C. *personatus* it is given between $1/2$ hour to $1^{1}/_{2}$ hour on average after sunrise (Kinzey and Becker, 1983).

A remarkable difference between C. *torquatus* and C. *cupreus ornatus* is the response to playbacks of duets (Kinzey and Robinson, 1983). The latter, in keeping with their manner of approach during boundary encounters, always approached duet playbacks of conspecifics (Robinson, 1979a,b, 1981a). C. *torquatus*, in contrast, in keeping with their having larger territories and their manner of duetting in mid-territory, did *not* approach duet playbacks. This difference superficially resembles the difference in spacing between the gibbon (*Hylobates* sp.) and the siamang (*Symphalangus*). If the difference between C. *torquatus* and C. *c. ornatus* in response to duet playbacks is directly related to location of calling during intergroup encounters, then C. *personatus* [which calls from mid-territory on a high open bough (Kinzey and Becker, 1983)] should respond in the same manner as C. *torquatus*. Responses to solo male calls also differed between the two *Callicebus* species (Kinzey and Robinson, 1983).

Play

Play behavior occurs by the juvenile alone, between the adult male and juvenile, and between juveniles 1 year apart in age. See summary in Kinzey (1981).

Posture and Locomotion

Utilization of the forest habitat, as defined by support types, forest strata, and vegetational densities, shows that C. *brunneus* demonstrated a strong preference for small supports (between 2 and 5 cm diameter). Supports between 10 and 20 cm were used less frequent, and supports greater than 20 cm in size were not used. They also used horizontal supports much more frequently than supports of oblique or vertical orientation (4% of the time), and were never observed to engage in vertical clinging and leaping behavior (Crandlemire-Sacco, 1988). In contrast, *C. torquatus* used large (20 cm) horizontal supports for sleeping, and preferred terminal branches for feeding and locomotion, but not for resting (Kinzey, 1977b). Nine percent of feeding took place on vertical supports, and vertical clinging and leaping were a frequent activity in suitable habitats. Further details of posture and locomotion are summarized in Kinzey (1981).

17

Callimico

Introduction

Callimico goeldii (Goeldi's monkey) is a small primate, about the size of a tamarin (*Saguinus*), and there is only one species in the genus. It is intermediate in several morphological and behavioral characteristics between the marmosets and tamarins (Callitrichini) and the larger New World monkeys (see below). It may be sexually monomorphic in body weight, but only male weights are recorded in the literature (Ford and Davis, 1992). The pelage is completely black; the species is described in more detail in Heltne et al. (1981) and Hershkovitz (1977). The diet is predominantly insects and fruit and does not include exudates (Moynihan, 1976a) as is found in the diets of marmosets and tamarins. Goeldi's monkey lives in small groups, but data are not yet available on social organization.

Goeldi's monkey is a rare species; relatively little is known about it in its natural environment. Its sparse distribution makes it of conservation concern, as it would be difficult to maintain viable populations in small reserves (Mittermeier, 1986b). On the other hand, breeding of *Callimico* in captivity has been very successful (Poole and Box, 1986). They have rapidly increased population sizes (Mace, 1986) and meet the criteria of Perry et al. (1972) for self-sustaining status (Flesness, 1986).

Taxonomy and Distribution

Because of its intermediate characteristics, *Callimico* has been placed by some investigators with the larger neotropical primates, primarily on the basis of having three molars that are quadritubercular and in giving birth to only a singleton young (Simpson, 1945; Cabrera, 1958; Martin, 1992). Others (e.g., Dolman, 1937; Hershkovitz, 1977; Goldizen, 1987a; Smuts et al., 1987; Mittermeier et al., 1988b) place it in its own family, the Callimiconidae. This intermediate position is also supported by features of the placenta (Soma and Kada, 1989), and by internal anatomy (Hill, 1959). Most now (e.g., Fleagle, 1988) include it in the same subfamily Callitrichinae (or family Callitrichidae) as the marmosets and tamarins, because of its small body size and the possession of clawlike nails on all digits except the hallux (big toe). Other data that support this position include cranial and dental characters (Rosenberger, 1979, 1981; Natori, 1988), postcrania (Ford, 1980a, 1986b), immunological data (Baba et al., 1975; Cronin and Sarich, 1978; Sarich and Cronin, 1980), hypocone gradient of upper molars (Natori, 1986, 1989), structural pattern of dental enamel (Nogami and Natori, 1986), karyological analysis (Dutrillaux, 1988a,b; Dutrillaux et al., 1988), and DNA sequencing (Seuánez et al., 1989, Crovella et al., 1992).

Callimico has a broad distributional range in the upper Amazon basin of Peru, Colombia, and the States of Acre and western Amazonas in Brazil (Hernández-Camacho and Cooper, 1976; Hershkovitz, 1977; Rylands, 1987; Peres, 1990), and the Department of Pando in Bolivia (Izawa, 1979; Brown and Rumiz, 1986) (Map 7). Its distribution is very patchy (Izawa, 1979).

Evolutionary History

Ford (1986a) astutely suggested that *Callimico* represents a stage in the evolution of Callitrichinae that had developed vertical clinging posture (using its claw-like nails) but had not yet developed exudativory. This concept was further developed by Garber (1992). Sarich and Cronin (1980) estimated that the ancestor of *Callimico* diverged from other Neotropical primates 7–9 million years ago (mya), based on immunological data. On the basis of satellite DNA sequencing, however, Fanning et al. (1987, 1989) estimated the divergence at more than 30 mya. Fanning's interpretation is supported by the presence of *Dolichocebus*, a cebine, in the late Oligocene, suggesting that the cebine/callitrichine split occurred before that (Rosenberger et al., 1990). The presence of *Micodon* (Setoguchi and Rosenberger, 1985) (cladistically close to the *Callithrix/Cebuella* clade) at 15 mya suggests that *Callimico* differentiated earlier than this date.

Kinzey (1982), in his discussion of Pleistocene refugia, postulated that recent differentiation of Goeldi's monkey occurred in the Napo center of endemism and spread extensively from there. The Napo center may represent the "shabby" forest of the early Quaternary referred to by Izawa (1979), in which both *Cebuella* and *Callimico* became ecologically segregated (see "Posture and Locomotion" below). Hershkovitz (1977) agrees that *Callimico* originated within its present geographical range, but he believes that traits shared by *Callimico* and tamarins are the result of parallel evolution.

Previous Studies

Callimico was only discovered in 1904 (Thomas, 1904) (see also early history in Hershkovitz, 1977), and relatively little is known about it.

Captive Studies

Early studies are reviewed by Heltne et al. (1981). There is a large reproductive colony at the Brookfield Zoo in Chicago (Warneke, 1988). Reproduction has been studied by Altmann et al. (1988), Carroll (1986), Carroll et al. (1989, 1990), Christen et al. (1989), and Ziegler et al. (1987, 1989). Infant care was studied by Heltne et al. (1973) and Carroll (1987a), social behavior by Carroll (1985, 1987b, 1988), marking displays by Wojcik and Heltne (1978), and vocalizations at the Japan Monkey Center by Masataka (1983, 1986) and by Masataka and Kohda (1988). Soini (cited in Hershkovitz, 1977:905) describes the daily rhythm of a pet in Iquitos, Peru.

Field Studies

Information about *Callimico* in its natural habitat comes primarily from studies in the Acre River basin of northern Bolivia by the Pooks (40 hours observation of *Callimico* over 3 months) (Pook and Pook, 1979, 1981, 1982) and by Masataka (1981, 1982) (205 hours over 5 months). These sites were identified during the ecological surveys of Izawa (1979). Masataka's study group was provisioned. There was also a brief study by Moynihan (1976a,b) in Colombia. The rarity of *Callimico* in the wild is illustrated by reports of Soini and Medem, cited in Hershkovitz (1977:904).

Habitat and Ecology

Callimico has been described as a habitat specialist but exactly why is not known (Rylands, 1987). Izawa (1979) found *Callimico* primarily in bamboo forest, and never in mature forest, and he concluded that its habitat is a variety of "shabby" forests. Nevertheless, these forests have considerable diversity and the Pook study site supports 13 sympatric primate species. Within the forest *Callimico* is unique among neotropical primates in inhabiting primarily the lowest levels of the forest canopy.

Pook and Pook (1981) reported that 88% of sightings were at 5 m or less! Much of the time at this level, including coming to the ground, is spent foraging for insects. The foraging pattern may be similar to that of *Saguinus fuscicollis* and *Leontopithecus*, in which vertical trunks serve as a primary platform from which to locate and capture insects (Garber, 1992). Additional early reports are cited in Heltne et al. (1981).

Diet

The preferred items in the diet appear to be insects that are obtained below the 5 m level in the forest or by jumping to the ground. (Sympatric tamarins have not been observed to go to the ground to catch insects.) They ascended to upper levels of the canopy to obtain fruit, but did not spend much time there, feeding only 5–10 minutes in each fruiting tree (Pook and Pook, 1979). There is no evidence of exudativory by *Callimico*, except the eating of the sticky seed pods of *Piptadenia* (Leguminosae) (Pook and Pook, 1981). They also fed on young leaves and epiphytes. Additional details of feeding are in Pook and Pook (1981), and earlier anecdotal feeding data are in Heltne et al. (1981).

In an attempt to relate differences in location of gastric glands and histochemistry of the gastric mucosubstances to differences in diet, Suzuki and Nagai (1986) examined 16 species of primates, and found *Callimico* to be of the same type as other species classified as frugivore-insectivores (*Saimiri, Cebus, Saguinus, Callithrix, Aotus, Galago,* and *Pan*) —the cardiac glands occupied a narrow area surrounding the cardia.

Predation

There is no evidence of humans hunting *Callimico* (Izawa, 1979). Their response to potential avian predators was to run down vertical trunks and hide in dense low shrubbery (Pook and Pook, 1979). At low levels of the forest they would be likely prey of the tayra (*Eira barbara*) or large snakes (Pook and Pook, 1981); thus, polyspecific association with tamarins (Pook and Pook, 1982) may well reduce the risk of predation.

Behavior

Goeldi's monkey lives in small groups. The study group of Pook and Pook (1979) consisted of eight animals (six adults and two juveniles). The study group of Masataka (1981) originally consisted of six animals (2 ♂♂ and 4 ♀♀) to which were later added three captive male (one subadult and two juveniles). The presence of two juveniles of the same age and two females in another group simultaneously carrying young (Masataka, 1982) implies either that two females were pregnant simultaneously, or that immigration of a (pregnant or lactating) female occurred. Most other reported groups were fewer in number, and monogamy is often assumed.

Social Behavior

During traveling, foraging, and feeding Pook and Pook (1981) observed very little social interaction. During resting in small subgroups, social interactions, most commonly grooming, occurred. Urine tail-marking was also frequent. Allogrooming is a frequent behavior in captivity as well (Heltne et al., 1981). In captivity, the most stable social groups seem to be monogamous pairs. In the wild, however, Masataka (1981) found no monogamous groups. Strong bonds exist between captive pairs, but they are not equal for male and female (Carroll, 1985). At the feeding site Masataka (1981) observed variation in grouping patterns. The entire group appeared together only during 3 out of 5 months. Based on displacements at the feeding site a dominance hierarchy existed in the group. More data are needed to determine whether the variable social system seen in marmosets and tamarins also occurs in *Callimico*.

Ranging Behavior

Individual groups (as is the case with *Cebuella*) are generally widely separated from one another (Izawa, 1979). No intergroup encounters were observed over their 3-month study (Pook and Pook, 1981). Izawa and Yoneda (1981) reported the range of one group at about 40 ha. Pook and Pook reported a single group's range at 30 ha over 3 months, but suggested it might be larger. The day range of the Pook and Pook (1981) study group was about 2 km/day. Ranging often utilized certain routes in the forest more than once per day. There was strong group cohesion while traveling—usually within 15 m of each other, seldom over more than 30–40 m. When resting, however, the group never rested as a single unit in contact, but rather in subgroups of one to four animals separated by several meters (Pook and Pook, 1981).

Pook and Pook's (1981) estimated population density is very low (1 group/4 km²) Goeldi's monkey is frequently found in association of one or both sympatric tamarins in Bolivia (Pook and Pook, 1982).

Activity Pattern

The activity pattern varied markedly between wet and dry seasons (Pook and Pook, 1981). In the wet season the group was active from about 6:15 to 17:00–17:45 (roughly just after dawn to just before dusk). About 23% of the active day was spent visiting feeding trees, 16% resting, and 60% traveling (Pook and Pook, 1981). "Traveling" included "foraging for insects." Sleeping sites were low and in dense tangles of vegetation, and widely distributed throughout the range. Resting sites were always within 4 m of the ground, in dense undergrowth, sometimes in bamboo thickets. Masataka (1981) observed that the study group, once sleeping, segregated in two sites that were separated by more than 100 m.

Mating and Reproduction

Most data on reproductive biology of *Callimico* come from captive studies. According to Carroll et al. (1989) they have a behavioral estrous cycle of 22–24 days, postpartum estrus, a gestation of 150–160 days, nonseasonal breeding (in captivity), and a single offspring that matures at approximately 1 year of age. Christen et al. (1989), based on urinary estradiol levels of 2 nonpregnant females, estimated ovarian cycle length at 27.8 ± 5.2 days. Ziegler et al. (1987) found postpartum conception occurring within 12 to 19 days after parturition; mean gestational length was 146 days. Ziegler et al. (1989) reported a postpartum conception rate of 83%, with a gestation length of 148.8 ± 5.2 days (range, 144–157 days). Stevenson (1986) reported an interbirth interval of 153 to 341 days (median = 169 days). Among platyrrhines, the cycle length of *Callimico* most closely resembles that of *Callithrix jacchus* (Christen et al., 1989).

Females are polyestrous and may have two births per year in the wild (Masataka, 1981). The two birth periods in northern Bolivia were in September/October and April/May (Masataka, 1981).

Some reproductive suppression is exhibited in *Callimico* males (Carroll, 1986). Atrophy of the testes in one individual followed a shift in dominance subsequent to aggression between him and his son, upon introduction of a new female into their group. The father's testes never redeveloped, even after his separation and pairing with another female.

The behavior of three groups housed together in trios of one male and two unrelated females was studied by Carroll (1987a). Of the two trios that were stable for a year, both females conceived, and in one the infants were raised to independence. There is a case of mother and daughter both becoming pregnant while housed together with a new breeding male. Analyzing these data, Carroll (1986) believes that the mechanism involved in reproductive suppression in *Callimico* is neither physiological suppression nor the result of dominance, but an "incest taboo" between mother and son, and between father and daughter.

The average age at first reproduction for *Callimico* females, at 1.3

years, is very low when compared with that of other primates; lower values are found only in *Galago* and *Microcebus* (Ross, 1988:217).

There are a few registered cases of twinning in *Callimico*. Erwin (personal communication in Shively and Mitchell, 1986) reported three sets of twins born at the Chicago Zoological Park. More recently, of five sets of twins born there, none was reared together successfully by the parents (Altmann et al., 1988).

There is an international studybook for *Callimico* (Warneke, 1992), which replaces the 1988 edition.

Growth and Development

In the wild *Callimico* infants are carried exclusively by the mother during the first 2 to 3 weeks (Pook, 1975; Masataka, 1981). In six groups studied by Carroll (1987b) from birth to independence at Jersey Wildlife Preservation Trust, the mother was the principal caregiver for the first 3 weeks of the infant's life, after which time the father and other group members began carrying, grooming, and sharing food at greater levels. These data are in marked contrast with those of tamarin and marmoset males (as well as monogamous *Callicebus* and *Aotus*), which typically carry and groom their infants within a few hours after birth. Animals become completely independent with full locomotor capability by 7–8 weeks (Masataka, 1981). Additional data on growth parameters in the wild are provided by Masataka (1981).

Communication

Callimico has a varied vocal repertoire (Epple, 1968) with 40 calls identified in the natural habitat (Masataka, 1981). Pook and Pook (1981) identified seven distinct calls, one of which was a loud and distinctive feeding call heard only when feeding high up in a fruit tree—presumably to locate the food resource for the remainder of the group traveling low in the forest. Heltne et al. (1981) provide an extensive list of communicative behaviors, largely from captive studies.

Play

Play behavior appears to be relatively infrequent, but data are sparse. There is a summary in Heltne et al. (1981).

Posture and Locomotion

Callimico's primary mode of posture and locomotion is vertical clinging and leaping at low levels in the forest (Moynihan, 1976b; Pook and Pook, 1981). When traveling or feeding at levels above 5 m they adopted quadrupedal walking and only occasionally leaping (Pook and Pook, 1981). The typical tamarin horizontal leaps from terminal branches of one tree to the next were rare. Instead they usually climbed down to the lower forest levels for travel (Pook and Pook, 1979).

If *Callimico* and *Cebuella* both differentiated in the Napo refuge, it may not be coincidental that they are the vertical clingers and leapers *par excellence* among the neotropical primates. This mode of locomotion would have been extremely well suited to the flooded habitat (*Cebuella* at "edges" of the forest and *Callimico* within the forest), which probably existed here as the large Tertiary lake receded into formation of the upper Amazon River basin (Izawa, 1979).

18

Callithrix

Introduction

The genus *Callithrix* comprises 3 to 12 species depending on the taxonomy (see below). The common name for all members of the genus is "marmosets." *Callithrix* is characterized by small body size, average weight 300–450 g, and lack of sexual dimorphism (Stevenson and Rylands, 1988). They are the smallest Neotropical primates except for *Cebuella*. Previously, these animals were described as primitive and squirrel-like (e.g., Hershkovitz, 1977), and even recent publications (e.g., Preston-Mafham, 1992) erroneously describe them as being primitive. Continuing work on their anatomy, ecology, and behavior has indicated, however, that they are highly derived and specialized primates (Ford, 1986, 1987; Sussman and Kinzey, 1984). Some of the specializations include the presence of claws instead of nails on all but the hallux (big toe) and twinning instead of single births (Stevenson and Rylands, 1988). The conservation status of each of the *Callithrix* species is reviewed by Rylands et al. (1993).

Taxonomy and Distribution

The genus *Callithrix* comprises three (Hershkovitz, 1977) to eight (Coimbra-Filho, 1984), nine (Mittermeier et al., 1988), or 12 species (Vivo, 1991), depending on the taxonomy [See Stevenson and Rylands (1988) and Rylands et al. (1993) for discussions of the problem].

Callithrix is found only south of the Amazon River and east of the Madeira River (Map 8). Except for possible minor areas of intergrade, all species of *Callithrix* are allopatric. With the exception of *C. argentata melanura*, whose range extends into northern Paraguay (Stallings and Mittermeier, 1983) and eastern Bolivia, all species and subspecies are confined to Brazil. There are two species groups (Hershkovitz, 1977). The "jacchus" group is found from the State of Ceara in the northeast, south along the Atlantic coastal forests and adjacent inland cerrado. The "argentata" group is found in the Amazon region only; there does not appear to be any area of contact between the two species groups (see Map 8). Rylands et al. (1993) prefer to subdivide the "jacchus" group into an "aurita" group (*C. aurita* and *C. flaviceps*) and a "jacchus" group (*C. jacchus, C. geoffroyi, C. penicillata,* and *C. kuhli*), based on phenotypical and ecological distinctions.

Variation in pelage color and pattern has hampered attempts to agree on species and subspecies of *Callithrix*. In some regions there are narrow areas of either intergrade or species overlap with hybridization. For example, in southern Bahia, there is a narrow area of either hybridization between *C. jacchus* and *C. penicillata* (Alonso et al., 1987), a subspecies of *C. jacchus* (*C. j. kuhli*) (Mittermeier and Coimbra-Filho, 1981), or a separate species, *C. kuhli* (Coimbra-Filho, 1984). Whether this situation is due to natural (historical) occurrences or is the product of habitat shrinkage (recent/man-made) is not known.

In 1992 two new species of *Callithrix* were described, *C. mauesi* (Mittermeier et al., 1992), and *C. nigriceps* (Ferrari and Lopes, 1992). The Rio Maués marmoset was found on the west bank of the Rio Maués-Açu, in eastern Amazonas, just south of the Rio Amazonas. The black-headed marmoset was found at Lago dos Reis (7°31'S, 62°52'W) in Rondônia. Both species are located between the Rios Madeira and Tapajós, are members of the "argentata" group, and have not been discussed in previous reviews cited above. The ranges of both species are apparently very small.

Evolutionary History

There is no known fossil evidence for *Callithrix*. However, Rosenberger et al. (1990) suggested that the Miocene La Venta *Micodon* and *Mohanamico* are members of the subfamily Callitrichinae. The callitrichines are probably a relatively rapid (Pook, 1978) and recent lineage, so one would not expect *Callithrix* in the 12–15 mya time range. Whatever the time frame, however, callitrichines are characterized by phyletic dwarfism relative to the presumed ancestor (Leutenegger, 1990; Ferrari, 1993).

The genus *Callithrix* probably evolved in the Atlantic coastal forests of Brazil from an ancestor in common with *Leontopithecus*, and then spread west and north into the Amazon basin (Ferrari, 1993). Subsequent speciation, at least of the eastern Brazilian species, probably took place in forest refugia during the Pleistocene. Contrary to Kinzey (1982), *C. penicillata* does not occur in the Bahia Center; rather, *C. kuhli* originated there. *C. penicillata* is related instead to Campo Cerrado Center (Müller, 1973), but since it is a savanna form, it may not have originated there. Its relationship to other savanna primates, such as *Alouatta caraya* and *Cebus apella libidinosus*, needs to be investigated.

Hershkovitz (1968) suggested that differences in pelage color among species and subspecies of marmosets are due to metachromism; however, this is only a description, not an explanation, of evolutionary change. The adaptive advantage of albinistic forms of *C. argentata* and *C. humeralifer*, east and west of the Rios Tapajós and Canumã, respectively, remains unexplained.

Terborgh (1992:170) suggested, on the basis of a comparison with small omnivorous primates elsewhere in the world (which are nocturnal), that the marmosets are anomalous because they are diurnal. I do not believe that this is an anomaly; rather the marmosets fill a niche analogous to that of the baboons on the Old World savannahs. Baboons have evolved large body size to thwart predation, whereas marmosets, "edge" primates that are small and cryptic, entered onto the Brazilian savannas where small body size, and probably even lighter pelage coloration, continued to provide an adaptive advantage (Brown and Kinzey, 1996). For example, *C. penicillate* (a savannah form) is lighter in coloration that the otherwise very similar *C. kuhli* (a forest species).

Previous Studies

See Stevenson and Rylands (1988) for an earlier review of previous studies.

Captive Studies

Callithrix jacchus has been the most frequently studied captive marmoset species (Stevenson and Rylands, 1988), and was the first species to be bred in captivity on a regular sustained basis, beginning in 1965 (Kingston, 1969, 1975). The first account of *Callithrix* behavior was by Epple (1967), and early accounts are reviewed there, in Stevenson and Poole (1976) and in Stevenson and Rylands (1988).

Captive studies have concentrated on maintenance of captive populations (dietary supplements: Coimbra-Filho and Silva, 1984; Silva, 1984; reproductive and pathological problems: Corillas et al., 1984; comparative behavior of callitrichid genera and species: Snowdon, 1983). Many studies have focused on social/physiological behaviors tied to reproductive patterns (suppression of ovulation in nondominant females within a group: Hubrecht, 1984; Evans and Hodges, 1984; pair-bonding and monogamy: Evans, 1983; parent–offspring attachments, imprinting, and development: Cebul et al., 1978). Other studies have included behavioral variability both in individuals and between groups in response to environmental change (Box, 1984), play, allogrooming, and scent marking (Box, 1978). When species other than common marmosets are included in reports, the context is generally one of comparative analysis between behaviors of some species other than *C. jacchus* and the more well-known behavior of *C. jacchus* (e.g., reintroduction of hand-reared infants into existing groups for *C. jacchus*, *C. argentata*, and members of *Saguinus* spp.: Pook, 1978).

Patkay (1992) thoroughly reviewed the incidence of diseases in captive callitrichines, including those caused by arthropods, organ pathologies, spontaneous neoplasms, nutritional diseases, helminths (common in marmosets and tamarins), mycoses (uncommon), protozoa (common but rarely causing illness), bacteria (major contributor to illness and death; usually found at time of importation, so may represent feral disease), and viruses.

Field Studies

The first long-term marmoset field study was conducted by Rylands from 1978 to 1979 on *C. humeralifer intermedius* at Aripuanã, MG (Rylands, 1979, 1981, 1982, 1984b, 1986a,b). *C. argentata melanura* was studied in Paraguay by Stallings (1985; Stallings and Mittermeier, 1983; Stallings et al., 1989). *C. jacchus* was studied at Tapacura, Pernambuco by Hubrecht (1984, 1985), Scanlon and Chalmers (1987), Scanlon et al. (1988, 1989), and Stevenson (1978; Stevenson and Rylands, 1988), and at João Pessoa, Paraíba by Maier et al. (1982), Alonso (1984), and Alonso and Langguth (1989). *C. penicillata* was studied in the savanna near Brasília by Lacher et al. (1984), Fonseca and Lacher (1984), and Faria (1984a,b, 1986, 1987, 1989). *C. kuhli* was studied at Una, Bahia by Rylands in 1980 (1982, 1984a, 1989). *C. flaviceps* was studied at the Caratinga Biological Station, MG, by Ferrari (1987a,b, 1988, 1992; Ferrari and Lopes Ferrari, 1990). *C. aurita* was studied at Fazenda Barreiro Rico, SP, by Torres de Assumpção (1983a), and in Minas Gerais by Muskin (1984a,b) and Stallings (1988; Stallings and Robinson, 1991).

Habitat and Ecology

Members of the genus *Callithrix* are generally described as having a "preference" for dry secondary or disturbed forests, or edge habitats (Sussman and Kinzey, 1984; Fleagle, 1988). While this is true, field studies have shown that the habitats occupied by marmosets are quite diverse. See Rylands and de Faria (1993) for a review of habitats of the different species.

Comparison between geographically separated Amazonian and Atlantic forest species shows that marmosets inhabit a variety of forest types, which include high primary forests, secondary forests, savanna patch forests, white sand, and disturbed forests (Rylands, 1987). *C. penicillata* and *C. jacchus*, the two most adaptable species in the genus, occupy habitat mosaics including open woodland with variously dense undergrowth or with large numbers of lianas and bromeliads, as well as savanna/dry forest formations, *cerradão, cerrado,* and *caatinga* (Ferrari and Lopes Ferrari, 1989; Scanlon et al., 1988; Stevenson and Rylands, 1988). The remaining species occupy secondary growth and tall forest with broken canopy of Amazonian and Atlantic forests. Marmosets are not found in várzea (flooded forest) or in extensive areas of primary forest with closed canopy and sparse understory. In contrast to *C. penicillata,* *C. kuhli* (contra Kinzey, 1982) has darker pelage, which is better adapted to the coastal forest than the paler pelage of *C. penicillata,* which is probably an adaptation to its typically more open woodland habitat.

Diet

Marmosets are frugivore–insectivores or, in the case of *C. penicillata* and *C. jacchus,* exudativore–insectivores. The type of forest inhabited by members of the genus appears to determine their foraging pattern and diet. These parameters, in turn, affect ranging behavior, group size, social organization and behavior, and activity pattern (Harrison and Tardif, 1989). In their excellent summary article Ferrari and Lopes Ferrari (1989) describe the difference in the diets of *C. flaviceps* (secondary forest) and *C. humeralifer* (primary and disturbed primary forests) as being based more heavily on exudates or fruit, respectively. Habitat-dependent changes in diet, from frugivory–insectivory to exudativory–insectivory, are evident throughout the genus (Scanlon et al., 1988; Rylands, 1987).

All marmosets (including *Cebuella*) eat plant exudates, but it is much more common in the "jacchus" group (especially *C. jacchus* and *C. penicillata*) than in the "argentata" group. The degree of development of the dental gouge (lower anterior teeth) in different species of marmosets is correlated with the degree of exudativory (Natori, 1986, 1990; Masahito and Shigehara, 1992). All marmosets lack enamel on the lingual side of lower incisors, which provides a self-sharpening mechanism for the dental gouge (Rosenberger, 1978). Exudativory in marmosets is fully discussed by Stevenson and Rylands (1988), and more generally by Nash (1986).

There is a trend among marmosets, especially those in more seasonal environments, to depend heavily on certain keystone resources. In a riverine forest bordering caatinga in Brazil, *C. penicillata* depends very heavily on exudates of *Tapirira guianensis* whose representatives make up 85% of the tree species in the area; remaining species provided fruit and flowers (de Faria, 1986; see also Seabra et al., 1987, for reliance on *Vochysia pyramidalis*). Even in less seasonal environments there is a tendency to depend on a few species to provide the bulk of plant food sources. For *C. humeralifer intermedius* the two highest ranking plant species made up 50% of the diet in almost every month of the year (Rylands, 1982). A list of more than 60 species of fruit eaten by *C. h. intermedius* is provided in Stevenson and Rylands (1988).

Much of the food eaten by marmosets is not plant material. Insects, small lizards, and frogs are important elements of the marmoset diet (Stevenson and Rylands, 1988). Foraging for insects and vertebrates takes the form of on the ground foliage-gleaning and exploitation of the activities of army ants as they "disturb" prey (Rylands, 1989). Secondary growth appears to be more attractive to folivorous insects, as plants in these environments tend to produce fewer secondary compounds (Rylands, 1987). This makes these environments more attractive to insect-foraging marmosets as well.

Predation

Another reason for choosing dense, viney, second growth or scrub/caatinga environments is that they provide hiding places from predators. Ferrari and Lopes Ferrari (1990) report a higher degree of vigilance behavior, calls, and open space avoidance behavior by buffy-headed marmosets when they are in higher canopy strata. They attribute this to marmosets' vulnerability to avian predators. Although predation of marmosets has not yet been observed (Stevenson and Rylands, 1988), raptors are presumably a threat.

Behavior

Social Behavior

The social organization of *Callithrix* in the wild is not so well known as that of *Saguinus*, in part because of the paucity of data on *Callithrix* with individually marked animals. In all species studied thus far there is only one breeding female per group at any one time. Often, however, there are additional adult males and/or females in the group that are probably offspring or siblings of the breeding pair. Except for immigration/emigration of adults, group composition remains quite stable (Ferrari and Lopes Ferrari, 1989). Only *C. jacchus jacchus* has been studied in the wild with individually marked animals (Scanlon et al., 1988; Dixson et al., 1992; Digby, 1994).

Marmoset groups are even more stable in captivity (Rothe and Darms, 1993). As in the case of *Saguinus, Callithrix* in captivity are usually kept in family groups in which a monogamous pair forms the breeding group. Cooperation, predominantly in the form of infant care, is a prominent aspect of callitrichid social organization (Rothe and Darms, 1993). Major differences between the social organizations of *Callithrix* and *Saguinus* are reviewed by Ferrari and Lopes Ferrari (1989).

Ranging Behavior

Habitat and type and distribution of primary food resources have predictable effects on average daily path lengths for marmosets. Recorded path lengths are *C. humeralifer*, 1.5 km; *C. flaviceps*, 1.2 km; *C. penicillata*, 1.0 km; and *C. jacchus*, 0.5–1.0 km (Ferrari and Lopes Ferrari, 1989). Path lengths are shorter than those of tamarins, which are more frugivorous (Ferrari and Lopes Ferrari, 1989). They also reflect similar trends within the genus: *C. penicillata* and *C. jacchus* are the least frugivorous of the marmosets, have greater dependence on more localized resources such as plant exudates, and hence have shorter path lengths.

Home range sizes of marmosets are from 0.5 ha (one *C. jacchus* group), to 16 ha (largest *C. aurita* group), to 28.25 ha (*C. humeralifer*) to 35.5 ha (for one *C. flaviceps* group) (Rylands and de Faria, 1993). With the exception of *C. flaviceps* (which exploits small, widely dispersed gum sources), smaller home ranges are correlated with higher degrees of exudativory (Rylands and de Faria, 1993).

Callithrix tend to live in larger groups than either *Saguinus* or *Leontopithecus* (Ferrari and Lopes Ferrari, 1989). Average group size for the five species studied ranges from 6.6 (*C. kuhli*) to 11.5 (*C. humeralifer*). The largest observed group was 15, including 5 adult females and 4 adult males (Rylands, 1982).

In captivity, common marmosets have been kept in "stable" groups of from 8 (Koenig and Rothe, 1991) to as many as 20 individuals (Darms, 1990), although intragroup aggression was reported for the largest group. Group stability is of great importance to animals where alloparenting is common across age–sex classes, and where the presence of extended families is fundamental to the development and well-being of offspring (Watson and Petto, 1988; Goldizen, 1990).

Scanlon et al. (1989) describe the environmental needs of common marmosets as being centered on exudate-producing trees and suitable

sleeping sites. Home ranges are small and contain specialized vegetation. Home ranges of defendable size containing limited resources are
parameters predicting territoriality (Oates, 1987); yet, intergroup territorial behavior is not described as typical of these animals.

Activity Pattern

Marmosets are active from about one-half hour after sunrise until
about one-half hour (*C. jacchus*) to 1½ hours (*C. humeralifer*) before sunset. *C. jacchus* spent 35% of their active day moving and foraging, 12%
feeding, 10% in social activities, and 53% stationary. *C. humeralifer* spent
38% locomoting, 30% feeding on animal prey and foraging, 18% feeding
on plant foods, and 15% resting and social activities (Stevenson and
Rylands, 1988). Marmosets normally sleep in dense vegetation or a vine
tangle. They may, or may not, use the same sleeping site on consecutive
nights.

Mating and Reproduction

In their natural habitat *Callithrix* live in extended family units in
which a single reproductive female may mate with more than one male
(Dixson, 1993). Whether species other than *C. humeralifer* (Rylands, 1982)
are polyandrous is not yet known. Ovulation is suppressed in all subordinate females (Snowdon, 1990).

Marmosets, as all callitrichids, normally have dizygotic twins; however, the twins have the same placental circulation, and thus have
immunological compatibility (Wislocki, 1939). In captive studies the
twinning rate is 80% (Goldizen, 1990).

Litter weight has been reported as representing 21–27% of the mother's weight for *Callithrix jacchus* (Goldizen, 1990). This is about 2.5 times
the average ratio for other, single birth, species of anthropoids (Goldizen, 1990). The evolution of monogamy or fraternal polyandry in *Callithrix* has been attributed, at least in part, to the production of twins
(Ferrari and Lopes Ferrari, 1989).

Groups of marmosets almost invariably show cooperation in infant
care (Goldizen, 1987a), reflecting dietary and locomotor constraints imposed by twinning on the breeding female (Sussman and Garber, 1987).
Both captive and field studies of marmosets indicate that pairs of infants
are the offspring of a single breeding pair within any group [for reproductive suppression of adult females by breeding female in captivity see
Rothe and Koenig, 1991 (*C. jacchus*); in the wild see Ferrari, 1987 (*C. flaviceps*) and Rylands, 1986 (*C. humeralifer*); for review, see Abbott et al.
(1993)]. Dominant males may also suppress endocrine function in unrelated subordinate males (Abbott, 1993).

Bimodal birth peaks, reported for marmosets in the wild (Ferrari and Lopes Ferrari, 1989), adds to the need for group support of developing infants; and while all age–sex classes contribute to infant survival and development, not all individuals contribute equally (Koenig and Rothe, 1991; Rothe, 1990). Darms (1987) observed flexibility in levels of alloparenting among captive common marmosets. In wild common marmosets there is no reason to believe that all group members are related (Sussman and Garber, 1987).

A successful initiation of pregnancy during postpartum estrus requires both chemical and behavioral cues (Dixson, 1992).

Growth and Development

Marmoset young develop quickly. Puberty for captive male common marmosets was estimated at 200 days (Dixson, 1986), reduced carrying by parents at 5–8 weeks (Tardif et al., 1986), and adolescence is achieved by 10 months (Stevenson and Rylands, 1988). Experience by juveniles in carrying infants increases their future reproductive success (Tardif et al., 1984). The degree to which callitrichids share food with younger group members is very high (Ferrari, 1987a).

Infant *Callithrix* are carried for a significantly shorter period of time than are infant *Saguinus*, reflecting their more rapid maturation (Tardif et al., 1993). Extended infant carrying by tamarins (cf. marmosets) may represent adaptation to a long distance foraging pattern (Tardif et al., 1993). Developmental stages for *C. jacchus* are given by Yamamoto (1993).

Communication

Callithrix has an extensive repertoire of visual, vocal, and olfactory signals. Visual signals include facial, body, and tail positional behaviors (for list, description, and function see Stevenson and Rylands, 1988). Vocal communications include calls typical of species where alloparenting is of tremendous importance (Masataka and Kohda, 1988). In comparative captive studies (Masataka, 1986) responses to one type of *Callithrix* and two types of *Saguinus* alarm calls elicited equal response types in the other species. Threat calls/responses were shared by *Cebuella*, *Callithrix*, *Leontopithecus*, and *Callimico*, while location calls of *Callithrix* acted as warning calls in the other three species. Masataka (1986) states that *C. jacchus* alarm calls were more tonal, hence less easily localizable, than *Callimico* and *Cebuella*. This difference is attributed to the higher canopy position occupied by marmosets, making them more vulnerable to avian predators.

Note Moynihan (1967) points that callitrichids all have high pitched sounds (related to evolution in a savanna-open habitat).

Olfactory signals have been observed to be mediators of intragroup behaviors (Box, 1988). They are also strongly related to feeding ecology. Bartecki and Heymann (1990) comment on the extensive use of suprapubic scent by *Callithrix humeralifer intermedius* to delimit home ranges and areas of intense use. Male olfactory input is important to coordinate reproductive behavior with female hormonal changes (Dixson, 1993). Successful initiation of pregnancy during postpartum estrus requires both chemical and behavioral cues (Dixson, 1992). Olfactory communication is reviewed by Epple et al. (1993).

Play

In captive *C. jacchus* infants engage in solitary play during weeks 2–4, and add social play during week 5 (Yamamoto, 1993). Other siblings were the main play partners; only if none was available was a twin a play partner. Cleveland and Snowdon (1984) suggested that by tolerating infants on their backs during play, juveniles learn carrying skills. Earlier references to play behavior are found in Stevenson and Rylands (1988).

Posture and Locomotion

The importance of tree exudates in the diets of marmosets and the postures associated with vertical clinging are considered to be mechanisms responsible for the development of claws (Sussman and Kinzey, 1984). Recent work by Garber (1989) supported the evidence that vertical clinging is most highly developed in the smallest of marmosets, *Cebuella* and *Callithrix jacchus*, who are also most dependent on tree exudates. He notes that vertical clinging is less frequently observed in *C. humeralifer* whose diet contains a higher percentage of fruit. When not engaged in exudate feeding locomotor behavior includes quadrupedal behaviors and leaping (Fleagle, 1988).

19

Cebuella

Introduction

The pygmy marmoset, *Cebuella pygmaea*, is the smallest living anthropoid primate. Among extant primates, only *Microcebus* ssp. and *Galago demidovii* are smaller (Fleagle, 1988). Soini (1988) gives a thorough description of the physical characteristics of a free-ranging *Cebuella* group. The dorsal pelage is predominantly tawny agouti. The lower back, rump, and legs of young adults are marked by a gray-ticked color. In adults this may extend over the mid-dorsum, often forming a dark gray or blackish semistriated pattern. The tail is faintly to conspicuously annulated and in adults the tail is longer than the body. The external genitalia of adult females are unpigmented, whereas the scrotum of adult males is heavily mottled by black pigmentation. The anogenital area of male and female adults is bordered by a thick frame of completely black pelage, which strongly enhances the visual effect of the genitals in genital displays. Of particular interest are the facial characteristics, which consist of a whitish spot above either mouth corner and a vertical, buffy to white nasal ridge stripe. These markings can be readily seen in adults, especially when viewed in the gloomy light conditions of the forest understorey. Soini (1988) believes that these distinct facial marks serve to enhance perception of facial expressions and head movements during visual communication.

Taxonomy and Distribution

The genus *Cebuella* Gray, 1866, is monotypic (Rosenberger, 1988), including the only species *Cebuella pygmaea* Spix, 1823 (Soini, 1988). Some investigators (Delson, 1988) include the pygmy marmoset within the genus *Callithrix*, a position once held by Rosenberger (1979). Most investigators, however, recognize *Cebuella* as a distinct genus (Rylands, 1984; Goldizen, 1987; Fleagle, 1988; Rosenberger et al., 1990).

The distribution of the pygmy marmoset is restricted to the upper Amazon Region, thus encompassing part of Brazil, southern Colombia, Ecuador, Peru, and northern Bolivia (Rylands, 1984; Fleagle, 1988). In the upper Amazon basin they are specifically limited to the tropical lowland forests ranging from the south bank of the Rio Caqueta in southern Colombia, and south through eastern Ecuador, eastern Peru, and western Brazil to the north banks of the Rio Orthon-Manupiri or the Rio Madre de Dios in northern Bolivia and southeastern Peru (Soini, 1988). The southernmost locality where *Cebuella* has been recorded is the National Park of Manu in the Rio Manu basin of Peru. The western boundary of *Cebuella* extends as far as the Rio Pastaza and the Rio Huallaga basins, while its eastern is the Rio Japura (north of the Amazon) and the Rio Purus (south of the main river).

Evolutionary History

There are no fossil *Cebuella*. See Chapter 18 for discussion of evolutionary history of the family Callitrichidae.

Previous Studies

Previous studies are reviewed by Soini (1988).

Captive Studies

Vocalization studies of the pygmy marmoset range from individual recognition (Snowdon and Cleveland, 1980; Snowdon, 1981) to positional orientation (Snowdon and Hodun, 1981), and behavioral responses to synthesized vocalizations (Snowdon and Pola, 1978). In the early 1970s Snowdon was one of the few to study the calls of the pygmy marmoset. Pola and Snowdon (1975) conducted a study on vocalizations and associated behavior patterns of a *Cebuella* colony at the Wisconsin Regional Primate Research Center. They described vocalizations made by adult pygmy marmosets when acoustically and visually separated from their family group. Although much attention was devoted to the "isolation calls" generated by adult *Cebuella*, very little attention was given to infants, even though they too were observed giving these same "isolation calls." Therefore, Newman et al. (1987) extended the Pola and Snowdon work by concentrating their study on the infant *Cebuella*. Newman and his colleagues characterized the acoustic structure of infant marmosets during the first weeks of life. A total of 20 infants were recorded from 1–6 times at ages of 1 day to 8 weeks. It was observed that the variability of individual notes as well as note sequences ("calls") was greatest during the first postnatal week. No adult-like pattern of isolation calling was observed even for the 8-week age group.

Field Studies

Only five long-term studies of callitrichids were published as of 1987 (Goldizen, 1987). Soini (1982) conducted one involving a long-term field study of *Cebuella* in Amazonian Peru that lasted from 1976 to 1979. The study involved three phases. The first phase consisted of a one and one-half-year intensive demographic and socioecological study of a *Cebuella* population in a 3-km^2 plot of forest and riverine environment within the Manití basin. A total of 19 social groups were recognized including several incipient but transitory individuals. An additional 3-month follow-up study concentrated on two groups in the Tahuayo river basin in 1977. The second phase, which lasted 6 months in 1978 (Soini, 1988), involved a series of short-term observations of groups and solitaries at several sites in the Tapiche river basin. During this second phase, members of three groups were live-captured for the purposes of assessing sex, weight, stature, age, and reproductive status. The third phase involved an intense follow-up study by capturing approximately 83% of the total population inhabiting the 3-km^2 rectangular plot to establish sex, weight, stature, age, and reproductive status. This phase lasted about 75 days during which 115 individuals were catalogued, whereas six wild-born *Cebuella*, three females and three males, were held and raised in captivity from early infancy to adulthood. Five offspring were born to these over the course of 3 years, thus providing additional information on behavior and ontogenetic development (including the dental pattern development) which Soini has in preparation (Soini, 1988). It should also be noted that before and throughout the study period, additional population surveys and cursory observation on *Cebuella* ecology and behavior were made at many localities along the Amazon river and its tributaries.

Habitat and Ecology

"The natural general habitat of *Cebuella* consists of heterogeneous, evergreen, mesophytic lowland forest" (Soini, 1988:82). It is within this general ecological pattern that *Cebuella* is found to be a habitat specialist that prefers the river edge and flood plain forests (Soini, 1982, 1988). This preference is well illustrated in Soini's study of *Cebuella* in Amazonian Peru, which saw 85% of the total population living at the edge of the Rio Manití and on other occasion about 83% of the population also living at the edges of the Rio Maniti's tributary stream, the Hatun Quebrada (Soini, 1988). According to Soini (1988) sparse populations of *Cebuella* may also be found in the high ground forest beyond the floodplain ecosystem. These populations live along the edges of small forest streams or brooks, which are subjected to occasional flooding. The general pattern observed in Soini's study is that *Cebuella*'s preference to floodplains is usually restricted to the areas where the general inundation level does not reach beyond 2–3 in depth and the forest floor does not remain flooded for more than 3 months of the year. Population density was found to be 51.5–59.0 animals per km^2, whereas at the river edges it reached 210.0–233.3 animals per km^2 (Soini, 1987a).

Diet

Pygmy marmosets are insectivore–exudativores. They feed principally on insects and exudates of trees and vines, and fruits are only a minor part of the diet. In Peru *Cebuella* fed on exudates of more than 57 plant species, especially from the families Vochysiaceae, Leguminosae, Anacardiaceae, and Meliaceae (Soini, 1993). A troop normally has a single primary exudate tree; when that source is exhausted, the troop moves to a new home range, centered on a new exudate source.

Cebuella is the most specialized (exudativorous) of the callitrichines (Garber, 1992). It is an exudate specialist in that transit time in the digestive tract is remarkably long, and the energy digestibility of gums is greater than that of any other callithrichine (Power, 1991). In its natural habitat, more than two-thirds of its feeding time is devoted to feeding on saps and guns (Ramirez et al., 1977; Soini, 1982).

Cebuella may spend up to 30% of its daily activity procuring and feeding on gums for which it is dentally and digitally adapted (Harrison and Tardif, 1991).

Predation

Because of their small size, pygmy marmosets are not normally hunted for food by humans. Diurnal raptors are the primary predators (Ferrari and Lopes Ferrari, 1990).

Behavior

Social Behavior

In each social group only a single female (the dominant female) bears young. If there is more than one adult male in the group, they may all attempt to copulate with the female, but it is not known whether more than one male actually fathers the young. Thus, the social system is that of at least a temporary pair bond, or possibly polyandry (Soini, 1988).

Ranging Behavior

The distribution and density population of the pygmy marmoset have been linked to the abundance of special exudate feeding trees (Fleagle, 1988). They have small home ranges usually centered around whatever the main food tree is at the time and wherever there is an adequate supply of insects. It appears that such a combination of resources is usually found only at vegetational edge situations. Since these primary exudate trees change from year to year, so do the home ranges of *Cebuella* groups. The normal day range of a group has been observed to be as short as 30 m but never surpassing 100 m (Soini, 1988). *Cebuella* have been seen to travel from tree to tree within their home range and usually along the same pathways. Soini measured the horizontal distance of two individuals, which averaged out to be 290 m, not taken into account the various vertical movements on the boles of the feeding tress. One particular solitary individual that was observed and tracked was seen to log a path length of about 850 m in one day.

Activity Pattern

Cebuella has been observed to be sympatric throughout its range with two tamarin species, *Saguinus fuscicollis* and *Saguinus mystax* (Soini, 1987a; Goldizen, 1987), but is allopatric with the marmoset *Callithrix jacchus* (Goldizen, 1987). When compared with the tamarin genera, *Cebuella* appears to be less coordinated and synchronized in their social groups. Some members of a group may be foraging for insects while some may be resting while still others may be gouging exudate holes. A general daily activity is as follows: The moment they get up in the morning they go to their primary exudate tree and begin to feed on the exudate exuded during the night; there is a shift to huddling, grooming, basking, and play; there is another stage of feeding with insect foraging and exudate-feeding again dominating their time; there is another interval of huddling, grooming, play, and resting, with some intermittent feeding and, finally, their day ends with an intensive period of feeding activity that continues throughout the remaining hours of the day (Soini, 1988). When the group has exhausted the primary exudate yield the group will either break apart or usually change home ranges.

Mating and Reproduction

Callitrichids are the only anthropoid primates that regularly twin (Fleagle, 1988; Goldizen, 1990). All populations of callitrichids studied as of 1987 have exhibited one or two distinct birth peaks during the year, but *Cebuella* is the only species for which this is documented in the wild (Goldizen, 1987). Only one female bears young for one particular group (Soini, 1987b, 1988). The reproductive female is socially dominant over all the other group members including her mate. The two mates will spend a considerable amount of time forming a pair bond between the two. Soini (1987b) also reports on the sociosexual behavior of a wild troop that contained two adult males during a postpartum periestrous period of its reproductive female. It is speculated that this period was brought about by male-initiated behavior, which in turn was brought about by a periestrous change in the odor of the genitalia and/or urine of the reproductive female (Soini, 1987b). During this period both males attempted to mate with the female but the dominant male restricted the mating access of the female by exhibiting guarding and aggressive behavior toward the subordinate male. The dominant male constantly followed her. There was an increase in huddling and grooming between the consorts.

There were two successful copulations. However, when the female did not tolerate the male's behavior she would present her genitalia to the male, in which case anogenital licking would be performed by the male with no mounting attempts. In other instances the male would vibrate his tongue and/or protrude it. Two aspects are of importance from this study: (1) *Cebuella* exhibited a behavior unique among Callitrichidae genera: genital presenting; and (2) in free-ranging troops with two or more adults, a functionally monogamous mating pattern is maintained.

Growth and Development

Live births have never been observed, but it appears from captivity and in wild populations studies that parturition takes place at nights. Observations on infant care and development showed that most group members participated in the carriage and caretaking of the neonates. Infants are usually carried constantly only for the first 1–2 weeks of life. From the third week on, parents begin to leave the infants in specific but protected places where they are left by themselves for gradually increasing time periods. This is done while the other group members would go foraging. The baby marmoset gets around by being diligently carried on the backs of many group members. The female carrier would be responsible for the back-carrying for approximately 19 days and then a male would switch with the female on around the 20th day (Soini, 1988). This switch of principal carriers might reflect the increase weight gained by the baby, which by then would presumably have grown too heavy for the female to carry. During the third week, exploratory play behavior was observed by Soini (1988). The two main types of play are solitary play and social play. The infants are forced into short-range locomotory independence by the seventh week, but are still often carried to and from the primary exudate tree (their home base) in the eighth week. The beginning of the weaning process starts at about 6 weeks of age, which seems to follow closely with the infants' first attempts at feeding independently from the exudate holes gouged by the adults. Around the fifth month the pygmy marmoset begins to develop the facial features of an adult (such as nasal stripe) and it is during this time frame where attempts to gouge for themselves have been reported.

Communication

Wild groups often mob nonthreatening animals that move into their home base areas; squirrels, snakes, and human observers were the most frequent mobbing targets (Soini, 1988). This mobbing behavior involves jerky movements and body swaying, accompanied by click vocalizations that become mixed with sharp-"tsk" notes. If they are not mobbing a human observer they would engage in genital displays. This is done by turning its rump toward the recipient, arching its back, and raising its tail in a stiff arch thereby exposing its genital area to the recipient. A dominant male may display his genitals in a similar fashion to a subordinate male when competing for a female during estrus (Soini, 1987b). See the "Captive Study" section for additional discussion of vocal behavior.

Play

Infants, juveniles, and subadults spend a considerable part of the day in play (Soini, 1988). During the third week, exploratory play behavior was observed by Soini (1988). The two main types of play are solitary play and social play. Social play consists of chasing and physical contact. Solitary play consists of exploring, acrobatic hanging, leaping, and imitative insect stalking and catching. Play groups are usually formed of 2–3 individuals. Littermates tend to play more together than with other group members. Most play behavior occurs during the major resting bouts, which occurred twice a day. The play behavior usually takes place near the crown of the group's principal feeding tree.

Posture and Locomotion

Cebuella engages in a squirrel-like life, running up and down tree trunks, supporting their weight on the tip of long and sharp clawlike tegulae. When feeding on a primary exudate tree, they frequently adopt clinging positions on large vertical supports (Kinzey et al., 1975). This vertical clinging may be an adaptation to feeding on exudate while vertical-leaping locomotory pattern may not be functionally related. *Cebuella* engages in branch and vine-running, which is their primary means of travel. This mode of locomotion includes walking or running along the underside surfaces of horizontal or diagonal branches and vines. They can completely turn around on a slender support, whether vertical or horizontal, and approach insect prey in a peculiar slow, creeping locomotion. *Cebuella* has the ability to turn its head in 180 degrees, which allows the marmoset to scan the environment for predators when in the vertical clinging position.

Postural behaviors include lying on the stomach, sitting, and vertical clinging, as mentioned previously. When either watching or resting, *Cebuella* positions itself on a diagonal to almost vertical substrate of small diameter. During vertical clinging position, the hindlimbs are flexed beneath the body, with the claws hooked to the bark and the entire plantar surface kept flat against the substrate. The forelimbs are kept near the body, flexed at the elbows, and several claws of one or both hands are hooked to the bark. The tail remains pressed against the substrate behind the body, serving as a prop (Soini, 1988).

20

Cebus

Introduction

The success of the genus *Cebus* (capuchin monkeys) is apparent from its vast distribution from Honduras to Argentina, second among neotropical primates only to that of *Alouatta*. The ability of *Cebus* to exploit a variety of habitats may stem from the great variability in behavior observed among species, neighboring groups, and even among members of a single group. *Cebus* is an opportunistic generalist (Fragaszy et al., 1990). Ability to ingest a wide spectrum of foods permits flexibility in diet and foraging pattern. Body size (2–4 kg) also falls within a range that allows dietary adaptability to both plant and animal foods. Males average 34% greater weight than females (Ford and Davis, 1992), but are only 16% (*C. capucinus*) to 22% (*C. apella*) greater in canine tooth size (Kay et al., 1988). Capuchin monkeys have a distinctive pelage head pattern with a dark "cap" that is tufted in *C. apella*, forms a broad "V" in *C. capucinus* (black) and *C. albifrons* (brown), and is narrow or wedge-shaped in *C. olivaceus*. [See Hershkovitz (1949) for diagrams of head patterns.] Overall pelage color is darkish brown (*C. apella*), pale to dark brown (*C. olivaceus*), to dark brown and buff (*C. albifrons*), silvery gray to agouti (*C. kaapori*), and black with white front (*C. capucinus*) (Napier and Napier, 1967). The tail is prehensile, although there is no bare skin area on the tip as in the Atelinae (Rosenberger, 1983). Multimale, multifemale groups are the norm, dominance hierarchies are linear, with an α-male and a α-female generally the two highest ranked animals, and most aggressive interactions involve one or both of them (Robinson and Janson, 1987). Thus, *Cebus* social organization parallels that of many Old World guenons more than that of any other neotropical primate. The genus *Cebus* has been particularly well studied both in captivity and in the field, shedding light on many aspects of ecology and social behavior. A recent issue of *Folia Primatologica* was devoted entirely to the genus (Fragaszy et al., 1990). Even though heavily hunted in some areas, no *Cebus* species is currently threatened (Robinson and Janson, 1987).

Taxonomy and Distribution

The genus *Cebus* consists of five species that are usually divided into two groups (Hershkovitz, 1949): (1) the "tufted" group: *C. apella*, the black-capped or brown capuchin; and (2) the "nontufted" group, consisting of four species: *C. capucinus*, the white-faced or white-throated capuchin; *C. albifrons*, the white-fronted or brown pale-fronted capuchin; *C. olivaceus* (= *nigrivittatus*), the wedge-capped or weeper capuchin; and *C. kaapori*, the Ka'apor capuchin. *C. kaapori* was only recently discovered. The Ka'apor capuchin is longer bodied and less robust than the other nontufted species (Queiroz, 1992). This taxonomy is supported by chromosomal sequence composition (Dutrillaux, 1988a). All members of the genus share a diploid complement of 54 chromosomes. Dickinson et al. (1990), however, came up with curiously different phyletic groups: (1) *C. apella* and *C. olivaceus* and (2) *C. albifrons* and *C. capucinus*.

Eisenberg (1989) reviewed the extensive distribution of the genus, which reaches Honduras to the north and northern Argentina to the south. *C. apella* is the most widespread species (probably the most widespread neotropical primate species), from southern Colombia east and south to northern Argentina (Map 9). Eleven subspecies are generally recognized (Wolfheim, 1983). The distribution of *C. capucinus* is almost entirely in Central America, from Honduras to northwestern Colombia (Maps 10 and 15). Previous reports of the species in Belize have not been confirmed (Hubrecht, 1986). Five subspecies are recognized (Wolfheim, 1983). *C. albifrons* is found from southern Colombia and Venezuela south to northern Bolivia (Map 10) and 11–13 subspecies are recognized (Wolfheim, 1983). *C. olivaceus* is found from Colombia east and south to the Amazon (Map 10) and also has five subspecies. *C. apella* is sympatric with the latter two species; *C. capucinus* occurs outside the range of *C. apella*. *C. kaapori* has a limited distribution in a restricted area between the Rios Gurupi and Pindaré of Brazil, in the same region as the endangered black saki, *Chiropotes satanas satanas*. The four nontufted species are allopatric, although Eisenberg (1989) suggests that *C. olivaceus* and *C. albifrons* in some areas could be strongly competing or interbreeding (see Map 10). J. Robinson (personal communication) surveyed the area between the two species in 1981 and found a pet animal southeast of Lake Maracaibo that appeared to be a hybrid. Thus, I believe that *C. olivaceus* and *C. albifrons* intergrade in western Venezuela and may be the same species.

The genus *Cebus* has the second largest distribution of all neotropical genera (see Maps 9, 10, and 15). The cosmopolitan distribution of howler monkeys may be attributed to their capacity to subsist largely on a more or less ubiquitous folivorous diet; that of capuchins is the result of their highly developed capacity to manipulate the environment such that they have become effective omnivores (Robinson and Janson, 1987). Thus, for different reasons both *Cebus* and *Alouatta* are "pioneer" primates (Eisenberg, 1979).

Evolutionary History

The only fossils possibly attributed to *Cebus* are Recent specimens from Hispaniola in the West Indies (Ford, 1990b). MacPhee and Woods (1982) suggested that the 3860-year-old *Saimiri bernensis* from the Dominican Republic, as well as the 9550 year-old mandibular fragment from Haiti (UF 28038) may be more closely related to *Cebus* than to *Saimiri*. Newer, as yet undescribed, material from the island may also belong to the same taxon (Ford, 1990b). Nothing is known of the older fossil history of the genus. (See also under "Posture and Locomotion.")

Previous Studies

Early studies are reviewed by Freese and Oppenheimer (1981) and are not reassessed here.

Captive Studies

Tool use has been studied by Visalberghi and others (Visalberghi, 1990; Antinucci and Visalberghi, 1986; Anderson, 1990; Gibson, 1990; Parker and Poti, 1990; Westergaard, 1991), including social influence as a determinant of performance in tool use (Fragaszy and Visalberghi, 1990), food-washing behavior (Visalberghi and Fragaszy, 1990), self-treatment of wounds (Westergaard, 1987), and tool use by a mother to clean her infant's wounds (Ritchie and Fragaszy, 1988).

A variety of problem-solving tasks has been studied (e.g., Simons and Holtkötter, 1986), as well as object manipulation (Natale et al., 1988). Social influences on performance in a foraging task were described by Jorgensen et al., (1991). Masataka (1990) studied handedness in captive *C. apella*, *C. albifrons*, and *C. capucinus*. Mirror-image stimulated responses of *C. apella* were recorded by Anderson and Roeder (1989). Responses to visual representation of individual group members were observed by Thompson and Vinci (1991). Social behavior over the menstrual cycle in *C. apella* was studied by Linn et al. (1991). Relationships among members of a captive group of *C. apella* were reviewed by Welker et al. (1990). Ross and Giller (1988) described the daily activity patterns and interactions among members of a group of *C. apella*. Roeder and Anderson (1991) and Ueno (1991) studied urine washing. D'Amato, in a series of papers, studied cognitive processes (D'Amato and Van Sant, 1988), processing of auditory stimuli (D'Amato and Salmon, 1984), and learning (D'Amato and Colombo, 1989, 1990) in *C. apella*. Vilensky and Moore (1991) studied locomotion in *C. apella* and *C. capucinus*.

Capuchins (*Cebus apella*) have also been trained as aides for quadriplegics (Willard et al., 1982; Willard and Young, 1988).

Field Studies

Recent studies include a comparative ecological study by Terborgh (1983) of *C. apella* and *C. albifrons* in Manu National Park, Peru, their association with *Saimiri* by Mitchell (1987) and by Podolsky (1990), with *C. apella* further studied by Janson (1984, 1985, 1986a,b,c, 1988b, 1990a,b). Defler (1982) compared intergroup behavior in the same two species in Colombia; Soini (1986) studied the same two species as part of a synecological study in Peru; and the same two species also occur sympatrically in Bolivia (Izawa and Bejarano, 1981, Izawa and Yoneda, 1981). The dietary flexibility of *C. apella* in northwest Argentina was described by Brown et al. (1984, 1986) and by Brown and Zunino (1990). Range size in *C. apella* was studied north of Manaus by Spironelo (1987). At Fazenda Montes Claros, Brazil, choice of feeding trees by *C. apella* was studied by Cristoffer (1987). At Fazenda Barreiro Rico, Brazil the ecology of *C. apella* was studied by Torres de Assumpção (1981, 1983a). Izawa (1988, 1990; Izawa and Nishimura, 1988) studied social behavior, Escobar-Paramo (1989a,b) studied infant development, and Izawa (1975, 1978) studied feeding behavior of *C. apella* in Colombia. Brief studies of *C. apella* were conducted in Suriname as part of a larger synecological study (Fleagle et al., 1981; Mittermeier and van Roosmalen, 1981).

At Hato Masaguaral, in the llanos of Venezuela, the demography, social structure and interactions, vocalizations, and foraging activity of *C. olivaceus* have been described by Robinson (1981b, 1982, 1984a,b, 1986, 1988a,b), O'Brien and Robinson (1987), O'Brien (1988), and Srikosamatara (1987); responses to alarm calls were studied by Norris (1990a,b); the effect of group size on foraging behavior was studied by de Ruiter (1986); and the effect of diet on dental wear was studied by Teaford and Robinson (1989) and Ungar (1990). *C. olivaceus* are also being studied at a newly developed site in the llanos, Hato Piñero (Harding and Miller, 1991; Miller, 1991).

Observation of *C. capucinus* in Santa Rosa National Park, Costa Rica (Fedigan et al., 1985) included studies of demography (Fedigan, 1986), diet (Chapman, 1987, 1988a; Chapman and Fedigan, 1990; Fedigan, 1990), predation (Newcomer and De Farcy, 1985; Chapman, 1986), seed dispersal (Chapman, 1989a), tool use (Chevalier-Skolnikoff, 1990), and locomotor and postural behavior (Gebo, 1992). *C. capucinus* have been observed in Corcovado National Park, Costa Rica where they rarely associate with squirrel monkeys (Boinski, 1989b), and they have been surveyed (Massey, 1987) and their diet studied (Gilbert et al., 1991) in Palo Verde National Park, Costa Rica. At La Selva Biological Station, Costa Rica (Wilson, 1990), studies of *C. capucinus* have just begun (Fishkind and Sussman, 1987). *C. capucinus* have also been studied in Honduras (Buckley, 1983). Seed dispersal by *C. capucinus* was studied on Barro Colorado Island, Panama (Rowell and Mitchell, 1991).

Habitat and Ecology

The genus *Cebus* is adaptable to a variety of forest types but prefers canopy-covered forest, and uses the middle strata. Janson (1986a, b) notes the tall canopy of the forest in Manu National Park, Peru where he studied both *C. apella* and *C. albifrons*. The two species prefer to occupy different size trees. *C. apella* normally occupy understory trees that are 5–20 m high, with crowns less than 10 m in diameter. *C. albifrons* prefer canopy trees over 30 m high, with crowns up to 55 m in diameter. Brown and Zunino (1990) note that *C. apella* in northwest Argentina occupy mostly montane forest, with an elevation of 200 to 1100 m.

Cebus capucinus have been studied extensively in Santa Rosa National Park, Costa Rica. This park is mostly tropical dry forest that extends from volcanic foothills to coastal plain (Chapman and Fedigan, 1990). The capuchins here use both old successional forest 75–100 years old and nearly pristine semievergreen forest (Chapman and Fedigan, 1990). Fedigan (1990) also reports *C. capucinus* as an able exploiter of young successional forest.

Cebus olivaceus have been well studied in a semideciduous tropical forest in the llanos of Venezuela. The capuchins here occupy gallery forest and shrub woodland. The gallery forest averages 4–5 km in width, and has an incomplete canopy with trees only occasionally reaching 20 m high (Robinson, 1986). A variety of microhabitats are listed in Robinson (1986).

Diet

Cebus monkeys are frugivore–insectivores, whose diet includes both fruit and invertebrates. They are manipulative foragers that often pry

insects from the bases of palm fronds. They are the closest approxima-
tion to an omnivore among the neotropical primates since they eat such
a variety of food items. They function as both seed dispersers (Chapman, 1989a; Rowell and Mitchell, 1991) and seed predators (van Roosmalen, 1987b; Peres, 1991).

The diet of *C. apella* is very adaptable. Brown and Zunino (1990)
describe the flexibility in their diet in northern Argentina. At El Rey
National Park the bulk of feeding records (72%) is made up of the foliage
base of 6 bromeliad species, an important resource particularly in the
dry season, whereas at Iguazu National Park bromeliads make up only
2% of the total diet.

Janson (1985, 1986b) describes the diet of *C. apella* at Manu Park in
Peru. Over 90 plant species are exploited on a year-round basis. Insects
make up 17% of total energy intake, but require much greater foraging
time to exploit. *C. albifrons*, like *C. apella*, favor *Scheelea* palm seeds and
Ficus (fig) fruit. *Scheelea*, however, is much more difficult for *C. albifrons*
to manipulate. The robustness of the jaws and overall body size of
C. apella are associated with their proficient manipulation of tough
foods.

Robinson (1986) records that plant parts make up 55% of the diet of
C. olivaceus, while invertebrates and vertebrates make up 33%, and
drinking 6%. Most plant parts are fruits from over 50 species. Buds,
leaves, shoots, and roots (from uprooted saplings) are also taken. The
more important animal material includes snails, caterpillars, grasshop-
pers, ants, wasps (adult and larvae), and millipedes. Vertebrate prey
include adult squirrels (decapitated before consumed), iguanas, frogs,
ground doves, and bird eggs. O'Brien (1988) observed dominant females
suckling from lower ranked, but not always lactating females. He referred to this behavior as parasitic nursing.

The diet of *C. capucinus* has been well recorded at Santa Rosa National
Park, Costa Rica. Chapman (1987) notes that fruit make up 78% of the
diet but leaves or flowers are rarely taken. Invertebrates make up 17% of
the total feeding time and include wasps, stinging ants, and scorpions
(Chapman and Fedigan, 1990; Fedigan, 1990). Fedigan (1990) has added
to the list of vertebrate prey, to include lizards, many nestling bird
species and eggs, adult parrots and magpie jays, young and adult squirrels, bats, and nestling coatis. Most vertebrate prey are still alive when
consumed. Chapman and Fedigan (1990) discuss the large variability in
dietary intake among neighboring groups.

Visalberghi (1990) gives an excellent review of studies on tool use by
captive capuchins including the use of stones as hammers to open nuts
(Antinucci and Visalberghi, 1986; Anderson, 1990).

Predation

Predators of capuchins include large felids, boas, and eagles and hawks (Robinson and Janson, 1987; van Schaik and van Noordwijk, 1989).

Behavior

Robinson and Janson (1987) suggest a socioecological convergence between *Cebus* and several species of African *Cercopithecus*, including size, diet, and many aspects of behavior. The *Cercopithecus* species have only one male, but when *Cebus* groups have more than one male, the dominant male apparently does all the breeding.

Social Behavior

Cebus groups have stable matrilineal dominance hierarchies, which is atypical for neotropical primates (see Anderson, this volume). Rank reversals are rare. The highest ranking female is dominant to all but the α-male (Robinson, 1981b). Most agonistic interactions involve either or both of these two highest ranking animals (Robinson and Janson, 1987). Grooming is the most frequent activity during rest periods, and dominant animals receive the most grooming (Robinson and Janson, 1987). Females are phylopatric, and males migrate from their natal groups.

Ranging Behavior

For *Cebus* generally, the more patchy the distribution of fruit resources, the larger the home range. When *C. apella* and *C. albifrons* are sympatric, the latter has a much larger range (150 ha) (Terborgh, 1983). Terborgh (1983) described *C. apella* in Peru as consistently using the central portion of its home range, which fluctuates up to threefold seasonally. The small size of the range allows constant monitoring. The home range of *C. olivaceus* groups at Hato Masaguaral, in the llanos of Venezuela, vary from 200 to 300 ha and overlap extensively (Robinson, 1988a), as they do at Hato Piñero (Miller, 1991). *C. capucinus* has the smallest home range, about 32–80 ha (Eisenberg, 1989). Regarding the degree of range overlap, there is greater intraspecific variation than species differences, because of varying distribution of resources, population density, and group size. Day ranges average 1.8–2.1 km (Robinson and Janson, 1987).

Activity Pattern

Although there is considerable seasonal variation (Robinson, 1986), all species of *Cebus* devote a significant portion of their activity budgets foraging for immobile prey; however, they are not as successful at insect-foraging as are squirrel monkeys (Janson and Boinski, 1992). Even though less time is spent feeding on plant materials, the latter comprise the bulk of the diet (Robinson, 1986).

Cebus have a long active day, frequently begin moving before dawn, and often do not reach their sleeping trees until after dusk (Freese and Oppenheimer, 1981). There are marked differences in activity pattern between wet and dry seasons (Freese and Oppenheimer, 1981; Robinson, 1986). They usually sleep in forks near the ends of branches, and often use the same sleeping tree more than once (Freese and Oppenheimer, 1981).

Mating and Reproduction

All *Cebus* species are polygamous and have multimale, multifemale groups (Robinson and Janson, 1987; Chapman, 1990; Chapman and Fedigan, 1990; Fedigan, 1990). There is a single dominant male who is probably responsible for siring most offspring in the group, although other adult males are frequently observed to copulate with adult females. *C. apella* in captivity have an average gestation length of 155 days (Fragaszy, 1990); other species are similar. Interbirth interval is normally 2 years provided the infant survives; average intervals range from 18 months in *C. albifrons* to 26 months in *C. olivaceus* (Robinson and Janson, 1987).

Female *Cebus* are reproductively mature earlier (4–6 years) than males (7–12 years) (Eisenberg, 1989; Robinson and Janson, 1987). Female *C. apella* in captivity mature at 6 years (Fragaszy, 1990). Female *C. olivaceus* are reproductively mature at the age of 6 years, although most do not conceive for the first time until age 7.

In Peru the birth season lasts from October to January for *C. apella*, and August to January for *C. albifrons* (Robinson and Janson, 1987). North of the equator a birth peak in *C. albifrons* occurs from February to July in Colombia, from December to April in *C. capucinus*, and from May to August in *C. olivaceus* (Robinson, 1988a; Robinson and Janson, 1987).

Robinson (1988a) describes *C. olivaceus* at Hato Masaguaral, Venezuela, with one dominant male in each group that tolerates other adult males. The dominant male sires the offspring in his group. Females in large groups (over 15 individuals) have greater reproductive success than females in small groups (Robinson and O'Brien, 1991). O'Brien and Robinson (1987) report a population birth sex ratio at 1.5:1 (female:male), a significant difference. The ratio differs, however, in large groups (2:1) and small groups (0.75:1) and does not differ between high and low-ranking females. Males have a greater reproductive success rate the longer they enjoy breeding tenure and the greater the number of adult females in the group (Robinson, 1988b). Infanticide has been reported in the same population (Valderrama et al., 1990); among neotropical primates infanticide is known only for *Cebus* and *Alouatta* (*q.v.*).

Growth and Development

Growth parameters are better known for *Cebus* than for most other neotropical primates. Fragaszy (1990) provides an excellent review of the major early developmental features in captive *C. apella*. The weight of a neonate's brain is 49% of the maternal brain weight and is proportionately small. Nine weeks pass before the infant first climbs off the mother, and an additional 10 weeks pass before the infant spends at least 50% of its activities independently. Voluntary grasping begins at 5 weeks and a precision grip appears by week 13. Postural control is fully developed by 7 months, and prehensile control by 1 year. Hand-reared infants display motor skills sooner than mother-reared infants. Skeletal development takes 6 years for completion.

Craniofacial growth in *C. apella* has been analyzed with finite element scaling (Corner and Richtsmeier, 1991). Overall, craniofacial growth in size is greater than shape change in both sexes. The greatest size change is in the muzzle of both sexes. The greatest shape change is in the upper face of both sexes. The basicranium and neurocranium change little but constantly with time. Euclidean distance matrix analysis by the same authors demonstrates, however, that the basicranium changes greatly with growth. Matrix analysis shows that male crania are 3–6% larger than female crania. The data do not support sexual bimaturism despite dimorphism in body size and weight. The dimorphism is instead explained by a faster growth rate for males than for females. Leigh (1991), however, supports bimaturism as the primary cause of body size dimorphism. Galliari (1985) provides a series of tables on teeth eruption in captive-born *C. apella*. Leutenegger and Larson (1985) show that 19 of 24 postcranial dimensions in *C. albifrons* are greater in males than females, with four significant differences.

Robinson (1988a) provides a table of age classes and their corresponding developmental stages in *C. olivaceus*.

Cebus apella has the greatest longevity of any primate (except the hominoids) (Harvey et al., 1987), but contrary to popular opinion primates are not the longest lived order of mammals (Austad and Fischer, 1992). Abate (1984) gives the maximum life span as 48 years.

Communication

Communication in *Cebus* is described in Freese and Oppenheimer (1981). More recent work on vocalizations has been done primarily on

C. olivaceus (Robinson, 1982, 1984a). Robinson, (1988b) describes countercalling preceding intergroup interactions. Norris (1990a,b) has described differences in responses of *C. olivaceus* to distinct alarm calls. Harding and Miller (1991) hypothesize that *C. olivaceus* groups recognize specific other groups by their vocalizations and react according to their respective group size. Boinski (1993) discusses trill vocalizations in relation to troop movement and coordination in *C. capucinus*.

Play

Social play among infants, juveniles, and subadults is common, and has been summarized in Freese and Oppenheimer (1981). Captive capuchins are the only neotropical primates reported to play with objects (Candland et al., 1978).

Posture and Locomotion

Cebus are basically quadrupedal (Freese and Oppenheimer, 1981). Gebo (1989) records locomotor activity of *C. apella* as 86% quadrupedalism, 7% leaping, and 6.5% climbing; and for *C. capucinus* (Gebo, 1992) 54% quadrupedalism, 15% leaping, and 26% climbing.

Kaufman (1987) claims that *Cebus* and *Saimiri* share primitive postcranial traits, although the crania are highly derived. Gebo et al. (1990) also state that quadrupedalism and leaping are primitive in platyrrhines. Capuchins also possess a prehensile tail (without bare skin on the tip) that adds postural support, particularly during foraging and feeding. The prehensile tail has evolved independently from that in the Atelinae (Rosenberger, 1983).

21

Chiropotes

Introduction

Chiropotes, the bearded saki monkey, is one of the three genera of saki-oukaries. See Introduction to Chapter 24 on *Pithecia* for a description of this group. The name comes from the Greek and means "hand-drinker," because Humboldt saw one dipping its hand into water, allowing water to drip off into its mouth (Hershkovitz, 1985). *Pithecia*, as well as other primates, have also been observed to drink in this way.

The average weight of bearded sakis is 2.86 kg (Ford and Davis, 1992) and males average 20% greater weight than females. Canine teeth, however, have very low (about 7%) dimorphism (Kay et al., 1988), presumably because the robust canines are used by both males and females to open hard fruit (Kinzey and Norconk, 1990). Bearded sakis are predominantly black or dark brown and have distinct beards, bulging swellings of temporal hair, a long bushy tail, but otherwise short body hair. *Chiropotes albinasus* has a white nose. All *Chiropotes* males have a distinct pink scrotum. *Chiropotes* are highly frugivorous, feed in large part on relatively hard fruit, and are seed-predators. They live in large groups of 8–30. The southern bearded saki (*C. satanas satanas*) may be the most endangered primate subspecies in Amazonia (Johns and Ayres, 1987).

Taxonomy and Distribution

Chiropotes are found only in the lower Amazon and Orinoco basins and the Guyanas. There are two species of bearded saki: *C. satanas* found north of the Amazon, and between the Rios Xingu and Gurupi south of the Amazon; and *C. albinasus*, only in Brazil south of the Amazon between the Rios Madeira and Xingu (Hershkovitz, 1985) (Map 5). The former species has three subspecies, distinct karyologically (Seuánez et al., 1992). The distribution of *C. satanas chiropotes* in Bolívar State, Venezuela, not previously recognized (Bodini and Perez-Hernandez, 1987), was first described by Alvarez (Alvarez et al., 1986; Alvarez-Cordero, 1987). Although bearded sakis and *Cacajao* are generally thought to be allopatric in their distributions, there may be a small area of overlap in southern Amazonas, Venezuela (Map 5). See Hershkovitz (1985) for a detailed taxonomy of the genus.

Evolutionary History

There are no known fossils of *Chiropotes*, but the tribe Pitheciini is represented by *Cebupithecia* about 15 mya (MacFadden, 1990). See Chapter 24 on *Pithecia* for further descriptions of fossils.

Previous Studies

Captive Studies

Because of the difficulty of keeping bearded saki monkeys in captivity, there have been few captive studies. They are kept and bred at the Centro Nacional de Primatas, Belém, Brazil (Fernandes, 1987; Kingston, 1986, 1987), and the Cologne Zoo, Germany (Hick, 1968). Previous anecdotal accounts are given in Hill (1960).

Field Studies

Van Roosmalen, Mittermeier, and Fleagle studied *Chiropotes satanas chiropotes* in Surinam as part of a larger synecological study of eight primate species (Fleagle and Meldrum, 1988; Fleagle and Mittermeier, 1980; Mittermeier, 1977; Mittermeier et al., 1983; Mittermeier and van Roosmalen, 1981; van Roosmalen et al., 1988). In Brazil, Ayres studied *Chiropotes albinasus* for 17 months on the Rio Aripuanã (Ayres, 1981), and briefly studied *C. satanas satanas* north of Manaus (Ayres, 1978, 1981). Kinzey and Norconk began a study of *C. s. chiropotes* in Surinam in 1986 (Kinzey, 1987; Kinzey and Norconk, 1990), left as a result of the civil war, and have continued the study in Venezuela (Kinzey and Norconk, 1992, 1993; Norconk and Kinzey, 1992, 1993). *C. satanas* was briefly surveyed in Guyana (Vessey et al., 1978).

Habitat and Ecology

Bearded saki monkeys are restricted to terra firme forest (Ayres, 1989), high rainforest, high mountain savanna forest, and high moist forest. They have not been observed in lowland, disturbed, secondary, flooded, or gallery forest. Within the forest, *Chiropotes* prefer the upper canopy and have not been observed on the ground. A quantitative assessment of habitat preferences and forest strata utilization in Surinam is given in Mittermeier and van Roosmalen (1981).

Diet

Chiropotes is a frugivore, a seed predator, and a sclerocarpic forager (Kinzey and Norconk, 1990) (see also description in Chapter 24 on *Pithecia*). They frequently feed either on unripe fruit that are eaten by other primates only when ripe, or on seeds when other primates only eat the mesocarp (Kinzey and Norconk, 1990). In published studies fruit comprises 72–97% of the diet, and seeds make up a substantial portion (Kinzey, 1992). van Roosmalen et al. (1988) summarize data on 85 food species obtained from 217 feeding observations in Surinam. In Guri, over 9 months (45 complete days of feeding data) *Chiropotes* seeds comprised 27–96% of the diet in any given month; flowers were eaten in only 4 months; leaves (in marked contrast to *Pithecia*) were eaten in only 3 months. Insects were eaten virtually every month, and comprised as much as 24% (mostly soft-bodied caterpillars) of the diet in May (early wet season), when the variety and abundance of fruit were relatively low (Kinzey and Norconk, unpublished data). Kinzey and Norconk (1990) measured the resistance of fruit pericarp to puncturing and found that very hard fruit were punctured by *Chiropotes* canine teeth, whereas, the seeds that were chewed were softer than those ingested by *Ateles* or *Pithecia* (Kinzey 1987c, 1992). It is significant that most insects eaten were soft-bodied and malleable rather than rigid and chitinous. Ayres and Nessimian (1982) and Frazão (1991) reported on insect consumption by *C. satanas* in Brazil; most were soft larvae. Additional information on feeding is in van Roosmalen et al. (1981) and Ayres (1989).

Predation

Actual predation of bearded sakis has not been reported, but *Chiropotes* were relatively infrequent victims of the harpy eagle as evidenced by nest remains (Rettig, 1978).

Behavior

Social Behavior

Bearded sakis live in multimale groups of 19 to 30 (*C. albinasus*) or 8–27 (*C. satanas*) (Ayres, 1989). The suggestion that *Chiropotes* groups might be relatively permanent aggregations of monogamous subunits (Robinson et al., 1987) is almost certainly not true; most groups do not consist of equal numbers of males and females (Kinzey and Norconk, personal observation).

Grooming has been observed both between adults and between adults and younger animals (van Roosmalen et al., 1981; Norconk, personal observation), but no detailed account has been reported.

Ranging Behavior

For *C. satanas* van Roosmalen et al. (1981) reported an average (5 days) daily travel distance of about 2.5 km in Surinam. Norconk and Kinzey (1994) report an average at the same site of 3.2 km (range 1.8–4.5 km; N = 9 days). The overall average (N = 70; 14 months, 5 days/month) in Venezuela was 1051 m; the range of monthly averages was 424–1780 m; the range of daily travel distances was 140–2480 m (Kinzey and Norconk, unpublished data). The lower values in Venezuela are certainly the result of the constraint of a 385-ha island habitat. *C. albinasus* groups moved 2500–3500 m daily during the wet season and 4500 m during the dry season (Ayres, 1981). van Roosmalen et al. (1981) estimated the home range of *C. satanas* in Surinam at 200–250 ha. Home ranges of three groups of *C. albinasus* were between 80 and 200 ha (Ayres, 1981).

Activity Pattern

Bearded sakis are active from dawn to dusk, and sleep spread out in one or more neighboring emergent trees, usually on medium sized boughs (Kinzey and Norconk, unpublished data). Although they may occasionally use the same tree more than once, they have not been observed to sleep in the same tree on consecutive nights. They are upper canopy dwellers that rarely move into the lower canopy and have not been reported on the ground. In Surinam they spent 52% of their time in the upper part of the canopy (Mittermeier and van Roosmalen, 1981).

Mating and Reproduction

In Surinam (van Roosmalen et al., 1981) and in Venezuela (Kinzey and Norconk, unpublished data) *C. satanas* has a definite birth season from late December to early March—the dry season. One to three newborn infants were seen every year for 4 years in the group studied by Kinzey and Norconk. Occasional copulations were seen in virtually every month of the year, but peaked in July, August, and September (Norconk, unpublished data). This suggests a gestation period of about 4.5–5.5 months, which agrees with the captive data of Uta Hick (see van Roosmalen et al., 1981).

Growth and Development

A summary of the literature on previous captive studies is provided in van Roosmalen et al. (1981).

Peetz (in preparation) provides data on infant development of *Chiropotes satanas* under natural conditions. During the first month (data on six infants) infants are carried ventrally by the mother. There is little activity by the infant except suckling, and no interest was shown in the infant by other group members. During the second month (data on five infants) the infant is still carried ventrally most of the time, but crawls onto the back of the mother when awake. The infant is groomed by other group members. During the third month the infant is carried dorsally most of the time, and may move up to a meter away from the mother. The first contact play with juvenile animals occurs. During the fourth month (data on four infants) the infant is carried exclusively dorsally by the mother. The infant travels independently for short distances for the first time. First feeding on soft fruit occurs, mostly taken away from the mother. During the fifth month the infant travels up to 10 m away from the mother. During the sixth month (data on three infants) the infant travels short distances by itself, but is carried dorsally by the mother during long distance travel. By the ninth month the infant is fully independent except for long difficult leaps. From the tenth to the thirteenth month the animal is fully independent, but still follows the mother closely. Although it continues to take food from the mother, this now provokes aggression by the mother. By the fourteenth month the anogential area may begin to turn pink.

Communication

The only study as yet of communicative behavior was a 10-month study of *C. s. utahicki* in captivity (Fernandes, 1991). Previous anecdotal reports are summarized in van Roosmalen et al. (1981).

Play

Play behavior has been described in free-ranging *Chiropotes satanas* by Peetz (in preparation). Play was observed throughout the day, but was most frequent between 9:30 and 12:30 hours, with another peak around 14:30 hours. Peetz divides play behavior into solitary play and social play. In the annual activity budget 6% (range 1–16% per month) is spent in play, of which 21% is solitary play and 79% is social play. All age and sex classes take part in social play, but juveniles and subadults were most actively engaged in play activities.

Posture and Locomotion

Postural and locomotor behaviors of *C. satanas chiropotes* have recently been the subject of a long-term study in Venezuela by Walker (1992, 1993). For all activities (feed, travel, rest) *Chiropotes* utilize primarily the main crowns of forest trees (Walker, 1992). Leaping is markedly different from that in *Pithecia*, which generally take off from a vertical clinging position; *Chiropotes*, which leap much less often, take off from a pronograde position and land almost exclusively in the terminal branches of a neighboring tree (Walker, 1993). During feeding bearded sakis occasionally used a distinctive pedal suspension to reach for food (see Figure 6 in van Roosmalen et al., 1981). Additional data on locomotor behavior are provided in Fleagle and Mittermeier (1980) and Fleagle and Meldrum (1988).

22

Lagothrix

Introduction

The woolly monkey is a large robust animal with a prehensile tail. The overall color is gray to brown to black, and one species (*L. flavicauda*) has yellow coloration on the ventral surface of the tail. Males are heavier than females, and male canines are significantly larger than those of females (Kay et al., 1988). Woolly monkeys are primarily frugivorous (Ramirez, 1988). They range in group size from 4 to 45, are polygamous, and have a dominance hierarchy (at least among males). They are noted for their nonbellicose social behavior. They have the highest vulnerability to hunting and habitat destruction of the neotropical primates (Fonseca et al., 1987), being hunted for meat, the pet trade, and for their pelts. Despite its large size, it is probably the least known platyrrhine genus.

Taxonomy and Distribution

There are two species of woolly monkey, *L. lagotricha* and *L. flavicauda* (Mittermeier et al., 1988b). Geoffroy named the genus in 1812, and the two allopatric species were designated by Humboldt the same year. *L. flavicauda*, the yellow-tailed woolly monkey, was known only from museum specimens, and thought to be extinct until 1974 when the species was rediscovered in the wild (Mittermeier et al., 1977). *L. flavicauda* is restricted to montane rainforest in northern Peru, between the Rios Huallaga and Marañon in the Departments of Amazonas, San Martin and La Libertad (Mittermeier et al., 1984; Leo Luna 1987) (Map 11). There is one individual of *L. flavicauda* in captivity, at the Parque de las Leyendas in Lima, Peru (Leo Luna, 1982).

L. lagotricha are found only in the upper Amazon basin, west of the Rio Tapajos in Brazil, eastern Peru, and eastern Ecuador, and the Andean headwaters of the Orinoco in eastern Colombia and western Venezuela (Ramirez, 1988) (see Map 11). As a result of hunting pressure they are apparently no longer found in Bolivia (Izawa and Bejarano, 1981; Brown and Rumiz, 1986). The northernmost populations, possibly isolated, are in the Departments of Cordoba and Bolívar in Colombia (Ramirez, 1988). There are four subspecies of *L. lagotricha* (Fooden, 1963) distinguished by coat color.

Evolutionary History

The four subspecies of *L. lagotricha* probably differentiated in Pleistocene refugia, such as those reconstructed by Simpson and Haffer (1978). *L. l. poeppigii*, for example, appears to have differentiated in the Napo Refuge (Kinzey, 1982). *L. flavicauda* was isolated in the Huallaga center of endemism (Lamas, 1982) together with *Callicebus oenanthe* and *Aotus miconax*, and an endemic hummingbird (Mittermeier et al., 1977) [see Brown (1975) for refugia terminology].

There are no known fossils related to living woolly monkeys, but the earliest ateline was probably much like *Lagothrix* (Rosenberger and Strier, 1989).

Previous Studies

Captive Studies

Leonard Williams (1974) established a semi-free-ranging colony of *L. lagotricha* in Murrayton, England in 1964 to study reproduction and behavior [see Ramirez (1988) for description and additional references]. Communication (Eisenberg, 1976) and reproduction (Mack and Kafka, 1978) were studied in a group at the National Zoological Park, Washington, DC. More recently White et al. (1988a,b, 1989a,b) conducted research on marking behavior and on urinary steroids and Stearns et al. (1988) studied predation on small birds by woolly monkeys at the Louisville Zoological Park.

There is a regional woolly monkey studbook for North America (Stearns, 1993).

Field Studies

One of the least known, though hardly cryptic, platyrrhines, *Lagothrix lagotricha* has most often been described anecdotally by researchers observing other animals, or during general surveys (e.g., Hill, 1962).

In Colombia *L. lagotricha lugens* was observed by Kavanagh and Dresdale (1975) for 48 hours in 1973, by Bernstein et al. (1976) while studying forest degradation, by Hernandez-Camacho and Cooper (1976) who studied distribution and ecology, by Klein and Klein (1976) in La Macarena National Park while they were studying *Ateles* in 1967, by Defler (1987, 1989) who studied a 20-member group near the Apaporis River from 1984 to 1987, by Izawa (1975; Nishimura and Izawa, 1975) and by Nishimura who studied woolly monkeys in the upper Rio Caquetá from 1973 to 1976 (Nishimura, 1987, 1990b), and has been studying woolly monkeys intermittently at La Macarena since 1986 (Nishimura, 1988, 1990a; Nishimura et al., 1990). In Peru *L. lagotricha cana* was studied briefly by Durham (1975) at three survey sites in southeastern Peru, and by Ramirez (1980, 1988) in Manu; *L. lagotricha poeppigii* was surveyed by Freese (1975) and Neville et al. (1976), and studied more extensively by Soini (1986, 1987) between 1979 and 1987 on the Rio Pacaya in northeastern Peru.

Information on *Lagothrix flavicauda* comes largely from Mariella Leo Luna (1980, 1982, 1987; Leo Luna and Ortiz, 1981) from a population studied between 1978 and 1980 on the rugged eastern slopes of the Peruvian Andes.

Habitat and Ecology

Lagothrix are found between sea level and 2500 m altitude (Mittermeier et al., 1977) in mature, continuous, undisturbed humid forests. Though they have been observed in somewhat degraded primary forest, they have not been observed in secondary growth forest areas and they seem to be particularly vulnerable to forest destruction or disturbance (Ramirez, 1988). They have been observed crossing through disturbed growth, however, to reach another area of tall forest (Ramirez, 1988). Woolly monkeys appear even more restricted to "continuous forest" than spider monkeys. They did not survive in remnant forests smaller than 2 km² in Colombia, even though spider monkeys did (Bernstein et al., 1976; Kavanagh and Dresdale, 1975). They are almost totally arboreal and spent about 80% of their time in the upper canopy (Ramirez, 1988).

L. flavicauda are found only in montane cloud forest of the eastern Andes of Peru, from roughly 500 to 2500 m. Further details about habitat preferences may be found in Ramirez (1988).

Diet

Among the atelines, the diet of *Lagothrix* is similar to that of *Ateles*, but the two genera have not yet been studied in sympatry (Peres, 1994). The diet of *Brachyteles* contains substantially more leaves than that of *Lago-*

thrix. Peres (1994) suggests that these differences are due primarily to differences in ecology. Rosenberger and Strier (1989), however, attribute the differences to morphological adaptations to folivory in *Brachyteles*.

L. lagotricha are basically frugivorous, eating fruit, leaves, and other plant parts (Ramirez, 1988). Nishimura (1990b) reported that 80% of feeding time was spent on ripe fruit by woolly monkeys at the Caquetá site. On the Rio Pacaya in Peru fruit accounted for 69–79% of observed feeding (Soini, 1987); values on an annual basis were fruit (including immature) 74%, seeds 17%, leaves 6%, and flowers 3%, plus a small amount of invertebrates (Soini, 1986). Seeds were most important during the early rainy season, coinciding with the seasonal low in availability of ripe fruits, when seeds comprised 26% of the diet and the most important species was a member of the Lecythidaceae (Soini, 1986). Peres (1994) also reported that *Lagothrix* (180 km southwest of Tefé, Amazonas, Brazil) included substantial amounts of seeds in their diet. Ramirez (1988) suggests that woolly monkeys may be important seed dispersal agents; but the above observation suggests that woolly monkeys may also be significant seed predators, taking up distributionally where bearded sakis (*Chiropotes*) leave off. They are allopatric with *Chiropotes* (compare Maps 5 and 11) and may compete for the same hard seeds in the Lecythidaceae and Sapotaceae. Milton and Nessimian (1984) found a large number and variety of insects in the stomach of a single animal (*L. l. lagotricha*) collected in July in Amazonian Brazil and suggested that insects may comprise an important part of the diet, at least at this time of the year.

The diet of *L. flavicauda*, in its cloud forest habitat, is apparently similar to that of *L. lagotricha*, with the addition of pseudobulbs of epiphytes (Leo Luna, 1987). A list of food items eaten by both species in the wild is provided by Ramirez (1988).

In captivity, during a 12-month study in the Louisville Zoo female woolly monkeys were observed to prey on small birds (sparrows) 15 times (Stearns et al., 1988). While the monkeys were observed to share some of the prey, higher ranking animals frequently took food from lower ranking animals.

Predation

Humans are significant predators on woolly monkeys (Ramirez, 1988) as they are a prized food. They suffer from hunting for the market over much of Peruvian Amazonia (Marsh et al., 1987). In Colombia and Peru, their skins are used for saddle covers and wall hangings, and to fabricate a "cuica," which is used by jaguar hunters to imitate the vocalizations of these cats (Marsh et al., 1987). Peres (1987) found that primate body mass was negatively correlated with population density in hunted but not in nonhunted sites in western Amazonia, thus defining *Lagothrix* as a very attractive game animal. Lactating females are in particular danger, being shot for easy access to an infant for the pet trade (Peres, 1990).

Nonhuman predators of *Lagothrix* include the larger birds of prey, and possibly jaguar (Ramirez, 1988).

Behavior

Lagothrix lagotricha live in large heterosexual groups of 10 to 45 individuals, with at least one (usually more than one) adult male per group (Ramirez, 1988; Nishimura, 1990b). *L. flavicauda* appear to live in smaller groups of 4 to 14, with 1 to 3 adult males per group (Leo Luna, 1982). Ramirez (1988) provides a table of group sizes.

Social Behavior

The social structure of *L. lagotricha* is still poorly known (Kay et al., 1988). Nishimura recognized a dominance hierarchy, but only among males, at both La Macarena (Nishimura, 1990a) and the Caquetá site (Nishimura, 1990b). As is the case for *Brachyteles*, agonistic interactions between males are rare (Kavanagh and Dresdale, 1975; Nishimura, 1990a). Troops observed at the Caquetá site moved in proximity for hours, without any agonistic interactions (Nishimura, 1987). *L. flavicauda* groups have also been described as having low levels of competition (Leo Luna, 1980).

Grooming took place frequently (Kavanagh and Dresdale, 1975) and was done mostly by less- to more-dominant animals; adult males received the most grooming; adult females received relatively little grooming, mostly from their juvenile daughters (Nishimura, 1990b).

Stearns et al. (1988) found that higher ranking woolly monkeys in the Louisville Zoo were more likely to steal avian prey from other individuals. An index based on agonistic interactions and urinary cortisol levels reflected social influence on space utilization, food sharing, maturation, and stress. Urinary cortisol was positively correlated with the previous day's hostile social encounters, suggesting that the relationship between social standing and cortisol is consistent with the poor survival of females in captivity and with the high incidence of stress-related pathology in this species (White et al., 1988a,b).

Ranging Behavior

Nishimura (1990b) estimated the home range of two adjacent groups at the Rio Caquetá to be 3.5 and 4.5 km² with 2.3 km² overlap. He suggested that because of other groups in the area there was no area of exclusive range use. All observers report overlap of home ranges, and woolly monkeys probably do not defend home range against conspecifics (Ramirez, 1988). The island group studied by Soini (1986) had a home range of 3.5 km². The two groups studied by Nishimura and Izawa (1975) were estimated to have ranges of 4 and 11 km².

At the Caquetá site groups appeared to be more cohesive than those of *Ateles* or *Brachyteles* and traveled in a stable foraging group (Nishimura and Izawa, 1975; Nishimura, 1990b). However, at other sites the typical atelin pattern of separating (fissioning) into smaller foraging groups seems to occur (Ramirez, l988).

Daily travel distance at the Pacaya site averaged 540 m ($N = 14$) with a range of 100–950 m (Soini, 1986). Woolly monkeys studied by Defler (1987) traveled an average of 3 km per day.

Activity Pattern

Daily activity period was about 12.5 to 13 hours, from before dawn to after sunset (Soini, 1986). Most daily activity takes place in the upper strata of the canopy (Soini, 1986; Ramirez, 1988). Sleeping occurs in the upper canopy, often in the crown of a tall feeding tree, usually on the middle and distal parts of stout branches or boughs (Soini, 1986). Additional information may be found in Ramirez (1988).

Mating and Reproduction

The period of receptivity of females is rather long: copulations took place over 6–11 days (Nishimura, 1990a). When receptive, the females make sexual overtures to a male and copulations occur frequently throughout estrus (Ramirez, 1988). In the wild *Lagothrix* has a discrete birth season from late dry to mid-wet season; births at La Macarena all took place between August and December (Nishimura et al., 1990); at the Pacaya site 60% of births took place between September and November (Soini, 1986). In captivity, however, Williams (1974) indicated there was no seasonal birth season in the Murrayton Sanctuary.

Gestation length is between 7 and 7½ months (Ramirez, 1988). (A figure of 139 days reported by Brizzee and Dunlap (1986) is almost certainly in error.] The interbirth interval has variously been reported at 3 years ($N = 5$) (Nishimura et al., 1990) and 12–24 months (Ramirez, 1988).

Growth and Development

L. lagotricha lagotricha (the largest subspecies) has an average body weight of 7.5 kg (Peres, 1990). Males are about 45% heavier than females (Ford and Davis, 1992). *Lagothrix* is monomorphic in body length (Rosenberger and Strier, 1989). Canine dimorphism is very low in *L. flavicauda* and about 22% in *L. lagotricha peoppigii* (Kay et al., 1988).

Nishimura (1990b) provides behavioral descriptions of age–sex classes but does not give actual age in years for any category. Additional data on infant development are provided by Soini (1986). Infants in the Rio Pacaya area are carried until they reach 6–8 months of age (Soini, 1987). Neonates have light straw-colored fur, which contrasts with the dark color of the adult (Kavanagh and Dresdale, 1975; Soini, 1986). Puberty is reached at 4 years in the female (Ramirez, 1988). A female is easily recognized by her enlarged pink clitoris. Male genitalia are rather hidden from view in the perineal hair (Hill, 1953).

Females emigrate from the group (Nishimura, 1990a) as for *Brachyteles* and *Ateles*. The ratio of reproductive females to male in groups ranges from 1.1 to 1.4 (Nishimura, 1990b), consistently lower than that for spider monkeys.

Lagothrix have the highest encephalization index of any New World primate studied by Stephan et al. (1988), comparable to that of gibbons and chimpanzees.

Communication

Ramirez (1988) provides a table of over 50 patterns (actions) of the communication repertoire of woolly monkeys. *Lagothrix* faces are very expressive and can show very subtle changes in intentions and moods. Ramirez (1988) emphasizes the importance of olfactory cues, especially chest-rubbing, which also occurs in captivity (White et al., 1989b). Mouth-to-mouth contact ("kissing") is a frequent greeting behavior (Kavanagh and Dresdale, 1975; Nishimura, 1990b).

Stearns et al. (1988) found frequent scream encounters in a group of captive woolly monkeys at the Louisville Zoo. They concluded that screaming is a component of social communication.

Play

Play is the most common social interaction among woolly monkeys (Nishimura, 1990b). Juveniles and subadults are the usual participants, frequently involving more than two animals, but virtually all age–sex classes, except adult males, may be involved (Kavanagh and Dresdale,

1975). Body contact play often takes place while hanging by the tail, as occurs in howlers and spider monkeys. Play bouts increased in duration with increase in size of group, up to an average of 11.5 minutes/bout for groups of four animals. Play is often accompanied by characteristic vocalizations.

Posture and Locomotion

Woolly monkeys are basically quadrupedal walkers and runners, sometimes using their prehensile tails to locomote, but not with the agility of *Ateles* or *Brachyteles* (Ramirez, 1988). They do not leap, and gaps are crossed by dropping (as much as 20 feet). Durham (1975) estimated that only 8% of locomotor movements involved assistance of the tail ("semibrachiation"). The tail is used for anchoring while moving, but seems to be employed mostly in positioning. The tail is also used in the investigation of novel objects and for hanging and gripping when feeding and playing. Woolly monkeys at the Murrayton Sanctuary use their tails even more frequently than their hands to pick up objects (Ramirez, 1988).

23

Leontopithecus

Introduction

The lion marmoset, *Leontopithecus*, is one of the rarest primate genera (Mittermeier et al., 1989). *Leontopithecus* exhibits typical features of the subfamily Callitrichinae, including clawlike nails on all digits except the hallux, small body size, absence of the third molar, tritubercular upper molars, tendency for producing twins, and cooperative breeding (Rosenberger and Coimbra-Filho, 1984; Sussman and Kinzey, 1984). *Leontopithecus* is the largest member of the subfamily with an average body weight of 600 g (Ford and Davis, 1992) and is distinguished by its impressive "golden" coat color, and long hairs that form a mane on cheeks, throat, ears, and head surrounding a bare face. It is a manipulative forager that uses elongated fingers to probe and extract concealed prey such as insects and small vertebrates (Garber, 1992). Like other members of the Callitrichinae, *Leontopithecus* groups are frequently polyandrous (Baker et al., 1992) or polygynous (Dietz and Baker, 1992). Despite its uniqueness, *Leontopithecus* has only recently been the focus of study in the wild, largely due to recent advances in radiotelemetry allowing habituation; the use of local, trained observers allowing systematic observation of many groups, and the endangered status of lion marmosets and their habitats (Dietz, personal communication). Most previous research efforts were concentrated on reintroducing captive populations of *L. rosalia* back into small refuges, the last remnants of the Atlantic Coastal forest habitat in Brazil (Goldizen, 1988; Kleiman et al., 1988).

Taxonomy and Distribution

The genus *Leontopithecus* consists of four distinct forms, which are generally considered separate species (Rosenberger and Coimbra-Filho, 1984; Snowdon et al., 1986; Mittermeier et al., 1988b), or subspecies

(Hershkovitz, 1977; Kleiman, 1981; Mittermeier and Coimbra-Filho, 1981; Forman et al., 1986). The fourth species, *L. caissara*, was only discovered in 1990 (Lorini and Persson, 1990; Anon., 1991). All forms have some golden pelage, though all except *L. rosalia* are predominantly black. *L. rosalia* (golden lion marmoset) is golden throughout its pelage; *L. chrysomelas* (golden headed lion marmoset) has a golden face and forelimbs; *L. chrysopygus* (golden rumped, or black lion marmoset) is the largest species, and has golden coloration only on its rump and tail base; the entire back of *L. caissara* (black-faced lion marmoset) is golden in color. Forman et al. (1986) found the small amount of genetic variation among three members of the genus was similar to expected values for populations that had been briefly isolated, and less than values reported for many mammalian subspecies. They suggested that the morphological variation among taxa could simply be the result of adaptation to their respective microenvironments.

Leontopithecus occupies four disjunct areas in lowland Atlantic Coastal Forest in southeastern Brazil within the states of Bahia, Rio de Janeiro, São Paulo, and Paraná (Map 12). *L. chrysomelas*, the most abundant species, numbers less than 600 individuals in the state of Bahia between the Rio Jequitinhonha and Rio das Contas. Its distribution outside of the Una Biological Reserve is poorly known (Kleiman et al., 1988; Rylands, 1989). *L. rosalia* is found in the State of Rio de Janeiro, in remnant forest patches and in the Poço das Antas Biological Reserve. *L. chrysopygus* used to occupy a large area between the Rios Paranapanema and Tieté, but deforestation has reduced its range to two reserves within the state of São Paulo: Morro do Diabo State Park and Caetetus Ecological Station (22°18'S, 49°34'W), and a few remnant forest patches. *L. caissara* is known only from Superagui Island (25°18'S, 48°11'W), off the coast of southern São Paulo and northern Paraná, with only about 5000 ha of suitable habitat (Seal et al., 1990). *L. caissara* has the most southern distribution of any marmoset or tamarin.

Descriptions of each of the four species may be found in Rylands et al. (1993).

Evolutionary History

There are no known fossils of *Leontopithecus*. See "*Callithrix*" for discussion of fossil history of the subfamily. *Leontopithecus* are referred to as "lion tamarins" by some (e.g., Dietz and Baker, 1992) and as "lion marmosets" by others (e.g., Garber, 1992; Rosenberger, 1992). The difference in terminology reflects a conceptual difference of opinion. The former term focuses on the morphological similarity between *Saguinus* ("tamarins") and *Leontopithecus* in large size of the canine teeth. This is probably a case of convergent evolution. The latter, I believe more appropriate, term reflects the cladist view that *Leontopithecus* are more closely related to *Callithrix* ("marmosets") than to *Saguinus* (see Figures 6 and 7 in Ford and Davis, 1992).

The extant distribution of three taxa of lion marmosets was explained on the basis of the Pleistocene refuge model (Kinzey, 1982), with one species each found in the Bahia, Rio Doce, and Paulista Centers of endemism. Since the distribution of the recently discovered *L. caissara* cannot be explained by the refuge model, it may simply be the result of island endemism.

Previous Studies

Captive Studies

The leading location for long-term captive studies of *Leontopithecus* has been the National Zoological Park, Washington, DC, largely related to its position as the center of the reintroduction program (Kleiman et al., 1988). Studies have involved social behavior including parental care (Mack and Kleiman, 1978; Rathbun, 1979; Hoage, 1982; Cameron, 1988), social dynamics (Kleiman, 1980), social reintegration (Inglett et al., 1989), vocalizations (Green, 1979), reproduction (Kleiman, 1977b; Wilson, 1977), endocrinology (Kleiman et al., 1978), and feeding (Power, 1991). Other studies have been conducted at Monkey Jungle, Miami, Florida (Snyder, 1974; DuMond et al., 1979; Snowdon, et al. 1986), the Oklahoma City Zoo (Wilson, 1976), Rio de Janeiro Primate Center (Coimbra-Filho et al., 1981; Snowdon et al., 1986; Pissinatti, 1992), the Los Angeles Zoo (Baker and Woods, 1992), and in Omaha at the University of Nebraska (Inglett and French, 1987a,b,c; Inglett et al., 1989; Benz, 1990; Benz et al., 1992; French and Inglett, 1989; Santos et al., 1992; Forman and French, 1992; French, 1992a,b). See Kleiman et al. (1988) for further discussion.

The results of captive studies have enhanced the breeding program for *L. rosalia*. Kleiman et al. (1988) report that captive studies have improved the infant survival rate to 60% and enhanced the reproductive success of captive females, producing more litters and more multiple births per year than wild conspecifics. The most significant consideration learned from relocating captive-bred animals is that without proper training and postrelease support, they do not locomote properly, do not respond appropriately to calls of conspecifics, and do not know how to survive in the wild (Kleiman et al., 1991).

International studbooks are available for *L. rosalia* (Ballou, 1991), *L. chrysomelas* (Mace, 1991), and *L. chrysopygus* (Padua, 1990).

Field Studies

Only recently have lion marmosets been studied in their natural habitat. *L. rosalia* has been studied at Poço das Antas (RJ) and is currently under long-term study there by Dietz et al. (Baker, 1987, 1990, 1991; Peres, 1986a,b, 1989a,b; Dietz et al., 1992; Dietz and Baker, 1993). A total of 465 animals, including 129 in 19 continuously monitored groups, were individually marked (as of 1990). *L. chrysomelas* was studied at Una (Bahia) by Rylands (1980, 1982, 1983, 1984a, 1989) and by Dietz (1993). *L. chrysopygus* has been studied at Morro do Diabo State Park (SP) by Albernaz et al. (1988), de Carvalho and de Carvalho (1989), de Carvalho et al. (1989), and Padua (1992). Kleiman et al. (1988) list a few short-term studies that were conducted prior to 1982. The reintroduction of captive-born *L. rosalia* has also been studied (Beck et al., 1991; Kleiman et al., 1986, 1991; Pinder, 1986; Dietz et al., 1992).

Habitat and Ecology

L. chrysomelas (and possibly other species as well) prefer tall primary forest in stream valley bottoms (Rylands, 1989). Swamp forest or riparian forest also seems to be an important habitat for the three northern species (Dietz, personal communication). Lion marmosets utilize secondary forest, especially for feeding on fruit, but Rylands (1989) felt that primary forest was required for their sleeping holes; however, Poço das Antas has no primary forest and all groups there use nest holes for sleeping (Dietz, personal communication).

Most of the lower montane Atlantic Coastal forest habitat of *Leontopithecus* has been decimated by agriculture and logging, and therefore much of the range occupied by members of the genus is filled with secondary forest. Annual rainfall in these refuges varies between 1100 to 2000 mm, with average temperatures of 22°C. Within the ranges of *L. rosalia* and *L. chrysopygus* the wet season occurs between September and March; within the range of *L. chrysomelas* it is wettest between March and June, but temperature and precipitation are much less seasonal. A recent description of the ecology of the *Leontopithecus* species may be found in Rylands (1993).

Diet

Leontopithecus feed on a variety of fruits, flowers, insects, small vertebrates, and exudates, and soft, sweet fruits are preferred (Kleiman et al.,

1988). In its constant search for insects and small vertebrates, *Leontopithecus* frequent bromeliads, palm crowns, and dead palm fronds. They are manipulative foragers that use their elongated fingers to probe and extract concealed prey such as insects and small vertebrates (Garber, 1992).

All members of the Callitrichinae, including *Leontopithecus*, are known to feed on exudates. Tamarins are opportunistic feeders, but Peres (1989a) has shown that *L. rosalia* adults, like *Callithrix*, occasionally stimulate exudate flow, "by actively biting the base of *Machaerium* (Papilionaceae) lianas," although juvenile animals were unable to do so. Exudate feeding on *Parkia pendula* seed pods has been reported for *L. chrysomelas* (Rylands, 1983), and *L. chrysopygus* has been seen feeding on exudates of *Euterpe* palms (Peres, 1989a).

Predation

Dietz (personal communication) observed a boa constrictor approach, attack, kill, and eat a wild adult female golden lion marmoset. He has also seen an adult reintroduced animal killed by an ocelot, and seen animals chased by *Cebus* and by raptors. The greatest threat to *Leontopithecus*, however, is probably destruction of their habitat by humans.

Behavior

Summaries of early studies on captive animals may be found in Kleiman et al. (1988).

Social Behavior

Since the conceptualization of social organization in callitrichines has been undergoing change in recent years (e.g., Sussman and Garber, 1987; Goldizen, 1988; Ferrari and Lopes Ferrari, 1989), data on *Leontopithecus* are of particular interest. At Poço das Antas *L. rosalia* groups consist of 2–11 individuals (mode = 5). Groups contain 1–2 adult females, 1–3 adult males, plus subadults, juveniles, and infants. Considering only adults, a group may contain 1♂ + 1♀ (50% of groups), 2–3♂♂ + 1♀ (40%), or 1♂ + 2♀♀ (10%) (Baker et al., 1992; Dietz and Baker, 1992). *Leontopithecus* has a variable mating system that does not necessarily correlate with group composition. Monogamy appears to be the most frequent system, and polyandry and polygyny also occur. In all known groups there are social dominance hierarchies among males and females, and cooperative raising of the young by other members of the social group. Feeding aggregations of 15–16 *L. rosalia* have been reported

(Coimbra-Filho and Mittermeier, 1973), and as many as four groups have intersected during group encounters (Dietz, personal communication).

Ranging Behavior

L. rosalia has an average territory size of 41.4 ha (SD = 21.1; range 6.7–116.8; N = 47) (Dietz and Baker, 1996). *L. chrysomelas* has a territory size averaging 79 ha (SD = 23.4; N = 4) (Dietz, 1993), with day ranges between 1410 and 2175 m/day (Rylands, 1989). Rylands (1989) reported only 7% range overlap between two groups of *L. chrysomelas*, but overlap is considerable in Poço das Antas (Dietz, 1993). Home range is fiercely defended. The one group studied by Peres (1986a) traveled and foraged consistently along the fringes of its home range, engaging in face-to-face and long-range interactions with neighboring groups. Censuses of *L. rosalia* at Poço das Antas estimated the population density at 0.05 individuals/km^2 (Green, 1980) and more recently at 1 individual/8–12 ha (Seal et al., 1990); for *L. chrysomelas*, 0.9–3.0 groups/km^2 (Rylands, 1989).

Activity Pattern

Lion marmosets usually sleep in tree holes, which they vacate at dawn. The number of tree holes/group at Poço das Antas varies from 1 or 2 to 20 (Dietz, personal communication). The same hole is used for up to six consecutive nights by *L. chrysomelas* (Rylands, 1989). Major activities (in captivity) take place in early to mid-morning (Green, 1979). Activity budgets for *L. chrysomelas* were locomotion 43%, feeding on plant foods 24%, foraging for insects 13%, and resting and social behaviors 20% (Rylands, 1989). *L. chrysomelas* generally used higher canopy levels than sympatric *Callithrix kuhli*; 80% of activity records were above 12 m, at heights where the greatest abundance of large bromeliads occurred (Rylands, 1989).

Mating and Reproduction

Most data have been obtained from captive studies. In the northern hemisphere most *Leontopithecus* births occur between February and August; the first litter, if a second litter occurs, occurs between February and March; births occur between September and February in the Southern Hemisphere. Twinning is common, with triplets and even quadruplets being reported. Normal gestation length is 129 days with a range of 125 to 132 days, but it is difficult to measure, since pregnancy is hard to detect, and sexual activity is frequent midway through the gestation period (Kleiman et al., 1988). Birth data from Poço das Antas have recently been reported, with a birth peak in late October (Dietz et al., in preparation).

Estrus is also difficult to detect. The length of the cycle is estimated at 19.6 days with a range of 14–31 days, based on hormonal data (Stribley et al., 1987). Kleiman et al. (1988) suggest that copulation frequencies appear to increase around mid-cycle, but Stribley et al. (1987) dispute this pattern. Both authors agree, however, that females tend to be continuously receptive. Copulatory behavior involves sniffing, allogrooming, and on occasion tongue flicking with copulation occurring in a ventrodorsal position. Several intromissions precede ejaculation, which is discernible only by prolonged thrust activity.

Unlike most other callitrichines all females in a *L. rosalia* group usually exhibit ovulatory cycles (Küderling et al., 1992) and there does not appear to be suppression of ovulation by a single breeding female. In captivity, however, breeding by daughters housed with their mothers does not take place (Dietz, personal communication). In multimale groups, males usually include a father and his son(s), but replacement of the dominant male may result in unrelated adult males in a group (Baker, 1991).

Growth and Development

A mother does not usually transfer her offspring to the father or to other care-givers until the second week postpartum, with transfer occurring later in the case of singleton births (Hoage, 1977). This is in contrast with other members of the Callitrichinae in which paternal care usually begins on the first or second day after birth (Baker and Woods, 1992). Care of infant(s) is given by all members of the group. A group member may actively grab an infant, and Baker (1991) suggests that such behavior by juvenile *L. rosalia* may have long-range consequences, for example, he notes that males tend to carry male infants more often, possibly to elicit reciprocal cooperation during coemigration. Females show no sex preference in carrying infants. Their involvement in infant carrying may enhance maternal skills and/or reaffirm affiliative interactions from other group members.

Infants tend to travel on their own after 12 weeks of age, although they take solid food beginning at about 3 weeks. Food sharing in *Leontopithecus* appears to be important to the social structure of the group, and food is shared among group members more often than by any other callitrichid (Brown and Mack, 1978). However, the period of eliciting food from group members by infants is short, and soon after they begin to forage independently, group members begin to solicit food from infants (Kleiman et al., 1988).

Infants are frequently groomed by all group members, but juveniles groom adults more than the reverse. Sexual behaviors, including circumgenital and sternal marking and mounting behavior, occur about 13 weeks after birth; scent marking does not occur until approximately 12 months of age (Kleiman et al., 1988).

As individuals grow older, exposure to aggressive encounters increases. Juvenile males tend to be the recipients of most aggressive encounters (Inglett and French, 1987b). Although all group members engage in aggressive encounters, most agonistic behavior is toward individuals of the same sex (Baker, 1987); however, adult males rarely exhibit any aggression toward other group members (Inglett and French, 1987b). At Poço da Antas almost all individuals left their natal groups by age 4 years; patterns of aggression indicated that natal females were evicted through intrasexual competition, but dispersal of natal males was probably driven by inbreeding avoidance (Baker, 1991).

Communication

Leontopithecus exhibit a limited repertoire of visual signals. These consist of postural and facial expressions, including tongue protrusion in sexual and social contexts; arch display, tail thrashing, and piloerection during aggression; and rump display in sexual contexts, to name a few. Scent marking is the dominant form of olfactory communication, accomplished through rubbing of the sternal and/or circumgenital gland. *Leontopithecus* appear to engage in such behavior at a much higher frequency than other members of the Callitrichinae; adult males are the most active (Kleiman et al., 1988).

Vocalizations of *Leontopithecus* consist of high frequency and ultrasonic whines, clucks, trills, and nontonal sounds. Food calls are generally given by individuals willing to share (Kleiman et al., 1988) and can be divided into categories possibly indicating food preference and type (Benz, 1990). Long calls serve a dual purpose. They act to maintain pair bonds, resulting in duets, as well as to advertise a group's presence in its territory. Snowdon et al. (1986) demonstrated that the calls of *L. rosalia* and *L. chrysopygus* are more similar to each other than are either to the calls of *L. chrysomelas*, reinforcing this distinction based on morphological criteria (Rosenberger and Coimbra-Filho, 1984).

Play

Play behavior is one of the most common behaviors of juveniles and infants and occurs more frequently than grooming. In addition, play behavior is more common among juveniles and infants than with adults. More detailed descriptions are found in Kleiman et al. (1988).

Posture and Locomotion

Leontopithecus is characterized as an arboreal quadruped that engages primarily in walking and running in captivity (Kleiman et al., 1988;

Stafford et al., 1992). Climbing, suspension, and jumps make up the rest of the locomotor repertoire. *Leontopithecus* may engage either in a head down posture or tail down posture when descending vertical supports (Kleiman et al., 1988), depending on trunk diameter.

Leontopithecus appear to possess limited grasping capabilities of the hallux and instead use pedal digits 4 and 5 against the sole to enhance the grasping action of the foot. This feature combined with novel forearm and hand proportions may be responsible for its unique bounding behavior. During bounding the hands are placed on the same side of the branch, as are the feet, and the foot is advanced ahead of the ipsilateral hand (Stafford et al., 1992). This feature has been noted only in captive populations thus far.

In foraging for food *Leontopithecus* assume a variety of postures, including suspension, and they constantly probe holes, leaves, bark, etc. with their slender, elongated hands. *Leontopithecus* also spend a significant amount of time resting, exhibiting a variety of postures, including sitting, crouching, and lying sprawled across a support (Kleiman et al., 1988).

24

Pithecia

Introduction

The saki monkey (*Pithecia*), a medium-sized arboreal primate, is one of three living genera of saki-oukaries, and one of the least known South American monkeys. Until recently, *Pithecia* had frustrated field workers with its furtive habit of hiding, then fleeing and suddenly stopping in a tangle of dense foliage, making it exceedingly difficult to follow. However, recent studies of *Pithecia* have been conducted on islands and in forest fragments where they have been easier to habituate.

The saki-oukaries, including *Pithecia*, *Chiropotes*, and *Cacajao*, have always been recognized as a group of closely related taxa (Ford, 1986b), now referred to as the Tribe Pitheciini. They share a suite of derived characters, including robust splayed canine teeth, styliform lower incisors, and very flat molar teeth almost without relief (Kinzey, 1992).

Pithecia are the smallest of the Pitheciini. Average adult weight is 1.763 kg (♀♀) and 2.384 kg (♂♂) (Ford and Davis, 1992). They have long, coarse dark hair, although the chin and underside of the neck is bare; tails are long (generally longer than head plus body length), bushy, and not prehensile (Mittermeier et al., 1988b). Sex differences in pelage color (sexual dichromatism) are found in *P. pithecia* and *P. aequatorialis* (Hershkovitz, 1987). Facial pelage is sexually dimorphic in all species, but is most marked in *P. pithecia* in which the face of the male is white, and that of the female is dark except for a white line on either side of the nose and mouth. In all species juveniles of both sexes resemble females (Robinson et al., 1987). *Pithecia* are highly frugivorous, feed in large part on relatively hard fruit, and are seed-predators. *Pithecia* live in relatively small, often family groups. Although none of the species of *Pithecia* is listed as endangered (Wolfheim, 1983), they are hunted and the bushy tail is sought as a duster.

Taxonomy and Distribution

Until 1979 two species of *Pithecia* were recognized: *P. pithecia* north of the Amazon and *P. monachus* south of the Amazon. In that year Hershkovitz (1979) suggested that there were three species south of the Amazon and that two of them were sympatric; in 1987 he added another species. This taxonomy appears to be accepted by most authors (e.g., Smuts et al., 1987; Mittermeier et al., 1988b; Rosenberger et al., 1990), but it should be noted that "consistent size, and cranial or dental differences between species of the *P. monachus* group have not been found" (Hershkovitz, 1987:410). Hershkovitz (1987) should be consulted for changes in taxonomic terminology of the genus; *P. hirsuta* is no longer a valid taxon.

P. pithecia is found in the Guianan region and the *P. monachus* group is found in the Amazon region, west of the Rio Tapajos (Map 13). The two groups are separated by the Amazon River. Species are apparently allopatric except, based on the location of collecting localities in Loreto, Peru where *P. m. monachus* is sympatric with *P. aequatorialis* (Hershkovitz, 1987). It remains to be seen how these two similar species, if sympatric, have different ecological niches. Ayres (1990) explored the ecological factors influencing distribution and population densities of the Pitheciini.

Evolutionary History

Although there are no known fossils of *Pithecia*, the tribe is represented in the fossil record as early as the Friasian, 15 mya (MacFadden, 1990). Kay (1990) recognizes one or two pitheciin fossils, *Cebupithecia* and possibly *Mohanamico*; and Rosenberger et al. (1990) also recognize *Soriacebus*. Some (e.g., Stirton, 1951; Meldrum and Lemelen, 1990) believe that *Cebupithecia sarmientoi* is more similar to *Pithecia* than to any other pitheciin.

Previous Studies

Captive Studies

Pithecia have not been the subject of many captive studies. Dugmore (1986) described play, grooming, aggression, and olfaction over a 5-month period in a captive pair of *P. pithecia*. Homburg (1987, 1989) observed the social behavior of six groups of *P. pithecia* consisting of 22 individuals (3 in zoo cages; 3 in a semifree environment) over a 5-month period. Bartecki and Heymann (1988) described predatory behavior of a single *Pithecia monachus* (=*P. hirsuta*). The reproductive biology of a group of *P. pithecia* is currently being studied at the Roger Williams Park Zoo, Providence, Rhode Island (Savage et al., 1992a,b). Previous anecdotal studies are reviewed in Buchanan et al. (1981).

Field Studies

Of the 3 species, *Pithecia pithecia* has been studied most often. A brief survey was conducted in Guyana (Vessey et al., 1978). As part of a larger synecological study, it was studied at Raleighvallen-Voltzberg Park in Surinam (Mittermeier, 1977; Buchanan, 1978; Fleagle and Meldrum, 1988; Fleagle and Mittermeier, 1980; Mittermeier and van Roosmalen, 1981), and briefly in Brazil (Bonvicino and Lima, 1982; Oliveira and Lima, 1981; Oliveira et al., 1985). A long-term study, emphasizing feeding behavior, is under way in Brazil (Setz, 1985, 1987, 1988a,b); at Guri Lake in Venezuela, Kinzey and Norconk are conducting an intensive field study of four sympatric species (Kinzey et al., 1988), emphasizing the mechanical and chemical properties of foods eaten (Kinzey and Norconk, 1993; Kinzey et al., 1990; Norconk and Kinzey, 1990), and locomotion (Walker, 1992, 1993).

Species in the *P. monachus* group are less well known, mostly from surveys and brief studies. Studies of the buffy saki (*P. albicans*) include Johns' (1985, 1986b) study of patterns of social organization, and Peres' (1993a) 20-month study of diet and ecology (though none of the groups he studied was habituated). Happel (1982) and Soini (1986) briefly studied the ecology of *P. monachus* (=*P. hirsuta*). Freese et al. (1982) surveyed several areas in northeastern Peru and lowland Bolivia, and observed *P. monachus* in Peru, but did not observe sakis in Bolivia. Early naturalists' reports on *Pithecia* are summarized in Buchanan et al. (1981).

Habitat and Ecology

Pithecia seem to be the most versatile of the pitheciins in terms of their ability to persist in a variety of habitats (Johns, 1986b; Fleagle and Meldrum, 1988). They inhabit both highland and lowland forests (Robinson et al., 1987), seasonally flooded *igapó* (Peres, 1993a), secondary forests (Kinzey, 1992), and disturbed habitats (Johns, 1986b; Lovejoy et al., 1986). *P. irrorata* were reported only in high forest in Bolivia, (Izawa and Bejarano, 1981; Izawa and Yoneda, 1981). Their ability to survive in a wide variety of habitats apparently allows *Pithecia* to occur sympatrically with either *Chiropotes* or *Cacajao*, whereas the latter two genera do not co-ccur. An additional contributing factor may be that where sympatric, *Pithecia* occupies the middle and lower levels of canopy and *Chiropotes*

occupies the upper level. They may come to the ground to obtain preferred food items (Kinzey and Norconk, 1993).

Diet

Like most other New World primates, *Pithecia* are frugivorous with at
least 70% fruit in their diet (Kinzey, 1992). What distinguishes *Pithecia*
(and the other pitheciins) is that whereas most primate frugivores eat
soft, ripe fruit pulp, the pitheciins are predators on seeds of fruit with
hard pericarps (Kinzey, 1992). The ability to prepare and ingest food
with a hard pericarp—sclerocarpic harvesting—allows *Pithecia* to obtain
nutritious seeds with reduced tannins that are softer than those ingested
by other primates. Although *Pithecia* are apparently not capable of biting
through pericarps as hard as those punctured by *Chiropotes* and *Cacajao*,
they may compensate by eating foods, such as *Licania* (Chrysobalana-
ceae), with moderate tannin levels in combination with high levels of
lipids (Kinzey and Norconk, 1993). Whether *Pithecia* are capable of de-
toxifying tannins is not known. *Pithecia* routinely include leaves and
insects (Heymann and Bartecki, 1990) in their diet, employing a mixed
feeding strategy. Results from a 16-month study conducted by Kinzey
and Norconk (1993; Norconk and Kinzey, 1990) on *P. pithecia* indicate
that 85–93% of feeding time is spent on fruit (seeds are masticated in 95–
99% of the fruit eaten); young leaves are eaten every day, making up
between 1 and 13% of monthly samples; insects, especially ants, account
for 1 to 6% of the monthly samples; flowers are eaten seasonally. Data
for shorter-term studies on *P. pithecia*, *P. albicans*, and *P. monachus* (=
P. hirsuta) are in keeping with Kinzey and Norconk's findings (Kinzey,
1992), although rates of fruit consumption are somewhat lower (Johns,
1986b; Peres, 1993a; Soini, 1986; Robinson et al., 1987; Setz, 1987, 1988b).
Kinzey and Norconk (1990) and Setz (1987, 1988b) have both found
seasonal changes in fruit consumption, which they attribute to seasonal
differences in fruiting phenology.

Predation

P. pithecia were relatively infrequent victims of the harpy eagle (Rettig,
1978), but due to their cryptic nature they probably are not easy prey for
raptors or for large felids.

Behavior

Knowledge of *Pithecia* social behavior is rather scant, especially in the
wild.

Social Behavior

Using brief field accounts (Buchanan et al., 1981) and data from captive studies (e.g., Homburg, 1987, 1989), most researchers have regarded *Pithecia* as a monogamous primate (Buchanan et al., 1981; Shively and Mitchell, 1986; Robinson et al., 1987). Recent field data, however, call this assumption into question. Groups of *P. hirsuta* (= *P. monachus*) ranged in size from 2 to 8 with 2–5 adults (1–3 adult ♂♂ and 1–2 adult ♀♀); group composition was extremely stable, indicating that groups were essentially closed social units (Soini, 1986). A group of *P. pithecia* currently under study at Guri Lake, Venezuela, consists of 3 adult ♂♂, 3 adult ♀♀, and 3 subadults and juveniles. Copulations have been observed between one adult male and two females in the group. *Pithecia* may live in small groups that come together to form larger congregations (Fleagle and Meldrum, 1988). As is the case with callitrichine primates, the form of social organization most compatible with captive breeding may not be the typical form under natural conditions.

Dugmore's 1986 study on captive *P. pithecia* showed low levels of aggression. Low levels of aggression and competition for females are also reported for free-ranging *P. pithecia* (Kay et al., 1988).

Allogrooming is a common behavior in *Pithecia* (Buchanan et al., 1981). They use both their fingers and toes to groom. Females groom males 68% of the time, and grooming may be actively solicited. Most allogrooming took place in the afternoon (Dugmore, 1986).

Ranging Behavior

Pithecia has usually been reported to have a small home range [e.g., 4–10 ha (Buchanan, 1978)], and this is probably true for *P. pithecia*. Species in the *P. monachus* group probably have larger ranges; for *P. hirsuta* (=*P. monachus*) home ranges varied from 9.7 to 40–42 ha, with large troops having larger ranges than small troops (Soini, 1986); Peres (1993a) reported 172 ha as the minimum range size for *P. albicans*. Day ranges are less than 1 km (Robinson et al., 1987). There are conflicting reports about territoriality in *Pithecia*. They have been thought to have relatively exclusive areas with defined boundaries and little overlap (Robinson et al., 1987). However, Soini believes that *P. hirsuta* (=*P. monachus*) is not territorial. He found that overlap between ranges of two troops was 70%. Among eight troops that were monitored for 2–6 years, there were only 3 cases of immigration into the troops; all three migrants were males (Soini, 1986). Temporary aggregations between troops did form occasionally, but more commonly contiguous troops did not associate with each other even when near each other. When they did come in contact, they would enter into vigorous vocal displays and sometimes visually threatened each other.

Activity Pattern

White-faced sakis have a relatively short active day, frequently rising after dusk, and generally moving into their sleeping tree 1–1½ hours before dusk (Norconk and Kinzey, unpublished observations).

P. pithecia is seen most frequently in the understory and lower canopy levels (Fleagle and Meldrum, 1988) and does come to the ground to obtain preferred food items (Kinzey and Norconk, 1993). All members of the *P. monachus* group observed thus far appear to prefer the middle and upper canopy (Peres, 1993a). *P. albicans* were observed by both Johns (1986b) and Peres (1993a) most often in the middle and upper canopy, rarely in the emergents or in the lower canopy, and never on the ground. *P. monachus* (=*P. hirsuta*) also confined its activities largely to middle and upper canopy levels (Soini, 1986).

P. pithecia regularly sleep in well camouflaged vine tangles (Norconk and Kinzey, unpublished observations). Sleeping sites for *P. monachus* (=*P. hirsuta*) were observed only three times. In each case the site was a stout horizontal branch located in the upper canopy (Soini, 1986). Johns (1986b) did not see evidence of regular sleeping sites for *P. albicans*.

P. monachus (=*P. hirsuta*) did not associate actively with other species, although they occasionally fed in the same tree as *Alouatta, Saimiri, Saguinus, Lagothrix,* or *Cebus apella*; of these, only *C. apella* displaced *Pithecia* (Soini, 1986). Other occasional associations of *Pithecia* were with *Chiropotes* and *Cacajao* (Buchanan et al., 1981).

Mating and Reproduction

Soini (1986) reported a well-defined birth season for *P. monachus* (= *P. hirsuta*) from September to December (late dry season to early wet season); copulation and sexual play were observed only in April and May (Soini, 1986). *P. pithecia* in Surinam has a birth peak in January and February (Robinson et al., 1987), as does *P. pithecia* in Venezuela (Norconk and Kinzey, unpublished observations). Johns (1986b) did not observe any seasonality in *P. albicans* births. The youngest confirmed age for a *P. pithecia* female to give birth is 37 months (Dugmore, 1986). The estrous cycle of *P. pithecia* is 16–17 days (Savage et al., 1992a).

P. monachus (=*P. hirsuta*) troops had only one reproductive female; offspring were born at 2–3 year intervals. A group of *P. pithecia* in Venezuela has two breeding females (Norconk and Kinzey, unpublished observations).

Growth and Development

Available field data on *P. albicans* (Johns, 1986b), *P. monachus* (=*P. hirsuta*) (Soini, 1986), and *P. pithecia* (Norconk and Kinzey, unpublished observations) indicate that mothers are the primary care takers of saki monkey young, although fathers do groom infants. Soini (1986) reports that for the first week of *P. monachus* (=*P. hirsuta*) the daughter may help carry the infant. They begin to leave the mother's back in the second month and are self-locomoting at 3–4 months [4–5 months in *P. pithecia* (Savage et al., 1992a)], although they still seek safety on the mother's back up to 7 or 8 months. In *P. monachus* (=*P. hirsuta*) weaning begins at 8 months and is completed at about 1 year of age. Mothers share fruit into their second year, although infants begin feeding on fruit at 2–3 months. A case was observed in which the mother opened tough pods for her 26-month-old daughter who could not open them by herself.

In threatening situations young juveniles often hid, remaining silent and motionless while the mother and troop fled a short distance, vocalizing loudly while they moved away from the hiding juvenile (Soini, 1986). *P. pithecia* (Norconk, personal communication) and *P. albicans* (Johns, 1986b) males have a distraction display that they use to lead observers away from the rest of the group.

P. monachus (=*P. hirsuta*) are born with short, sparse, dark pelage. As the coat becomes denser and longer, it turns lighter. At 2–8 weeks infants have lighter coats than adults. It is not until they are $2^{1}/_{2}$–3 years old that the adult facial hair pattern becomes evident. Johns (1986b) reports more rapid maturation for *P. albicans*: adult coloration is attained at 3 weeks, total independence is achieved at 6 months, and at 11 months it is hard to distinguish between adults and juveniles.

Communication

A description of visual and vocal patterns of communication is provided in Buchanan et al. (1981). *P. pithecia* has low pitched, atonal roars that are contagious between groups. *P. monachus* has a "hoo-hoo" that is also used as an intergroup call. It is still not clear how *Pithecia* calls are used in spacing (Robinson et al., 1987).

Dugmore (1986) reports the use of four behaviors that may be related to olfactory communication: rubbing the bare ventral area of skin that runs down from the throat against objects, rubbing urine on the one sole and one palm and leaving foot and hand prints, nasobuccal rubbing, and anogenital rubbing.

Play

Dugmore (1986) observed *P. pithecia* playing alone and with partners. Males and older siblings are the primary play partners (Savage et al., 1992a). Solitary play included exaggerated forms of leaping and swinging from branches and an element of object manipulation, pulling, chewing, and jumping on small branches and ropes. Partner play usually consisted of chasing and rough and tumble. Social play usually occurred in the morning (Dugmore, 1986).

Posture and Locomotion

P. pithecia locomotion is dominated by leaping and clinging behaviors. Leaping accounts for 70% of locomotor behavior during travel; quadrupedal walking and running just over 25% and climbing less than 5%. Less is known about postural and locomotor behavior during feeding (Fleagle and Meldrum, 1988). A frequent posture while resting and feeding is clinging to vertical tree trunks (Walker, 1992). Vertical clinging is also the most frequent take-off position for *P. pithecia* leaps (Walker, 1993). *P. albicans* is also quadrupedal and a leaper. They have been observed bouncing (propulsion is supplied by all four limbs) from one bough to another, but the pitheciin bipedal hop that is associated with *P. pithecia* was not observed in *P. albicans* (Johns, 1986b), and vertical clinging was very rare (Peres, 1993a), probably due to their larger body size. Due to its rapid locomotion through the arboreal habitat, Moynihan (1976b) says that in Colombia *Pithecia* are often referred to as *"volador"* (flier). Like other pitheciins *Pithecia* frequently grasps small branches between its second and third digits (Erikson, 1957).

25

Saguinus

Introduction

The genus *Saguinus*, the tamarins, is a group of small, largely insectivorous monkeys. Together with the marmosets they comprise the Callitrichinae, all of which have claw-like nails that they use for vertical
clinging postures on large tree trunks. A major argument about the
evolution of New World primates has centered on whether the callitrichines are primitive or derived. Some (e.g., Eisenberg, 1977; Hershkovitz, 1977) have argued that marmosets and tamarins are among the
most primitive of living primates, but there is a growing consensus that
numerous callitrichine features represent a set of derived and highly
specialized character states (e.g., Sussman and Kinzey, 1984; Garber,
1993a). It has been suggested by Leutenegger (1973, 1980), Rosenberger
(1977, 1979), and Ford (1980b, 1986b) that traits such as claw-like nails,
an elongated hind foot, loss or reduction in size of the hypocone on
maxillary molars, loss of third molar, and reproductive twinning are the
result of phyletic dwarfism and the constraints of small body size.
Saguinus ranges in body size from 310 g (*S. fuscicollis*) to about 650 g in
the largest species (Garber and Teaford, 1986).

Taxonomy and Distribution

There are 12 species of *Saguinus* that are distributed largely in the
Amazon basin (see Map 12), but extend into Central America as far north
as Costa Rica (see Map 15). All species are allopatric except *S. fuscicollis*,
the smallest tamarin and the most widely distributed, which is virtually
always sympatric with another larger species of *Saguinus*: *S. mystax*,
S. imperator, or *S. labiatus*.

Of the platyrrhine genera, *Saguinus* is the most numerous and most
diverse. There are three primary groupings (hairy-face taxa, mottled-face
taxa, and bare-face taxa), 11–13 species, and 29 recognized subspecies

(Hershkovitz, 1977; Mittermeier and Coimbra-Filho, 1981; Mittermeier et al., 1988b; Garber, 1993a). The taxonomic arrangement of *Saguinus* species is supported by data on long call vocalizations (Snowdon, 1993).

Saguinus and *Callithrix* have generally been thought to be allopatric, with the latter genus south of the Rio Amazonas and east of the Rio Madeira. Recently, however, *Saguinus fuscicollis* and *Callithrix emiliae* have been shown to be sympatric in a small area of Rondônia east of the Rio Madeira (Ferrari and Martins, 1995; Lopes and Ferrari, 1994), and *S. midas* is sympatric with *C. argentata* south of the Amazon, between the Tocantins and Xingu rivers (Ferrari, 1993). *Saguinus fuscicollis* and *S. mystax* are sympatric with *Cebuella*.

Evolutionary History

The evolutionary history of the tamarins, as for all callitrichines, is a subject of speculation. There is no fossil evidence for *Saguinus*, so that reconstruction of their evolutionary history is difficult. However, from the evidence available, it has been suggested that differentiation of most of the extant species occurred during the Pleistocene (Garber, 1993a), but the postulated mechanisms differ greatly (cf. Hershkovitz, 1977; Kinzey, 1982).

Comparative morphology has led to the suggestion that *S. nigricollis* is the prototype for all tamarin species and that from *S. nigricollis* two phylogenetic lines based on *S. mystax* and *S. midas* diverged (Hershkovitz, 1966, 1972, 1977; Hanihara and Natori, 1989). The facial morphology of the 13 subspecies of saddle-back tamarins was compared to trace their phylogenetic and geographic relationships (Cheveraud and Moore, 1990), resulting in the proposition that morphological variation among subspecies is not geographic.

Although several have suggested that phyletic dwarfism is the cause of many callitrichine adaptations (see above), Garber (1993a) argues that dwarfing provides only a partial explanation for the existence of these traits, and that the unique combination of features more likely reflects a series of ecological adaptations.

Previous Studies

Captive Studies

Studies concerning the social behavior of tamarins have included reproduction (Hampton et al., 1966; Epple, 1975, 1978; Malaga, 1985; Price, 1990, 1992a; Baker and Woods, 1992), parental and family care

(Price, 1991, 1992b), play (Caine and O'Boyle, 1989), vocalizations (Masataka, 1988; Caine and Stevens, 1990), social dynamics (Heistermann et al., 1989), and endocrinology (Widowski et al., 1990). Earlier studies were reviewed in Snowdon and Soini (1988).

Field Studies

Seven of the 11 species of tamarins have been the subjects of detailed and systematic investigations lasting 5 months or more. *Saguinus fuscicollis* has been the most frequently studied species to date, attributed to its widespread distribution and to its tendency to form stable associations with other species of *Saguinus*.

S. geoffroyi has been studied in Panama by Dawson (1976, 1978, 1979) and Garber (1980a,b, 1984a,b). *S. oedipus* and *S. nigricollis* were both studied in Colombia by Neyman (1977) and Izawa (1978b), respectively. *S. labiatus* has been investigated in Bolivia (Izawa and Yoneda, 1981; Yoneda, 1981, 1984; Pook and Pook, 1982; Buchanan-Smith, 1988, 1990b, 1991a,b; Buchanan-Smith and Jordon, 1992) and *S. mystax* in Peru (Garber, 1986, 1988a,b, 1993b; Garber et al., 1993, 1995; Garber and Pruetz, 1995; Garber and Teaford, 1986; Pruetz and Garber, 1991; Norconk, 1986; Ramirez, 1990). *S. fuscicollis* was extensively observed both in Peru (Terborgh, 1983, 1985, 1986a; Crandlemire-Sacco, 1986; Garber 1986, 1988a,b, 1993b; Garber and Teaford, 1986; Norconk, 1986; Goldizen, 1987a,b, 1990; Goldizen et al., 1988) and Bolivia (Pook and Pook, 1981, 1982; Izawa and Yoneda, 1981; Yoneda, 1981, 1984a,b; Buchanan-Smith, 1988, 1990a). *S. bicolor* was studied in Brazil (Egler, 1983, 1986, 1992). *S. midas* was studied in Amapá, Brazil (Thorington, 1968b) and in Surinam by Mittermeier (1977) and Mittermeier and van Roosmalen (1981). For a comprehensive list of studies refer to Snowdon and Soini (1988) and of long-term studies, to Garber (1993a).

Habitat and Ecology

Except for *S. oedipus* and *S. geoffroyi*, which are frequently found in dry deciduous forest, *Saguinus* is found primarily in humid tropical lowland forest. It seems to prefer secondary forest and forest edge, occupying all canopy levels, but preferring middle and lower canopies. Species differences in habitat preference are described by Snowdon and Soini (1988).

Diet

The primary components of the tamarin diet are insects, ripe fruits, plant exudates, and nectar. This diet is high in both nutrient quality and caloric content, which are extremely important in a small-bodied animal that exhibits a limited gut volume and rapid rate of food passage (Garber, 1993a).

Recently, there has been a particular focus on the dietary behavior of *Saguinus* (Garber, 1992, 1993a). The callitrichines are unique among platyrrhine taxa in their ability to cling to large vertical trunks while exploiting resources such as plant exudates and bark-refuging insects (Garber, 1992). By implanting their claws into the bark of large tree trunks *Saguinus* are able to maintain stable vertical clinging postures. This feeding position has been observed as a common means of exudate acquisition. This posture was adapted in 64–97% of observed exudate feeding episodes of *S. geoffroyi*, *S. mystax*, and *S. fuscicollis*. Exudates provide an important source of minerals, water, and other nutrients.

Unlike *Callithrix*, *Saguinus* lacks specialized mandibular incisors, and is unable to stimulate exudate flow by gouging holes in tree bark. Instead, gum availability depends upon more indirect means such as the natural weathering of bark, parasitic activity of wood-boring insects, and/or removal of already hardened gum with their mandibular incisors (Garber, 1993a). Although gums may constitute only a small fraction of the yearly diet of *Saguinus*, during the dry and early wet seasons, exudates appear to be a significant dietary staple for many species: gum feeding may contribute up to 50% of total plant feeding time in a 2- to 3-month period (Garber, 1980a; Norconk, 1986). In other areas, such as Manu, during times of food scarcity such as the dry season, *Saguinus* resorts to keystone resources (Terborgh, 1986b) such as nectar (Terborgh, 1983).

Highly opportunistic foragers, *Saguinus* procure insects, fruits, and nectar by using various "forelimb and hindlimb prehensile positional behaviors" on all branch sizes and in all areas of the tree, including vines and fragile foliage (Garber, 1993a). Trunk foraging for insects has been observed occasionally in *Saguinus*; however species such as *S. mystax* and *S. geoffroyi* have demonstrated this foraging strategy only rarely, accounting for less than 3% of their total insect foraging time (Garber, 1993a).

Species-specific differences in foraging behavior have been demonstrated within the genus *Saguinus*. Garber (1993a) divided tamarin insect foraging into three distinct patterns. The characteristics that separate the patterns are based on "hunting techniques, substrates exploited, vertical stratification, and modes of positional behavior" (Garber, 1993a).

Pattern 1: *S. geoffroyi*—cryptic hunting for prey on leafy substrates of small flexible branches in lower canopy and shrub layer.

Pattern 2: *S. mystax*, *S. labiatus*, *S. imperator* (and possibly *S. midas*)—

exploit insects on leaves and branches in the lower and middle levels of the forest canopy.

Pattern 3: *S. fuscicollis* (and possibly *S. nigricollis* and *S. bicolor*)—most distinctive foraging pattern, which entails feeding on relatively large and cryptic invertebrate prey, in all levels of the canopy, especially on moderate and large supports, utilizing a high proportion (79%) of vertical clinging postures. Note that *S. fuscicollis* is usually sympatric with tamarins using pattern 2.

Predation

Tamarins are preyed upon by raptors, snakes, and humans. Antipredator response varies among species and geographic populations. In general, most wild *Saguinus* respond with loud excitation calls, mobbing displays, and/or escape behavior (Moynihan, 1970b; Terborgh, 1983; Heymann, 1987). In mixed species groups of tamarins, there appears to be coordination between *S. fuscicollis*, which are more vigilant in the lower canopy for terrestrial predators; the larger tamarin (e.g., *S. mystax*) provides greater vigilance and protection in the upper canopy against arboreal predators (Peres, 1993b).

Moving objects seem to trigger a fear response in captive cotton-top tamarins (Hayes and Snowdon, 1990).

Behavior

Social Behavior

Saguinus tend to live in small multimale–multifemale groups consisting of unrelated adults. They exhibit communal care of the young, which is provided mainly by unrelated group members, particularly adult males. Polyandrous matings seem to be dominant, however, cases of monogamy and polygyny have been reported. There is only one reproductively active female in a group at a time. The breeding female suppresses the ovulatory cycles of other females in the troop and she will mate with more than one male (Goldizen, 1987b). Sussman and Garber (1987) showed that there was a significant correlation between the number of males in a group and the total number of surviving young, demonstrating the value to the female of mating with more than one male.

Ranging Behavior

Tamarins have moderate to large home ranges and extremely large day ranges in comparison to other primates of similar body size (*Aotus*,

Callicebus, and various prosimian taxa). Home ranges of 20–40 ha are common for most *Saguinus* species; however mixed-species groups of *S. imperator* and *S. fuscicollis* in Peru are reported to have home ranges of over 100 ha. Mean day range exceeds 1200 m for all species for which quantitative data are available (Garber, 1993a). Plant resource distribution is a primary determinant of primate daily ranging patterns (Garber, 1993a).

Where *S. fuscicollis* do not occur with congeners (e.g., on the Rio Nanay in Peru, Castro and Soini, 1978), groups coalesce to feed in large supergroups. Thus, the formation of mixed-species groups may substitute for the formation of supergroups as an alternative antipreditor device (Kinzey and Cunningham, 1994). *S. fuscicollis* apparently do not fragment and coalesce when they form mixed-species groups (e.g., Goldizen, 1990).

Within the home ranges of tamarins, overlap may be extensive. The degree of overlap varies among species and location and may depend on the distribution pattern of major food patches (Garber, 1993a).

In a unique experimental study of free-ranging moustached tamarins Garber and Dolins (1994) studied perceptual cues in foraging behavior, and determined that visual and temporal cues, but not olfactory cues, were important in locating food. Temporal cues, however, did not carry over from one day to the next.

Activity Pattern

The members of a tamarin group usually sleep together in a clump. The roost is normally the crotch of a branch or a vine tangle. Relative to other primates, tamarins leave their roosts and initiate daily activities late, from several minutes to 1 hour after sunrise (Snowdon and Soini, 1988). The daily group activity pattern seems to be very similar across species. There appear to be two peaks for foraging for fruit and insects: the first occurs early to mid-morning and the second shortly before settling for the night (Snowdon and Soini, 1988).

Mating and Reproduction

Saguinus births generally coincide with maximum fruit availability and, therefore, occur in the early half of the rainy season and last for about 2–3 months. For species north of the Amazon basin the birth peak falls between April and June (Dawson and Dukelow, 1976; Hershkovitz, 1977; Moynihan, 1970b), while those in the Upper Amazonian basin have a peak from November to March (Izawa, 1978b; Soini, 1987d; Soini and Soini, 1982). A few litters are born nearly every month of the year (Dawson and Dukelow, 1976; Soini, 1987d; Soini and Soini, 1982). In the northern hemisphere, peak season in captive groups occurs between February and August.

Tamarins, like other callitrichines, have the potential for two litters per year, which is common in captivity (Goldizen et al., 1988). The link between birth season and maximum food availability is particularly relevant because females have only one litter per year in the wild. The demands placed on female tamarins in the wild are high in terms of infant care and reproduction: they have a high neonate/adult ratio, high frequency of twinning, and high energetic demand of pregnancy and lactation. Lactation is especially expensive energetically, as transport of 30-day-old infants is only 10% as expensive as lactation (Tardif and Harrison, 1990). Female tamarins are unable to rear young without the help of others; polyandry solves the problem of excessive energetic demands by providing extra helpers for infant care (Sussman and Kinzey, 1984).

Growth and Development

Birth and development of the young, for both wild and captive animals, including a table of developmental parameters for four species, are reviewed extensively by Snowdon and Soini (1988). Larger species develop more slowly. Newborns differ in coloration from adults. Mothers rarely carry young beyond the second day of life, after which males and siblings provide infant care. Adequate parental care requires previous experience early in life.

Communication

Scent marking in tamarins has been reported for *S. fuscicollis*, *S. geoffroyi*, and *S. nigricollis*. *S. fuscicollis* uses anogenital, suprapubic, and sternal modes of marking. *S. geoffroyi* and *S. nigricollis* have demonstrated only anogenital and suprapubic marking (Bartecki and Heymann, 1990). Field observations of *S. fuscicollis* show that marking is correlated with home range use and is concentrated in feeding and resting areas. Scent marking was not observed to increase during confrontations with neighboring groups and, therefore, differs from observations of *S. geoffroyi* where scent marking is most concentrated in areas of group overlap (Bartecki and Heymann, 1990; Dawson, 1979; Lindsay, 1980).

Masataka studied the response of both male and female captive *S. labiatus* to male long calls of natal and foreign populations. Female tamarins responded selectively to male long calls from their natal populations. Males did not respond differentially to male and female calls (Masataka, 1988).

Tamarins are able to discriminate subtle acoustic cues that have communicative significance (Bauers and Snowdon, 1990). The cotton-top tamarin has a vocal repertoire of eight chirp vocalizations.

Play

Play has been observed in many species of tamarins. It is most commonly initiated by juveniles. It is rare between infants and adults or between infants and subadults (Cleveland and Snowdon, 1984). It has been noted that play is much more frequent within captive populations; this is possibly due to the greater demands placed on wild tamarins to learn insect foraging and other skills (Snowdon and Soini, 1988).

Different forms of play are performed on different substrates. Flat horizontal surfaces serve as wrestling ground; chasing and grasping are more common among vines, branches, and tree trunks (Caine and O'Boyle, 1989).

Posture and Locomotion

The most common modes of locomotion for *Saguinus* consist of quadrupedal walking and running, bounding, or galloping along horizontal and diagonal supports, combined with leaping between terminal branches (Snowdon and Soini, 1988). *S. fuscicollis*, *S. geoffroyi*, and *S. mystax*, despite differences in body size, were generally similar in positional behavior: 43–52% of travel time was spent in quadrupedal bounding and running; leaping accounted for 31–41% of travel time (Garber, 1991). *S. fuscicollis*, however, more frequently leapt between vertical trunks (see pattern 3, above).

All callitrichines use a I–V opposable pedal grasp, as defined by Gebo (1985). The presence of claw-like nails enables callitrichines to retain vertical clinging postures using all five pedal digits; prosimians of similar size are unable to do so.

26

Saimiri

Introduction

Saimiri, the squirrel monkey, is one of the best known neotropical primates, in large part because of its extensive use in a captive context. It is the second most commonly used laboratory monkey. The average weight of all wild caught specimens is 836 g, but *S. oerstedii* weighs less (695 g ♀♀, 829 g ♂♂) than *S. sciureus* (Ford and Davis, 1992). Males average 30% heavier than females. There is some variation among subspecies. Males additionally gain up to 222 g (about 20% of body weight) during the mating season, producing what is called the "fatted" phenomenon (DuMond and Hutchinson, 1967). In *S. oerstedii citrinellus* (but not in *S. o. oerstedii*) there is sexual dimorphism in crown color, with males having gray crowns and females having black crowns (Hershkovitz, 1984). Squirrel monkeys are frugivorous–insectivorous, live in relatively large multimale groups, but engage in very little social activity. Because of their widespread distribution, the genus, *Saimiri*, is not endangered, although many local populations, especially in Central America (Boinski, 1994), are scarce.

Taxonomy and Distribution

In squirrel monkey taxonomy, from one (Cabrera, 1957) to seven (Elliot, 1912) species have been recognized (Costello et al., 1993). Other taxonomies are included in Hershkovitz (1984) [four species; two additional subspecies added in Hershkovitz (1985)], and Mittermeier et al. (1988b). Following Costello et al. (1993), two species are recognized, *S. sciureus* in South America and *S. oerstedii* in Central America. They summarize the taxonomic evidence from pelage variation, geographic distribution, biochemistry, chromosomes, skull and dentition, and behavior. Silva et al. (1993), based on electrophoresis of 15 proteins, regard the South American forms as a single polytypic ring species. The geographic distribution of *Saimiri* covers most of the Amazon basin (Map 14), and the population in western Panama and Costa Rica is disjunct (Map 17). The genus is widespread in tropical lowland forest along the river courses. They are found along the Amazon and its continuation, the Rio Solimões, and, for varying distances, up the tributaries of this river system. In the west *Saimiri* reaches the higher latitudes following the arc of the Andes reaching as far south as the Bolivia/Argentina border and Paraguay. In the east the animals are found in Guyana, Surinam, and French Guyana. *Saimiri* is absent from altitudes reaching higher then 300 m and the more arid regions surrounding the Amazon.

Ayres (1985) described a new species *S. vanzolinii* based on pelage coloration, and its restricted geographic distribution, but Costello et al. (1993) do not consider it specifically distinct.

Thorington (1985) gives a detailed overview of the geographic variation in skins and skulls. He recognizes two South American species *S. sciureus* and *S. madeirae*; *oerstedii* represents a subspecies of *sciureus*.

Squirrel monkeys are frequently divided into two forms, "Gothic" Arch and "Roman" Arch (MacLean, 1964), based on the pointed or round black arch over each eye [see Hershkovitz (1984) for figures of the two forms]. These designations, however, are not correlated with current taxonomy, and it is far preferable to give the precise location from which an animal comes to provide accurate subspecific taxonomy.

Evolutionary History

Squirrel monkeys are one of the oldest identifiable neotropical monkey lineages with the fossil record having yielded the clearly related *Neosaimiri fieldsi* (Rosenberger et al., 1991a; Setoguchi et al., 1991) from the middle Miocene of Colombia. Nevertheless, the relationship of squirrel monkeys to *Cebus* is not agreed upon. The classical view is that these two are closely related, the view currently espoused by Rosenberger (1981, 1992). However, Hershkovitz (1984) using morphological data, Sarich and Cronin (1980) using molecular data, and Ford (1986b) summarizing a variety of data all believe that *Cebus* and *Saimiri* are not closely related.

Fossil remains provisionally attributed to *Saimiri* have also been recovered from recent cave deposits on Hispañola (Ford, 1990b), where squirrel monkeys do not occur today.

Previous Studies

The genus *Saimiri* is thoroughly reviewed by Boinski (1995).

Captive Studies

In the United States the squirrel monkey is the most commonly used neotropical primate in laboratory research (Baldwin and Baldwin, 1981). Studies have focused on social organization (Anzenberger et al., 1986; Goldsmith, 1987; Perloe, 1986; Strayer and Noel, 1986; Wheeler, 1990; Wiener et al., 1986; Williams and Abee, 1988), mother–infant relations (Biben et al., 1989; Hennessy, 1986a,b), feeding ecology and behavior (Andrews, 1986, 1988a; Hopf and Ploog, 1991; King et al., 1986, 1987; Landau, 1987), communication (Biben et al., 1986, 1989; Harris and Newman, 1988; Matsuka and Symmes, 1986; Symmes and Biben, 1988; Symmes and Goedeking, 1988), play behavior (Biben, 1986, 1989; Biben et al., 1989), and reproduction (Colgrove et al., 1990; Mendoza and Mason, 1989, 1991; Saltzman et al., 1988). Studies prior to 1981 are reviewed in Baldwin and Baldwin (1981).

Captive studies have shown that there is geographic variation in the behavior of squirrel (Martau et al., 1985). Some of these variations seem to reflect population-specific, and inferred, genotypic differences such as aspects of social behavior (Mendoza and Mason, 1989) or vocalizations (Winter, 1969). The disadvantage of captive studies, however, is that there is no ecological context. Many of the behaviors demonstrated in the small confines of captive situations have not been demonstrated in the natural habitat.

Captive studies are available for two populations of *Saimiri*, from Bolivia and from the Guyanas. The former has a sexually segregated social organization in which males and females remain spatially separated outside the breeding season. Both males and females have independent linear hierarchies. Females are found to have higher cortisol titers and longer and more sustained adrenal response to stress; they are dominant over males and initiate more agonistic interactions compared to *S. sciureus* (Coe et al., 1985). In animals from the Guyanas, males and females are integrated throughout the year and a single linear hierarchy includes both males and females (Mendoza et al., 1978). Male and female social interactions in the two groups, therefore, appear to be regulated by two different mechanisms, dominance relationships in the latter and sexual segregation in the former (Costello et al., 1993).

Field Studies

Prior to 1981 there were no long-term field studies on squirrel monkeys (Baldwin and Baldwin, 1981). Since then a relatively large number of field studies have been conducted.

S. oerstedii have been studied in Costa Rica at Corcovado National park (Boinski, 1986, 1987a–d, 1988, 1989a,b, 1991; Boinski and Fragaszy,

1989; Boinski and Newman, 1988; Boinski and Scott, 1988; Boinski and Timm, 1985, Herzog and Meyer, 1987; Mitchell et al., 1991) and in Manual Antonio National Park, Costa Rica (Boinski, 1987b; Boinski and Newman, 1988). *Saimiri* does not occur at La Selva, Costa Rica (McDade et al., 1984; Wilson, 1990). *S. oerstedii* has been studied in Panama by Baldwin and Baldwin (1972, 1973).

S. sciureus were studied briefly in Surinam at Raleighvallen-Voltzburg National Park (Fleagle and Mittermeier, 1980; Fleagle et al., 1981), extensively in Peru at Manu National Park (Boinski and Mitchell, 1992; Mitchell, 1987, 1990; Mitchell et al., 1991; Podolsky, 1990; Terborgh, 1990), in Peru at Rio Pacaya (Soini, 1986), in Colombia at Monte Seco (Thorington, 1967, 1968a), and in Colombia at La Macarena (Izawa and Nishimura, 1988; Klein and Klein, 1973; Yoneda, 1988, 1990).

Winter (1972) studied a small group of animals in a free ranging condition on Vieques Island just off the southwest coast of Puerto Rico.

The only locations where squirrel monkeys have been studied extensively are in Manu National park, Peru and Corcovado and Manual Antonio National Parks, Costa Rica. Since the species at the former site is *S. sciureus* and that at the latter two sites is *S. oerstedii*, contrasts between the two species that will be made in the remainder of this chapter represent differences between these two locations. Data from other sites will have to be obtained to determine that these differences are not just the result of differences in ecology.

Habitat and Ecology

Squirrel monkeys are exclusively arboreal and use a wide variety of habitats. They use primarily lower and middle canopies of the forest. According to Boinski (1995), they appear to have the least restrictive requirements in regard to forest type of any neotropical primate. They are not known, however, to exploit primary forest except on a temporary basis. *S. oerstedii* is exceptional, however, in its preference for early successional stages of forest growth.

Diet

Squirrel monkeys are insectivorous and frugivorous, and rely on whichever is more readily available during any particular season. The major protein source is insects, which are gleaned from surfaces of leaves. The squirrel monkey's method of insect foraging contrasts strongly with that of *Cebus*, which tear open plants, especially palms, to retrieve concealed insects (Jason and Boinski, 1992). Insect-eating birds, such as double-toothed kites and woodcreepers, often follow squirrel monkeys to obtain insects flushed by them, especially during the wet season when arthropods are least abundant (Boinski and Scott, 1988). Most fruit eaten is ripe, and, where available, the most important fruit source is figs. Any small seeds that are ingested are probably dispersed and not digested. In times of fruit scarcity, squirrel monkeys resort to keystone resources such as nectar (Terborgh, 1983).

Squirrel monkeys have a unique method of searching for food, at least in Peru, where they follow troops of *Cebus* monkeys and utilize the latter's extensive knowledge of large fruiting trees to find new fruit sources (Podolsky, 1990). In Costa Rica, where squirrel monkeys only feed in small fruit trees, they do not associate with *Cebus* (Janson and Boinski, 1992). An extensive description of patterns of feeding, and differences among age groups, between sexes, and between species is found in Boinski (1995).

In Costa Rica the greater the food abundance, the smaller the range area and the smaller proportion of time spent foraging for food (Boinski, 1987c).

Predation

Squirrel monkeys give alarm calls in the presence of large flying raptors, and other large flying birds, and mob snakes and potentially predatory mammals (Baldwin and Baldwin, 1981). Predation on squirrel monkeys by several raptors, including the harpy eagle, the Guianan crested eagle, the ornate hawk-eagle, and the slate-colored hawk, have been observed (Boinski, 1995). Squirrel monkeys have also been the prey of ocelots (Emmons, 1987).

Behavior

Squirrel monkeys have the largest cohesive groups of any neotropical primate, and yet spend almost no time in social grooming (Janson, 1993). This is totally contrary to Dunbar's (1993) assertion that time devoted to grooming is positively correlated with group size.

Social Behavior

Field studies of *S. oerstedii* and *S. sciureus* show differences in behavior, especially for females. In *S. oerstedii* there are no female alliances and dominance hierarchies. in *S. sciureus*, however, there is differentiation in female relationships. Here females form alliances, probably among related females, and have dominance hierarchies. Females remain in their natal group. The males of *S. sciureus* migrate from their group before maturation and form all male groups, while the few males that remain within the group show strong agonistic behavior toward each other and the females. This contrasts strongly with *S. oerstedii* where the males remain in the group and no agonistic behavior between males in the same group has been observed.

Ranging Behavior

Squirrel monkeys typically have the largest group size of any neotropical primate, with sizes ranging between 23 and 54 troop members (Robinson and Janson, 1987) and up to 65 animals in Costa Rica (Boinski, 1986). There are also early reports of very large aggregations of up to 500 animals (Baldwin and Baldwin, 1981). There is also seasonal variation in troop size. Troop composition is very difficult to determine, because of the difficulty in distinguishing males from females, and the fact that troops are large, the animals move rapidly, and are widely dispersed.

Though secondary forest is the preferred habitat of *Saimiri* they have been observed in many types of wet forest. Secondary growth and the mid-canopy level are the preferred strata for the squirrel monkey, but they also make use of the highest levels of the forest and will even come to the ground. Boinski (1988) indicates that for *Saimiri* in Costa Rica the range area varied between seasons. Baldwin and Baldwin (1981) also indicate that range area is variable for *Saimiri* and is influenced not only by habitat type and season but also by group size. The greatest amount of time spent traveling occurs during seasons of low food abundance (Boinski, 1988).

Activity Pattern

Squirrel monkeys normally arise at dawn and settle down for the night at dusk. They spend 61% of their time feeding and/or foraging for food (Mitchell, 1990). Except when feeding in a fruit tree, very little of this activity is synchronized. Thus, it is difficult to define any "typical" daily pattern of activity for squirrel monkeys. Traveling (except that when foraging for insects) occupies only 27% of the day. This leaves only 12% of the day for resting and social interactions. Sleeping sites are variable, but for *S. sciureus* generally involve a group of neighboring trees where the animals sleep on the terminal branches of the canopy. They generally sleep in the vicinity of the *Cebus* troop with which they are associated during the day. *S. oerstedii*, in contrast, often sleep in the same tree or trees for many consecutive nights, in the isolated canopy of an emergent tree. Thus, in the former species, vigilance for predators is provided by the *Cebus* troop, whereas the latter species, which does not travel in mixed-species troops (Boinski, 1989b), avoids predators by sleeping in an isolated crown.

Mating and Reproduction

Squirrel monkeys do not show any behavioral patterns, displays, or vocalizations that appear crucial for reproductive isolation (Costello et al., 1993). Wild individuals appear to copulate without any evident preparatory interactions. Males and females, both in the field and captivity, seem to lack any premating affiliative behaviors except for some association and olfactory investigation prior to copulation (Mendoza and Mason, 1991). Most copulations, at least during the peak of estrus, are performed by a single male.

Saimiri oerstedii show a seasonal reproductive cycle and have the most restricted birth peak of any neotropical primate (Boinski, 1987a). For any given troop most births are generally synchronized to within a 1-week period. Males undergo a seasonal fattening, gaining at least 20% of body weight during the reproductive period; this is unique among primates. The breeding season lasts from August to early October; the gestation period averages 145 days; births occur from February to early April; most adult females give birth every year; weaning takes place after about 6 months (Boinski, 1987d). Weaning takes place earlier in the wild than in captivity (Boinski and Fragaszy, 1989).

In contrast, *S. sciureus* give birth every other year; birth synchrony is not pronounced, and weaning does not take place until about 18 months (Mitchell, 1990). It appears that predation pressure on the very young animal is reduced here, and maternal investment is prolonged; in contrast, *S. oerstedii* are under the greatest predation threat at a very early age (Boinski, 1995).

Growth and Development

After birth the infant immediately clings to the mother's body. During the first 2 weeks of life it alternates between sleeping and nursing. The infant's pelage color does not show any difference from that of the adult animal. During the third week the infant becomes more mobile and will cling to the mother's back. During the infant's fifth–seventh week it starts leaving the mother's back and explores the immediate environment. From the eighth week through the fourth month the interaction between mother and infant becomes looser and the infant starts to associate more with other young animals. During the fifth through tenth month the independence process continues and at 6 months they are usually weaned. From about 16 to 24 months the animal has reached average adult size but is not considered to be an adult female until 2.5

years of age or an adult male until 4 years of age (Baldwin and Baldwin, 1981). In the wild infants are weaned earlier than under captive conditions (Boinski and Fragaszy, 1989).

Communication

Squirrel monkey vocalizations are well known from a large number of laboratory studies. Their vocal repertoire was first described by Winter et al. (1966) and early on differences in vocal structure between "Gothic" and "Roman" types were established. Different investigators use different terminologies. Finally Newman (1985) established a consistent system for identifying different squirrel monkey vocalizations.

Saimiri have easily identifiable calls that consist mainly of a pure-tone frequency structure that ranges up to about 12 kHz. Most vocalizations fall into two categories, chucks and peeps. Vocalizations appear to be geographically differentiated, making it unlikely that different populations of squirrel monkeys would be able to interpret all of each others vocalizations (Snowdon et al., 1985; Boinski, 1991). Most populations of *Saimiri* studied so far have a number of population-specific vocalizations (Costello et al., 1993). Squirrel monkeys are capable of identifying individual animals by their vocalizations (Biben, 1993). There is no evidence for long distance territorial calls (as in *Callicebus* or *Alouatta*), or for syntax (as in *Cebus* or *Cebuella*).

Differences exist between *S. sciureus* (Mitchell et al., 1991; Boinski and Mitchell, 1992) and *S. oerstedii* (Boinski and Newman, 1988). For example, adult female *S. oerstedii* increase the rate of calling with increasing distance from the nearest adult female, whereas in *S. sciureus* rate of calling is not affected by distance from nearest neighbor.

Play

The most frequently observed social behavior in wild squirrel monkeys is play. It is frequent among juvenile animals during the first year of life. It drops out of the behavioral repertoire only when food is scarce (Boinski, 1995).

Posture and Locomotion

Quadrupedal walk or run is the primary mode of locomotion for *Saimiri* (Boinski, 1989a). The majority of foraging activities takes place on small branches and vines. The animals are very agile and can move very quickly. Though the tail is not prehensile, it is important for balance. Adults can make leaps of up to 5 m. Brachiation is never used. A study of *Saimiri* in Surinam (Fleagle et al., 1981) indicated that during travel 55% was quadrupedal movement and 45% leaping; only 3% of the observations indicated that the animal was climbing. During feeding these ratios changed somewhat, but the categories of locomotion did not; quadrupedal movement 87%, leaping 11%, and climbing 2%.

Saimiri have two main rest postures, each with variation (Baldwin and Baldwin, 1981). The huddle involves sitting or squatting with the head down and the tail tucked between the legs and swung over a shoulder. This is the posture in which animals sleep. The sprawl is a posture in which the animal lies prone on a branch with limbs extended. Vertical clinging postures are occasionally used (Boinski, 1989b). There are few differences in posture between males and females, except that adult males are more often found at higher elevations in the canopy; this is related to their greater vigilance in detecting predators (Boinski, 1989b). Perhaps the most significant aspect of squirrel monkey positional behavior is the high degree of flexibility that is related to seasonal differences in food abundance and related foraging behaviors.

Maps

Distribution Maps of South and Central America by Genus

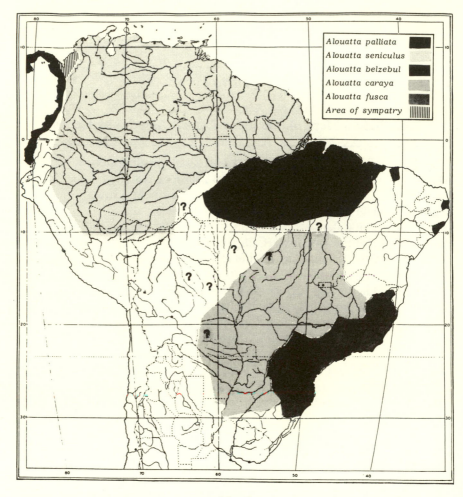

Map 1

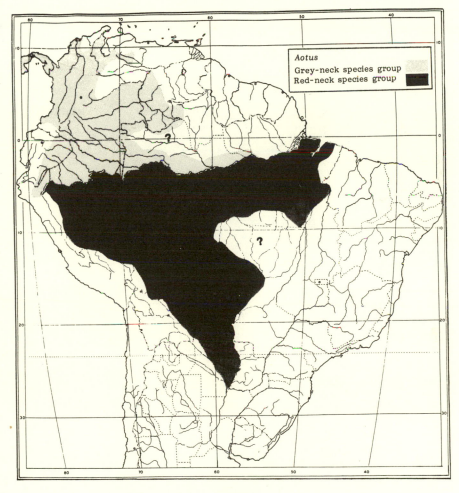

Map 2

Map 3

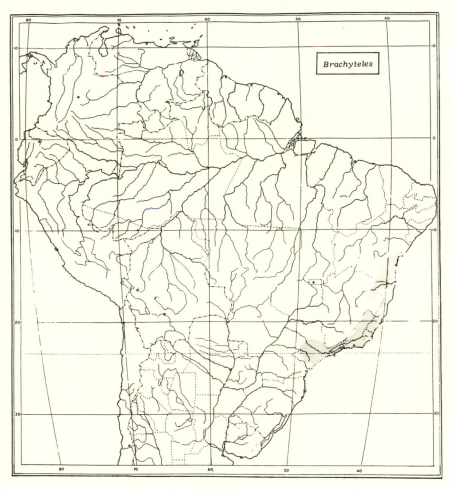

Map 4

Map 5

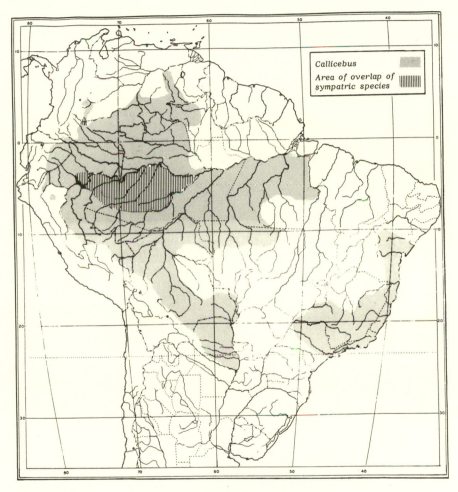

Map 6

Map 7

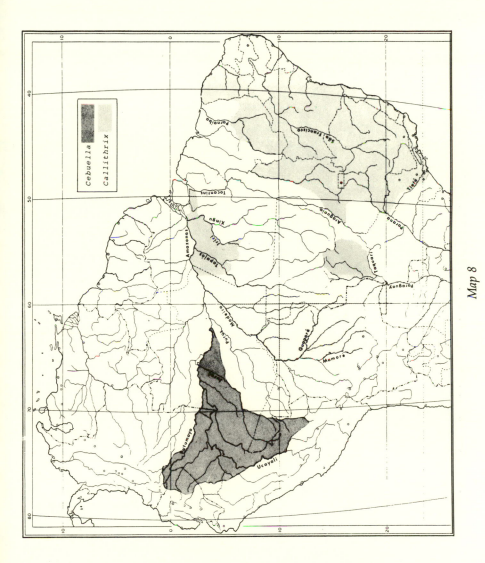

Cebuella

Callithrix

Map 8

Map 9

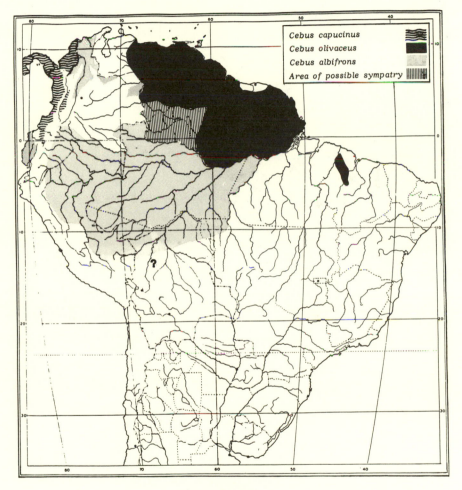

Map 10

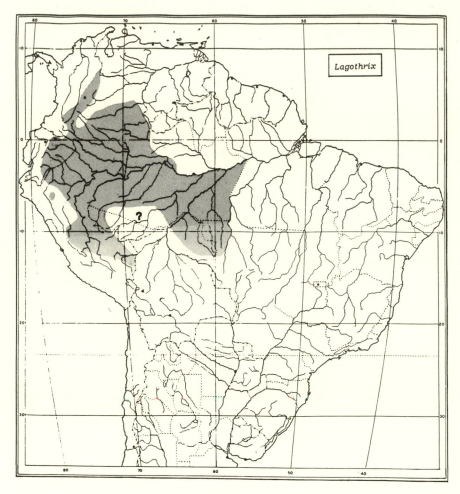

Map 11

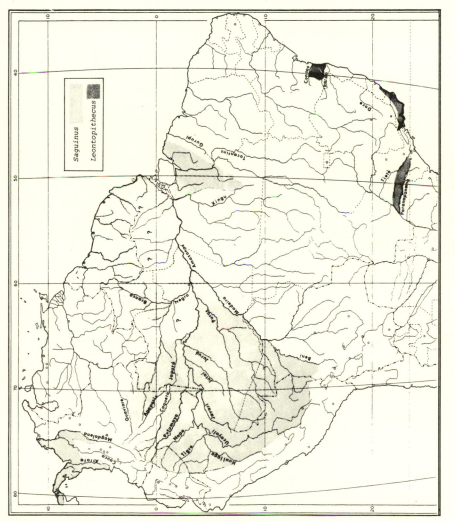

Map 12

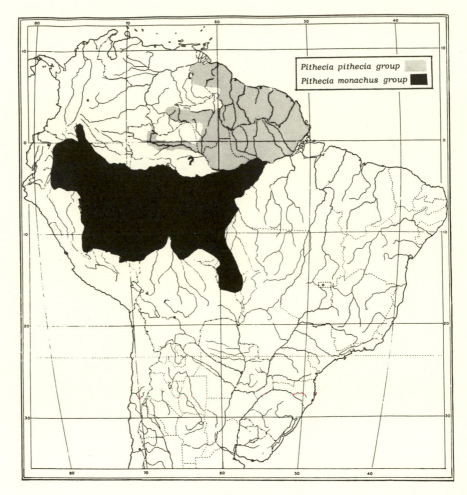

Map 13

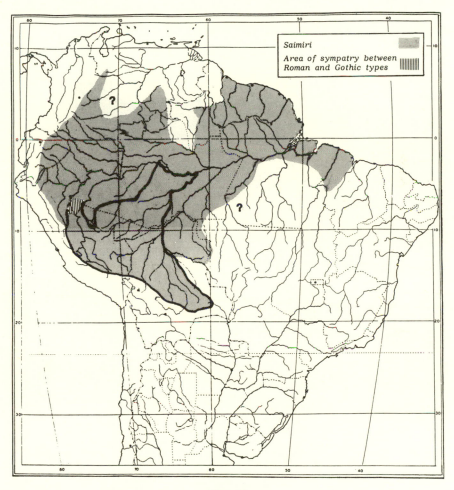

Map 14

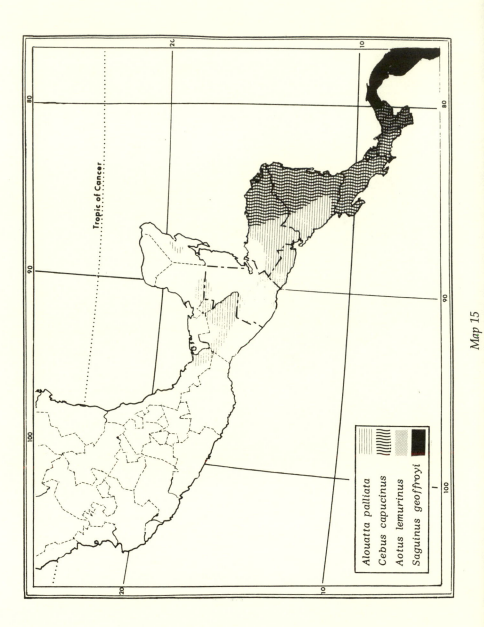

Map 15

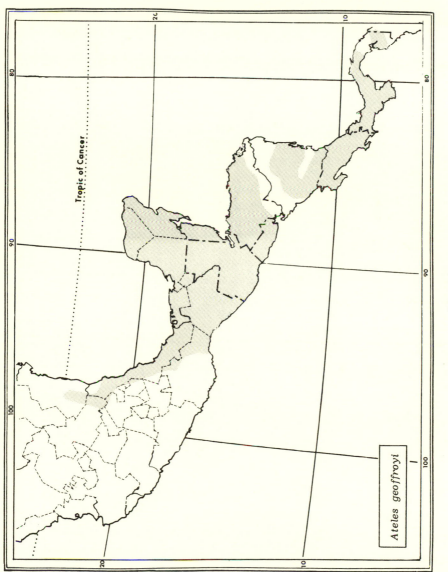

Map 16

Ateles geoffroyi

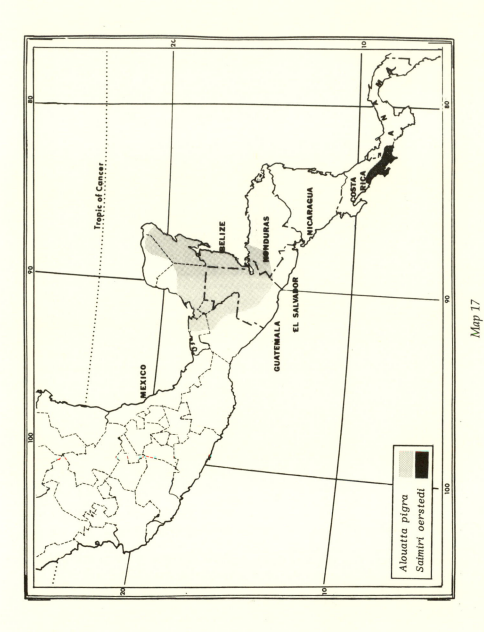

Map 17

References

Abate, V. (1984). [Bobo, 48-year-old laboratory-maintained cebus.] *Lab Animal* 13(3):cover, 5.

Abbott, D. H. (1993). Social conflict and reproductive suppression in marmoset and tamarin monkeys. In W. A. Mason and S. P. Mendoza (eds.), *Primate Social Conflict* (pp. 331–372). Albany: State University New York Press.

Abbott, D. H., Barrett, J., and George, L. M. (1993). Comparative aspects of the social suppression of reproduction in female marmosets and tamarins. In A. B. Rylands (ed.), *Marmosets and Tamarins; Systematics, Behaviour, and Ecology* (pp. 152–163). Oxford: Oxford University Press.

Adler, M. J. (1967). *The Difference of Man and the Difference It Makes*. New York: Holt, Rinehart, & Winston.

Aguirre, A. C.(1971). *O Mono Brachyteles arachnoides*. Rio de Janeiro: Academia Brasileira de Ciências.

Ahumedo, J. A. (1989). Behavior and social structure of free ranging spider monkeys (*Ateles belzebuth*). *La Macarena Field Studies of New World Monkeys, La Macarena, Colombia* 2:7–31.

Aiken, S. R., and Leigh, C. H. (1985). On the declining fauna of peninsular Malaysia in the post-colonial period. *Ambio* 14(1):15–22.

Albernaz, A., Ywane, M. S. S., and Carvalho, C. T. (1988). Study of the home range of a group of *Leontopithecus chrysopygus*. *XIIth Congress of the International Primatological Society, Abstracts Supplement*, Brasilia, p. 23.

Albignac, R. (1972). The carnivora of Madagascar. In R. Battistini and G. Richard-Vindard (eds.), *Biogeography and Ecology in Madagascar* (pp. 667–682). The Hague: Junk.

Albrecht, G. H., Jenkins, P. D., and Godfrey, L. R. (1990). Ecogeographic size variation among the living and subfossil prosimians of Madagascar. *American Journal of Primatology* 22:1.

Allman, J. (1982). Reconstructing the evolution of the brain in primates through the use of comparative neurophysiological and neuroanatomical data. In E. Armstrong and D. Falk (eds.), *Primate Brain Evolution: Methods and Concepts* (pp. 13–28). New York: Plenum.

Allman, J., and McGuinness, E. (1988). Visual cortex in primates. In H. Steklis and J. Erwin (eds.), *Comparative Primate Biology, Vol. 4. Neurosciences* (pp. 279–326). New York: A. R. Liss.

Alonso, C. (1984). Observações de campo sobre cuidado a prole e o desenvolvimentos dos filhotes de *Callithrix jacchus jacchus*. In M.T. de Mello (ed.), *A Primatologia no Brasil-1* (pp. 67–78). Brasilia: Sociedade Brasileira de Primatologia.

Alonso, C., and Langguth, A. (1989). Ecologia e comportamento de *Callithrix jacchus* (Primates: Callitrichidae) numa ilha de floresta Atlântica. *Revista Nordestina Biologia* 6:107–137.

Alonso, C., De Faria, D. S., Langguth, A., Santee, D. F. (1987). Variação do pelagem na area de intergradação entre *Callithrix jacchus* e *Callithrix penicillata*. *Revista Brasileira Biologia* 47(4):465–470.

Altmann, J. (1974). Observational study of behavior: Sampling methods. *Behavior* 49:227–267.

Altmann, J., Warneke, M., and Ramer, J. (1988). Twinning among *Callimico goeldi*. *International Journal of Primatology* 9(2):165–168.

Alvarez, E., Balbas, L., Massa, I., and Pacheco, J. (1986). Ecological aspects of the Guri Reservoir. *Interciencia* 11(6):325–333.

Alvarez-Cordero, E. (1987). Notas preliminares sobre el comportamiento del Mono Capuchino del Orinoco (*Chiropotes satanas chiropotes*) en la region del Embalse Guri, Estado Bolvar, Venezuela. *Proceedings of the 37th Annual Convention, Venezuelan Association for the Advancement of Science*, Maracaibo, Venezuela, p. 268.

Alves, M. C. (1986). Novas localizações do mono carvoeiro, *Brachyteles arachnoides* (Cebidae Primates) e situação atual do parque nacional do Caparaõ. In M Thiago de Mello (ed.), *A Primatologia no Brasil*, Vol. 2 (pp. 367–368). Brasilia: Instituto de Ciências Biológicas.

Ameghino, F. (1891). Nuevos restos de mamíferos fósiles descubiertos por Carlos Ameghino en el Eoceno inferior de la patagonia austral. *Revista Argentina de Historia Natural* 1:289–328.

Anapol, F., and Lee, S. (1994). Morphological adaptation to diet in platyrrhine primates. *American Journal of Physical Anthropology* 94:239–261.

Anderson, C. M. (1981). Subtrooping in a Chacma baboon (*Papio ursinus*) population. *Primates* 22:445–458.

Anderson, C. M. (1986). Predation and primate evolution. *Primates* 27(1):15–39.

Anderson, C. M. (1987). Female transfer in baboons. *American Journal of Physical Anthropology* 73:241–250.

Anderson, C. M. (1991). The hunting hypothesis revisited. Paper presented at the American Anthropological Association Annual Meeting, Chicago, IL.

Anderson, C. M. (1992). Male investment under changing conditions among chacma baboons at Suikerbosrand. *American Journal of Physical Anthropology* 87(4):479–496.

Anderson, J. R. (1990). Use of objects as hammers to open nuts by capuchin monkeys (*Cebus apella*). *Folia Primatologica* 54:138–145.

Anderson, J. R., and Roeder, J. J. (1989). Responses of capuchin monkeys (*Cebus apella*) to different conditions of mirror-image stimulation. *Primates* 30(4):581–587.

Andrews, M. W. (1986). Contrasting approaches to spatially distributed resources by *Saimiri* and *Callicebus*. In J. G. Else and P. C. Lee (eds.), *Primate Ontogeny, Cognition and Social Behaviour* (pp. 79–86). New York: Cambridge University Press.

Andrews, M. W. (1988a). Selection of food sites by *Callicebus moloch* and *Saimiri sciureus* under spatially and temporally varying food distribution. *Learning and Motivation* 19(3):254–268.

Andrews, M. W. (1988b). Spatial strategies of oriented travel in *Callicebus moloch* and *Saimiri sciureus*. *Animal Learning and Behavior* 16(4):429–435.

Andrews, M. W., and Rosenblum, L. A. (1988). Relationship between foraging and affiliative social referencing in primates. In J. E. Fa and C. H. Southwick (eds.), *Ecology and Behavior of Food-Enhanced Primate Groups* (pp. 247–268). New York: Alan R. Liss.

Anon. (1991). Stunning new primate species found in Brazil. *National Geographic* 180(4):152.

Antinucci, F., and Visalberghi, E. (1986). Tool use in *Cebus apella*: A case study. *International Journal of Primatology* 7(4):351–363.

Anzenberger, G. (1986). Social conflict in two monogamous New World primates: Pairs and rivals. *Primate Report* 14:143–144.

Anzenberger, G. (1988). The pairbond in the titi monkey (*Callicebus moloch*): Intrinsic versus extrinsic contributions of the pairmates. *Folia Primatologica* 50(3–4):188–203.

Anzenberger, G., Mendoza, S. P., and Mason, W. A. (1986). Comparative studies of social behavior in *Callicebus* and *Saimiri*: Behavioral and physiological responses of established pairs to unfamiliar pairs. *American Journal of Primatology* 11(1):37–51.

Aquilina, G. D. (1994). *North American Regional Cotton-Top Tamarin Studbook.* Buffalo: Buffalo Zoological Gardens.

Aquino, R. (1978). La fauna primatoligica en area de Janaro Herrera. *Proj. de Asent. Rural Int. Jararo Herrera, Bol. Tech.* 1:1–20.

Aquino, R., and Encarnacion, F. (1986a). Population structure of *Aotus nancymai* (Cebidae: Primates) in Peruvian Amazon lowland forest. *American Journal of Primatology* 11(1):1–7.

Aquino, R., and Encarnacion, F. (1986b). Characteristics and use of sleeping sites in *Aotus* (Cebidae: Primates) in the Amazon lowlands of Peru. *American Journal of Primatology* 11(4):319–331.

Aquino, R., and Encarnacion, F. (1988). Population densities and geographic distribution of night monkeys (*Aotus nancymai* and *Aotus vociferans*) (Cebidae: Primates) in northeastern Peru. *American Journal of Primatology* 14(4):375–381.

Aquino, R., Puertas, P., and Encarnacion, F. (1990). Supplemental notes on population parameters of northeastern Peruvian night monkeys, genus *Aotus* (Cebidae). *American Journal of Primatology* 21(3):215–221.

Ardito, G. (1975). *Checklist of the Data on the Gestation Length of Primates.* Torino: Instituto di Antropologia.

Ardito, G. (1979). Primate chromosome atlas. In B. Chiarelli, A. L. Koen, and G. Ardito (eds.), *Comparative Karyology of Primates* (pp. 213–281). The Hague: Mouton.

Armstrong, E. (1983). Relative brain size and metabolism in mammals. *Science* 220:1302–1304.

Armstrong, E. (1985a). Relative brain size in monkeys and prosimians. *American Journal of Physical Anthropology* 66:263–273.

Armstrong, E. (1985b). Allometric considerations of the adult mammalian brain, with special emphasis on primates. In W. L. Jungers (ed.), *Size and Scaling in Primate Biology* (pp. 115–146). New York: Plenum.

Armstrong, E. (1986). Enlarged limbic structures in the human brain: The anterior thalamus and medial mamillary body. *Brain Research* 362:394–397.

Armstrong, E. (1989). A comparative review of the primate motor system. *Journal of Motor Behavior* 21:493–517.

Armstrong, E. (1990). Evolution of the brain. In A. Paxinos (ed.), *The Human Nervous System* (pp. 1–23). New York: Academic Press.

Armstrong, E. (1995). Expansion and stasis in human brain evolution: Analyses of the limbic system, cortex and brain shape. 65th James Arthur Lecture on the Evolution of the Human Brain, American Museum of Natural History, in press.

Armstrong, E., and Bergeron, R. (1985). Relative brain size in birds. *Brain, Behavior and Evolution* 26:141–153.

Armstrong, E., and Frost, G. T. (1987). Evolution of the primate subiculum. *Society for Neurosciences Abstracts* 13:527.

Armstrong, E., Zilles, K., Schlaug, G., and Schleicher, A. (1986). Comparative aspects of the primate posterior cingulate cortex. *Journal of Comparative Neurology* 253:539–548.

Armstrong, E., Curtis, M., Buxhoeveden, D. P., Fregoe, C., Zilles, K., Casanova, M., and McCarthy, W. F. (1991). Cortical gyrification in the rhesus monkey: A test of the mechanical folding hypothesis. *Cerebral Cortex* 1:426–432.

Armstrong, E., Schleicher, A., and Zilles, K. (1993). Cortical folding and the evolution of the human brain. *Journal of Human Evolution* 25:387–392.

Armstrong, E., Schleicher, A., Omran, H., Curtis, M., and Zilles, K. (1995). The ontogeny of human gyrification. *Cerebral Cortex* 5:56–63.

Asdell, S. A. (1964). *Patterns of Mammalian Reproduction*, 2nd ed. Ithaca: Cornell University Press.

Austad, S. N., and Fischer, K. E. (1992). Primate longevity: Its place in the mammalian scheme. *American Journal of Primatology* 28:251–261.

Ayres, J. M. (1978). Situacão atual da area de ocorrência do cuxiu (*Chiropotes satanas satanas* Hoffmansegg, 1807). Unpublished report to the IUCN/SSC Primate Specialist Group.

Ayres, J. M. (1981). Observações sobre a ecologia e o comportamento dos cuxius (*Chiropotes albinasus* e *Chiropotes satanas*. Cebidae, Primates). Instituto Nacional de Pesquisas da Amazonia (INPA), Manaus, Brazil.

Ayres, J. M. (1985). On a new species of squirrel monkey, genus *Saimiri*, from Brazilian Amazonia (Primates, Cebidae). *Papeis Avulsos Zoologica* 36(14):147–164.

Ayres, J. M. (1986a). The conservation status of the white uakari. *Primate Conservation* 7:22–26.

Ayres, J. M. (1986b). Uakaris and Amazonian Flooded Forest. Unpublished Ph.D. dissertation, University of Cambridge.

Ayres, J. M. (1989). Comparative feeding ecology of the uakari and bearded saki, *Cacajao* and *Chiropotes*. *Journal of Human Evolution* 18(7):697–716.

Ayres, J. M. (1990a). Ecological factors influencing rarity in Pithecines. *Abstracts of the XIIIth Congress of the International Primatological Society*, Nagoya, p. 164.

Ayres, J. M. (1990b). Scarlet faces of the Amazon. *Natural History* March: 32–40.

Ayres, J. M., and Johns, A. D.(1987). Conservation of white uacaries in Amazonian Varzea. *Oryx* 21(2):74–80.

Ayres, J. M., and Nessimian, J. L. (1982). Evidence for insectivory in *Chiropotes satanas*. *Primates* 23(3):458–459.

Baba, M. L., Goodman, M., and Dene, H. (1975). Origins of Ceboidea viewed from an immunological perspective. *Journal of Human Evolution* 4:89–102.

Bailey, R. C., and Aunger, R. (1989). Significance of the social relationships of Efe pygmy men in the Ituri forest, Zaire. *American Journal of Physical Anthropology* 78:495–508.

Bailey, R. C., Head, G., Jenike, M., Owen, B., Rechtman, R., and Zechenter, E. (1989). Hunting and gathering in tropical rain forest: Is it possible? *American Anthropologist* 91(1):59–82.

Baker, A. J. (1987). Emigration in wild groups of golden lion tamarins (*Leontopithecus rosalia*). *International Journal of Primatology* 8(5):500.

Baker, A. J. (1990). Adaptive significance of infant care behavior by nonreproductive members of wild golden lion tamarin (*Leontopithecus rosalia*) groups. *XIIIth Congress of the International Primatological Society Abstracts*. Nagoya: International Primatological Society.

Baker, A. J. (1991). Evolution of the social system of the golden lion tamarin (*Leontopithecus rosalia*): Mating system, group dynamics and cooperative breeding. Unpublished. Ph.D. dissertation, University of Maryland, College Park.

Baker, A. J., and Woods, F. (1992). Reproduction of the emperor tamarin (*Saguinus imperator*) in captivity, with comparisons to cotton-top and golden lion tamarins. *American Journal of Primatology* 26(1):1–10.

Baker, A. J., Dietz, J., and Kleiman, D. (1992). Behavioral evidence for monopolization of paternity in multi-male groups of golden lion tamarins. *Abstracts of the XIVth Congress of the International Primatological Society*. Strasbourg, France, pp. 122–123.

Baldwin, J. D., and Baldwin, J. I. (1972). The ecology and behavior of squirrel monkeys (*Saimiri oerstedi*) in a natural forest in western Panama. *Folia Primatologica* 18:161–184.

Baldwin, J. D., and Baldwin, J. I. (1973). The role of play in social organization: Comparative observations on squirrel monkeys (*Saimiri*). *Primates* 14:369–381.

Baldwin, J. D., and Baldwin, J. I. (1978). Exploration and play in howler monkeys (*Alouatta palliata*). *Primates* 19:411–422.

Baldwin, J. D., and Baldwin, J. I. (1981). The squirrel monkeys, genus *Saimiri*. In A. F. Coimbra-Filho and R. A. Mittermeier (eds.), *Ecology and Behavior of Neotropical Primates* (pp. 277–330). Rio de Janeiro: Academia Brasileira de Ciências.

Baldwin, L. A., Patterson, T. L., and Teleki, G. (1977). Field research on callitrichid and cebid monkeys: An historical, geographical, and bibliographical listing. *Primates* 18:485–507.

Balée, W. (1989a). The culture of Amazonian forests. *Advances in Economic Botany* 7:1–21.

Balée, W. (1989b). *Footprints of the Forest: Ka'apor Ethnobotany—the Historical Ecol-*

ogy of Plant Utilization by an Amazonian People. New York: Columbia University Press.

Ballou, J. D. (1991). *1990 International Studbook—Golden Lion Tamarin (Leontopithecus rosalia rosalia).* Washington, DC: National Zoological Park.

Barash, D. (1977). *Sociobiology and Behavior.* New York: Elsevier.

Barnett, A., and da Cunha, A. C. (1991). The golden-backed uacari on the upper Rio Negro, Brazil. *Oryx* 25:80–89.

Bartecki, U., and Heymann, E. W. (1987). Sightings of Red Uakaris, *Cacajao calvus rubicundus*, at the Rio Blanco, Peruvian Amazonia. *Primate Conservation* 8:34–36.

Bartecki, U., and Heymann, E. W. (1988). Observations of predatory behavior of a young saki monkey, *Pithecia hirsuta* (Platyrrhini: Cebidae). *Primate Report* 20:29–33.

Bartecki, U., and Heymann, E. W. (1990). Field observations on scent-marking behaviour in saddle-back tamarins, *Saguinus fuscicollis* (Callitrichidae, Primates). *Journal of the Zoological Society of London* 220:87–99.

Bartlett, T. Q., Sussman, R. W., and Cheverud, J. M. (1993). Infant killing in primates: A review of observed cases with specific reference to the sexual selection hypothesis. *American Anthropologist* 95(4):958–990.

Bauchop, T. (1978). Digestion of leaves in vertebrate arboreal folivores. In G. G. Montgomery (ed.), *The Ecology of Arboreal Folivores* (pp. 1193–1204). Washington, DC: Smithsonian Institution Press.

Bauers, K. A., and Snowdon, C. T. (1990). Discrimination of chirp vocalizations in the cotton-top tamarin. *American Journal of Primatology* 21:53–60.

Beard, K. C., Dagosto ,M., Gebo, D. L., and Godinot, M. (1988). Interrelationships among primate higher taxa. *Nature (London)* 331:712–714.

Beard, K. C., Qi, T., Dawson, M.R., Wang, B., and Li, C. (1994). A diverse new primate fauna from middle Eocene fissure-fillings in southeastern China. *Nature (London)* 368:604–609.

Beard, K. C., Wang J., Dawson, M. R., Huang, X., and Tong, Y. (1996). Earliest complete dentition of an anthropoid primate from the late middle Eocene of Shanxi Province, China. *Science* (in press).

Bearder, S. K. (1987). Lorises, bushbabies, and tarsiers: Diverse societies in solitary foragers. In B. B. Smuts, D. L. Cheney, R. M. Seyfarth, R. W. Wrangham, and T. T. Struhsaker (eds.), *Primate Societies* (pp. 11–24). Chicago: University Chicago Press.

Bearder, S. K., and Martin, R. D. (1979). The social organization of a nocturnal primate revealed by radio tracking. In C. J. Amlaner, Jr. and D. W. Macdonald (eds.), *A Handbook on Biotelemetry and Radio Tracking* (pp. 633–646). Oxford: Pergamon Press.

Beck, B. B. (1980). *Animal Tool Behavior.* New York: Garland.

Beck, B. B., Kleiman, D. G., Dietz J. M., Castro, I., Carvalho, C., Martins, A., and Rettberg-Beck, B. (1991). Losses and reproduction in reintroduced golden lion tamarins *Leontopithecus rosalia. Dodo* 27:50–61.

Beckerman, S. (1979). The abundance of protein in Amazonia: A reply to Gross. *American Anthropologist* 81(3):533–560.

Beckerman, S., and Sussenbach, T. (1983). A quantitative assessment of the dietary contribution of game species to the subsistence of South American tropical forest tribal peoples. In J. Clutton-Brock and C. Grigson (eds.), *Animals and Archaeology. I. Hunters and Their Prey* (pp. 337–350). Oxford: British Archaeological Reports, International Series 163.

Benefit, B. R., and McCrossin, M. L. (1991). Ancestral facial morphology of Old World higher primates. *Proceedings of the National Academy of Science U.S.A.* 88:5267–5271.

Benz, J. J. (1990). Food-elicited vocalizations in golden lion tamarins (*Leontopithecus rosalia*): Design features for representational communication. Unpublished Ph.D. dissertation, The University of Nebraska, Lincoln.

Benz, J. J., Leger, D. W., and French, J. A. (1992). Relation between food preference and food-elicited vocalizations in golden lion tamarins (*Leontopithecus rosalia*). *Journal of Comparative Psychology* 106(2):142–149.

Bernardes, A. T., Rylands, A. B., Valle, C. M. C., Machado, R. B., Coimbra-Filho, A. F., and Fisher, L. R. B. (1988). *Primate Field Studies in Brazil: A Bibliography*. Belo Horizonte, Minas Gerais, Brazil: Sociedade Brasileira de Primatologia.

Bernstein, I. S., Balcaen, P., Dresdale, L., Gouzoules, H., Kavanagh, M., Patterson, T., and Neyman-Warner, P. (1976). Differential effects of forest degradation on primate populations. *Primates* 17(3):401–411.

Biben, M. (1986). Individual- and sex-related strategies of wrestling play in captive squirrel monkeys. *Ethology* 71(3):229–241.

Biben, M. (1989). Effects of social environment on play in squirrel monkeys: Resolving Harlequin's dilemma. *Ethology* 81(1):72–82.

Biben, M. (1993). Recognition of order effects in squirrel monkey antiphonal call sequences. *American Journal of Primatology* 29(2):109–124.

Biben, M., and Symmes, D. (1986). Play vocalizations of squirrel monkeys (*Saimiri sciureus*). *Folia Primatology* 46:173–182.

Biben, M., Symmes, D., and Masataka, N. (1986). Temporal and structural analysis of affiliative vocal exchanges in squirrel monkeys (*Saimiri sciureus*). *Behaviour* 98(1–4):259–273.

Biben, M., Symmes, D., and Bernhards, D. (1989). Contour variables in vocal communication between squirrel monkey mothers and infants. *Developmental Psychobiology* 22(6):617–631.

Bicca-Marques, J. C. (1990). A new southern limit for the distribution of *Alouatta caraya* in Rio Grande do Sul State, Brazil. *Primates* 31(2):449–451.

Bicca-Marques, J. C., and Calegaro-Marques, C. (1993). Feeding postures in the black howler monkey, *Alouatta caraya*. *Folia Primatologica* 60:169–172.

Bicchieri, M. G. (1988). *Hunters and Gatherers Today*. Prospect Heights, IL: Waveland.

Biedzicki de Marques, A. A. (1993). Field study of wild brown howling monkeys, *Alouatta fusca*. *Neotropical Primates* 1(4):22.

Bilgener, M. (1988). Chemical components of howler monkey (*Alouatta palliata*) food choice and kinetics of tannin binding with natural polymers. Unpublished Ph.D. dissertation, Boston University.

Blakeslee, B., and Jacobs, G. H. (1982). Color vision in the spider monkey (*Ateles*). *Folia Primatologica* 38:86–98.

Bluntschili, H. (1931). *Homunculus patagonicus* und die ihm zugereihten Fossil funde aus den Santa-cruz-Schichten Patagoniens. *Morphol. Jahr.* 67:811–892.

Bodini, R., and Perez-Hernandez, R. (1987). Distribution of the species and subspecies of cebids in Venezuela. *Fieldiana:* Zoology 39:231–244.

Boinski, S. (1986). *The Ecology of Squirrel Monkeys in Costa Rica.* Unpublished Ph.D. dissertation, University of Texas, Austin.

Boinski, S. (1987a). Mating patterns in squirrel monkeys (*Saimiri oerstedii*). Implications for seasonal sexual dimorphism. *Behavioral Ecology and Sociobiology* 21:13–21.

Boinski, S. (1987b). The status of *Saimiri oerstedii citrinellus* in Costa Rica. *Primate Conservation* 8:67–72.

Boinski, S. (1987c). Habitat use by squirrel monkeys (*Saimiri oerstedii*) in Costa Rica. *Folia Primatologica* 49(3):151–167.

Boinski, S. (1987d). Birth synchrony in squirrel monkeys (*Saimiri oerstedii*) in Costa Rica. *Behavioral Ecology and Sociobiology* 21(6):393–400.

Boinski, S. (1988). Sex differences in the foraging behavior of squirrel monkeys in a seasonal habitat. *Behavioral Ecology and Sociobiology* 23(3):177–186.

Boinski, S. (1989a). The positional behavior and substrate use of squirrel monkeys: Ecological implications. *Journal of Human Evolution* 18(7):659–677.

Boinski, S. (1989b). Why don't *Saimiri oerstedii* and *Cebus capucinus* form mixed-species groups? *International Journal of Primatology* 10(2):103–114.

Boinski, S. (1991). The coordination of spatial position: A field study of the vocal behaviour of adult female squirrel monkeys. *Animal Behaviour* 41(1):89–102.

Boinski, S. (1993). Vocal coordination of troop movement among white-faced capuchin monkeys, *Cebus capucinus*. *American Journal of Primatology* 30(2):85–100.

Boinski, S. (1994). The Costa Rican squirrel monkey: Waltzing to extinction. *American Journal of Primatology* 33(3):196–197.

Boinski, S. (1995). *Saimiri. Illustrated Monographs of Living Primates.* The Netherlands: Istitut voor Ontwikkelingsopdrachten. In press.

Boinski, S., and Fragaszy, D. M. (1989). The ontogeny of foraging in squirrel monkeys, *Saimiri oerstedii*. *Animal Behaviour* 37:415–428.

Boinski, S., and Mitchell, C. L. (1992). Ecological and social factors affecting vocal behavior of adult female squirrel monkeys. *Ethology* 92:316–330.

Boinski, S., and Newman, J. D. (1988). Preliminary observations on squirrel monkey (*Saimiri oerstedii*) vocalizations in Costa Rica. *American Journal of Primatology* 14(4):329–343.

Boinski, S., and Scott, P. E. (1988). Association of birds with monkeys in Costa Rica. *Biotropica* 20(2):136–143.

Boinski, S., and Timm, R. B. (1985). Predation by squirrel monkeys and double-toothed kites on tent-making bats. *American Journal of Primatology* 9(2):121–127.

Bolin, I. (1981). Male parental behavior in black howler monkeys (*Alouatta palliata pigra*) in Belize, Guatemala. *Primates* 22:349–360.

Bonvicino, C. R. (1989). Ecology and behavior of Alouatta belzebul (Primates: Cebidae) living near the Atlantic. *Revista Nordestina de Biologia* 6(2):149–179.

Bonvicino, C. R., and Lima, M. G. (1982). Notas preliminares sobre a ecologia e comportamento do parauacu (*Pithecia pithecia*, Linnaeus, 1766), Cebidae, Primata. Unpublished Bachelors Thesis, Instituto de Biocincias, Universidade Estadual de São Paulo, Rio Claro (SP).

Bonvicino, C. R., Langguth, A., and Mittermeier, R. A. (1989). A study of pelage color and geographic distribution in *Alouatta belzebul* (Primates: Cebidae). *Revista Nordestina Biologia* 6(2):139–148.

Boucher, D. H. (1981). The "real" disperser of *Swartzia cubensis*. *Biotropica* 13(Suppl. 2):77–78.

Bourliere, F. (1973). The comparative ecology of rain forest mammals in Africa and tropical America: Some introductory remarks. In B. J. Meggers, E. S. Ayensu, and W. D. Duckworth (eds.), *Tropical Forest Ecosystems in Africa and South America* (pp. 279–292). Washington, DC: Smithsonian Institution Press.

Bourliere, F. (1985). Primate communities: Their structure and role in tropical ecosystems. *International Journal of Primatology* 6(1):1–26.

Bourliere, F., Petter-Rousseaux, A., and Petter, J. J. (1961). Regular breeding in captivity of the lesser mouse lemur (*Microcebus murinus*). *International Zoo Yearbook* 3:24–25.

Bowmaker, J. K., and Mollon, J. D. (1980). Primate microspectrophotometry and its implications for colour deficiency. In G. Verriest (ed.), *Colour Deficiencies V* (pp. 61–64). Bristol: Hilger.

Bowmaker, J. K., Mollon, J. D., and Jacobs, G. H. (1983). Microspectrophotometric results for Old and New World primates. In J. D. Mollon and L. T. Sharpe (eds.), *Colour Vision Physiology and Psychophysics* (pp. 57–68). London: Academic Press.

Bowmaker, J. K., Jacobs, G. H., Spiegelhalter, D. J., and Mollon, J. D. (1985). Two types of trichromatic squirrel monkey share a pigment in the red-green spectral region. *Vision Research* 25:1937–1946.

Bowmaker, J. K., Jacobs, G. H., and Mollon, J. D. (1987). Polymorphism of photopigments in the squirrel monkey: A sixth phenotype. *Proceedings of the Royal Society of London Series B.* 231:383–390.

Bowmaker, J. K., Astell, S., Hunt, D. M., and Mollon, J. D. (1991). Photosensitive and photolabile pigments in the retinae of Old World monkeys. *Journal of Experimental Biology* 156:1–19.

Bown, T. M., and Larriestra, C. N. (1990). Sedimentary paleoenvironments of fossil platyrrhine localities, Miocene Pinturas Formation, Santa Cruz Province, Argentina. *Journal of Human Evolution* 19:87–119.

Box, H. O. (1978). Social interactions in family groups of captive marmosets (*Callithrix jacchus*). In D. G. Kleiman (ed.), *The Biology and Conservation of the Callitrichidae* (pp. 239–249. Washington, DC: Smithsonian Institution Press.

Box, H. O. (1984). Individual and intergroup differences in social behaviour among captive marmosets (*Callithrix jacchus*) and tamarins (*Saguinus mystax*). *Society of Biology and Human Affairs* 47(2):49–68.

Box, H. O. (1988). Behavioural responses to environmental change. Observations on captive marmosets and tamarins (Callitrichidae). *Animal Technology* 39(1):9–16.

Boynton, R. M. (1979). *Human Color Vision*. New York: Holt, Rinehart & Winston.

Bramblett, C. A. (1976). *Patterns of Primate Behavior*. Prospect Heights, IL: Waveland.

Bramblett, C. A., Bramblett, S. S., Coelho, A. M., and Quick, L. B. (1980). Party composition in spider monkeys of Tikal, Guatemala. A comparison of stationary vs. moving observer. *Primates* 21:123–127.

Braza, F. (1978). El Araguato Rojo (*Alouatta seniculus*). Unpublished doctoral thesis, University of Seville.

Braza, F., Alvarez, F., and Azcarate, T. (1981). Behavior of the red howler monkey (*Alouatta seniculus*) in the llanos of Venezuela. *Primates* 22:459–473.

Braza, F., Alvarez, F., and Azcarate, T. (1983). Feeding habits of the red howler monkey (*Alouatta seniculus*) in the llanos of Venezuela. *Mammalia* 47:205–214.

Brizzee, K. R., and Dunlap, W. P. (1986). Growth. In W. R. Dukelow and J. Erwin (eds.), *Comparative Primate Biology, Vol. 3: Reproduction and Development* (pp. 363–413). New York: Alan R. Liss.

Brokensha, D., Warren, D. M., and Werner, O. (1980). *Indigenous Knowledge Systems and Development*. Washington, DC: University Press of America.

Brown, A. D., and Rumiz, D. I. (1986). Distribucion y conservacion de los primates en Bolivia—Estado actual de su conocimiento.

Brown, A. D., and Zunino, G. E. (1990). Dietary variability in *Cebus apella* in extreme habitats: Evidence for adaptability. *Folia Primatologica* 54:187–195.

Brown, A. D., Chalukian, S., and Malmierca, L. (1984). Habitat y alimentacion de *Cebus apella* en el n.o. argentino y la disponibilidad de frutos en el dosel arboreo. *Revista del Museo Aegentino de Ciencias Naturales "Bernadino Rivadavia" e Instituto Nacional de Investigacion de las Ciencias Naturales (Zool.)* 13(28):273–280.

Brown, A. D., Chalukian, S., Malmierca, L., and Calillas, (1986). Habitat structure and feeding behavior of *Cebus apella* (Cebidae) in El Rey National Park, Argentina. In D. M. Taub and F. A. King (eds.), *Current Perspectives in Primate Social Dynamics* (pp. 137–151). New York: Van Nostrand Reinhold.

Brown, C. H. (1982). Auditory localization and primate vocal behavior. In C. T. Snowdon, C. H. Brown, and M. R. Petersen (eds.), *Primate Communication* (pp. 144–170). New York: Cambridge University Press.

Brown, K., and Mack, D. S. (1978). Food sharing among captive *Leontopithecus rosalia*. *Folia Primatologica* 29:268–290.

Brown, K. S., Jr. (1975). Geographical patterns of evolution in neotropical Lepidoptera. Systematics and derivation of known and new Hiliconiini (Nymphalidae: Nymphalinae). *Journal of Entomology (B)* 44(3):201–242.

Brown, L. M., and Kinzey, W. G. (1995). Why are there no terrestrial New World primates? In preparation.

Buchanan, D. B. (1978). Communication and ecology of Pithecine monkeys with special reference to *Pithecia pithecia*. Unpublished Ph.D. dissertation, Wayne State University.

Buchanan, D. B., Mittermeier, R. A., and van Roosmalen, M. G. M. (1981). The saki monkeys, genus *Pithecia*. In A. F. Coimbra-Filho and R. A. Mittermeier (eds.), *Ecology and Behavior of Neotropical Primates*, Vol. 1 (pp. 391–417). Rio de Janeiro: Academia Brasileira de Ciências.

Buchanan-Smith, H. M. (1988). A field study on the red-bellied tamarin and the saddle-back tamarin in Bolivia. *Primate Eye* 35:12.

Buchanan-Smith, H. M. (1990a). Polyspecific association of two tamarin species, *Saguinus labiatus* and *Saguinus fuscicollis*, in Bolivia. *American Journal of Primatology* 22(3):205–214.

Buchanan-Smith, H. M. (1990b). The social organization and mating-systems of the red bellied tamarin (*Saguinus labiatus labiatus*): Behavioural observations in captivity and in the wild. *Dissertation Abstracts International* B50(9):4275, 1990.

Buchanan-Smith, H. M. (1991a). A field study on the red-bellied tamarin, *Saguinus l. labiatus*, in Bolivia. *International Journal of Primatology* 12(3):259–276.

Buchanan-Smith, H. M. (1991b). Encounters between neighboring mixed species groups of tamarins in northern Bolivia. *Primate Report* 31:95–99.

Buchanan-Smith, H. M. (1991c). Field observations of Goeldi's monkey, *Callimico goeldii*, in northern Bolivia. *Folia Primatologica* 57:102–105.

Buchanan-Smith, H. M., and Jordon, T. R. (1992). An experimental investigation of the pair bond in the Callitrichid monkey, *Saguinus labiatus*. *International Journal of Primatology* 13(1):51–72.

Buckley, J. S. (1983). The feeding behavior, social behavior and ecology of the white-faced monkey, *Cebus capucinus*, at Trujillo, Northern Honduras, Central America. Unpublished Ph.D. dissertation, University of Texas, Austin.

Bugbee, N. M., and Goldman-Rakic, P. S. (1983). Columnar organization of corticocortical projections in squirrel and rhesus monkeys: Similarity of column width in species differing in cortical volume. *Journal of Comparative Neurology* 220:355–364.

Bunker, S. G. (1984). Modes of extraction, unequal exchange and the progressive underdevelopment of an extreme periphery: The Brazilian Amazon, 1600-1980. *American Journal of Sociology* 89(5):1017–1064.

Buss, D. H. (1971). Mammary glands and lactation. In E. S. E. Hafez (ed.), *Comparative Reproduction of Non-Human Primates* (pp. 315–333). Springfield, IL: Charles C Thomas.

Buss, D. H., and Cooper, R. W. (1972). Composition of squirrel monkey milk. *Folia Primatologica* 17:285–291.

Busse, C., and Hamilton, J. W., III (1981). Infant carrying by male chacma baboons. *Science* 212:1281–1283.

Bygott, J. D. (1979). Agonistic behavior, dominance, and social structure in wild chimpanzees of the Gombe National Park. In D. A. Hamburg and E. R. McCown (eds.), *The Great Apes* (pp. 405–427). Reading, MA: Benjamin-Cummings.

Cabrera, A. (1958). Catálogo de los mamíferos de America del Sur. *Revista del Museo Argentino Ciencias Naturales "Bernadino Rivadavia"* 4(1):1–307.

Caine, N., and O'Boyle, V. (1989). Forms of play as a function of substrate use in captive tamarins. *American Journal of Primatology* 18(2):138–139.

Caine, N., and Stevens, C. (1990). Evidence for a "monitoring call" in red-bellied tamarins. *American Journal of Primatology* 22(4):251–262.

Calegaro-Marques, C., and Bicca-Marques, J. C. (1994). Ecology and social relations of the black-chinned emperor tamarin. *Neotropical Primates* 2(2):20–21.

Cameron, E. (1988). Parental behaviour in captive primiparous golden lion tamarins (*Leontopithecus rosalia rosalia*). *Australian Primatology* 3(1):37.

Cameron, R., Wilshire, C., Foley, C., Dougherty, N., Aramayo, X., and Rea, L. (1989). Goeldi's monkey and other primates in northern Bolivia. *Primate Conservation* 10:62–70.

Campbell, A. F. (1994). Patterns of home range use by *Ateles geoffroyi* and *Cebus capucinus* at La Selva Biological Station, northeast Costa Rica. *American Journal of Primatology* 33(3):199–200.

Candland, D. K., French, J. A., and Johnson, C. N. (1978). Object-play: Test of a categorized model by the genesis of object-play in *Macaca fuscata*. In E. O. Smith (ed.), *Social Play in Primates* (pp. 259–296). New York: Academic Press.

Cant, J. G. H. (1977). Ecology, locomotion and social organization of spider monkeys (*Ateles geoffroyi*) Unpublished Ph.D. thesis, University of California, Davis.

Cant, J. G. H. (1978). Population survey of the spider monkey, *Ateles geoffroyi* at Tikal, Guatemala. *Primates* 19:525–535.

Cant, J. G. H. (1980). What Limits Primates? *Primates* 21(4):538–544.

Cant, J. G. H. (1986). Locomotion and feeding postures of spider and howling monkeys: Field study and evolutionary interpretation. *Folia Primatologica* 46:1–14.

Cant, J. G. H. (1990). Feeding ecology of spider monkeys (*Ateles geoffroyi*) at Tikal, Guatemala. *Human Evolution* 5(3):269–281.

Carlson, M., Huerta, M. F., Cusick, C. G., and Kaas, J. H. (1986). Studies on the evolution of multiple somatosensory representations in primates: The organization of anterior parietal cortex in the New World Callithrichid, *Saguinus*. *Journal of Comparative Neurology* 246:409–426.

Carpenter, C. R. (1934). A field study of the behavior and social relations of howling monkeys. *Comparative Psychology Monographs* 10(2):1–168.

Carpenter, C. R. (1965). The howlers of Barro Colorado Island. In I. DeVore (ed.), *Primate Behavior: Field Studies of Monkeys and Apes* (pp. 250–291). New York: Holt, Rinehart & Winston.

Carpenter, M. B. (1991). *Core Text of Neuroanatomy*. Baltimore: Williams & Wilkins.

Carroll, J. B. (1985). Pair bonding in Goeldi's monkey *Callimico goeldii* (Thomas 1904). *Dodo, Journal of the Jersey Wildlife Preservation Trust* 22:57–71.

Carroll, J. B. (1986). Social correlates of reproductive suppression in captive callitrichid family groups. *Dodo, the Journal of the Jersey Wildlife Preservation Trust* 23:80–85.

Carroll, J. B. (1987a). Caregiver behavior and the infant rearing strategy of captive *Callimico goeldii*. *International Journal of Primatology* 8(5):521.

Carroll, J. B. (1987b). A behavioural study of captive *Callimico goeldii* housed in polygynous social groups. *International Journal of Primatology* 8(5):522.

Carroll, J. B. (1988). The stability of multifemale groups of Goeldi's monkey *Callimico goeldii* in captivity. *Dodo, the Journal of the Jersey Wildlife Preservation Trust* 25:37–43.

Carroll, J. B. (1991). An investigation of multifemale breeding groups among captive Goeldi's monkeys, *Callimico goeldii*, at Jersey Wildlife Preservation Trust. In A. B. Rylands and A. T. Bernardes (eds.), *A Primatologia no Brasil*-3 (pp. 207–208). Brasilia: Sociedade Brasileira de Primatologia.

Carroll, J. B., Abbott, D. H., George, L. M., and Martin, R. D. (1989). Aspects of urinary oestrogen excretion during the ovarian cycle and pregnancy in Goeldi's monkey (*Callimico goeldii*). *Folia Primatologica* 52(3–4):201–205.

Carroll, J. B., Abbott, D. H., George, L. M., Hindle, J. E., and Martin, R. D. (1990). Urinary endocrine monitoring of the ovarian cycle and pregnancy in Goeldi's monkey (*Callimico goeldii*). *Journal of Reproductive Fertility* 89(1):149–161.

Cartelle, C. (1992). Achado de *Brachyteles* do Pleistoceno final. *Neotropical Primates* 1,8.

Cartmill, M. (1972). Arboreal adaptations and the origin of the order Primates. In R. Tuttle (ed.), *The Function and Evolutionary Biology of Primates* (pp. 97–122). Chicago: Aldine-Atherton.

Cartmill, M. (1974). *Daubentonia, Dactylopsila*, woodpeckers and klinorhynchy. In R. D. Martin, G. A. Doyle, and A. C. Walker (eds.), *Prosimian Biology* (pp. 655–670). London: Duckworth.

Cartmill, M. (1975). Strepsirhine basicranial structures and the affinities of the Cheirogalidae. In W. P. Luckett and F. S. Szalay (eds.), *Phylogeny of the Primates: A Multidisciplinary Approach* (pp. 313–356). New York: Plenum.

Cartmill, M. (1980). Morphology, function, and evolution of the anthropid post orbital septum. In R. Ciochon and A. B. Chiarelli (eds.), *Evolutionary Biology of New World Monkeys and Continental Drift* (pp. 243–274). New York: Plenum.

Cartmill, M., MacPhee, R. D. E., and Simons, E. L. (1981). Anatomy of the temporal bone in early anthropoids, with remarks on the problem of anthropoid origins. *American Journal of Physical Anthopology* 56:3–21.

Castro, R., and Soini, P. (1978). Field studies on *Saguinus mystax* and other callitrichids in Amazonian Peru. In D. G. Kleiman (ed.), *The Biology and Conservation of the Callitrichidae* (pp. 73–78). Washington DC: Smithsonian Institution Press.

Ceballos, C. (1989). Food change of a fourth month adopted infant in a wild group of *Alouatta seniculus*. *Field Studies of New World Monkeys, La Macarena, Colombia* 2:41–43.

Cebul, M. S., Alveario, M. C., and Epple, G. (1978). Odor recognition and attachment in infant marmosets. In H. Rothe, H. J. Wolters, and Hearn (eds.), *Biology and Behaviour of Marmosets* (pp. 141–146). Göttingen: Eigenverlag Rothe.

Chagnon, N. A. (1981). Terminological kinship, genealogical relatedness and village fissioning among the Yanomamo Indians. In R. D. Alexander and D. W. Tinkle (eds.), *Natural Selection and Social Behavior* (pp. 490–508). New York: Chiron.

Chagnon, N. A. (1988). Life histories, blood revenge, and warfare in a tribal population. *Science* 239:985–992.

Chagnon, N. A., and Bugos, P. (1979). Kin selection and conflict: An analysis of a Yanomamo ax fight. In N. A. Chagnon and W. Irons (eds.), *Evolutionary Biology and Human Social Behavior* (pp. 213–238). N. Scituate, MA: Duxbury.

Chamberlain, J., Nelson, G., and Milton, K. (1993). Fatty acid profiles of major food sources of howler monkeys (*Alouatta palliata*) in the neotropics. *Experientia* 49:820–824.

Chapman, C. A. (1986). Boa constrictor predation and group response in white-faced cebus monkeys. *Biotropica* 18(2):171–172.

Chapman, C. A. (1987). Flexibility in diets of three species of Costa Rican Primates. *Folia Primatologica* 49:90–105.

Chapman, C. A. (1988a). Patterns of foraging and range use by three species of neotropical primates. *Primates* 29(2):177–194.

Chapman, C. A. (1988b). Patch use and patch depletion by the spider and howling monkeys of Santa Rosa National Park, Costa Rica. *Behaviour* 105(1–2):99–116.

Chapman, C. A. (1989a). Primate seed dispersal: The fate of dispersed seeds. *Biotropica* 21(2):148–154.

Chapman, C. A. (1989b). Spider monkey sleeping sites: Use and availability. *American Journal of Primatology* 18:53–60.

Chapman, C. A. (1990). Ecological constraints on group size in three species of neotropical primates. *Folia Primatologica* 55:1–9.

Chapman, C. A., and Chapman, L. J. (1986). Behavioural development of howling monkey twins (*Alouatta palliata*) in Santa Rosa National Park, Costa Rica. *Primates* 27(3):377–381.

Chapman, C. A., and Chapman, L. J. (1987). Social responses to the traumatic injury of a juvenile spider monkey (*Ateles geoffroyi*). *Primates* 28(2):271–275.

Chapman, C. A., and Chapman, L. J. (1990). Dietary variability in primate populations. *Primates* 31(1):121–128.

Chapman, C. A., and Chapman, L. J. (1991). The foraging itinerary of spider monkeys: When to eat leaves? *Folia Primatologica* 56:162–166.

Chapman, C. A., and Fedigan, L. M. (1990). Dietary differences between neighboring *Cebus capucinus* groups: Local traditions, food availability or responses to food profitability? *Folia Primatologica* 54:177–186.

Chapman, C. A., and Lefebvre, L. (1990). Manipulating foraging group size: Spider monkey food calls at fruiting trees. *Animal Behaviour* 39:891–896.

Chapman, C. A., and Weary, D. (1990). Variability in spider monkeys' vocalizations may provide basis for individual recognition. *American Journal of Primatology* 22:279–284.

Chapman, C. A., Chapman, L. J., and McLaughlin, R. L. (1989a). Multiple central place foraging by spider monkeys: Travel consequences of using many sleeping sites. *Oecologia* 79:506–511.

Chapman, C. A., Chapman, L. J., and Richardson, K. S. (1989b). Sex ratio in primates: A test of the local resource competition hypothesis. *Oikos* 56:132–134.

Chapman, C. A., Fedigan, L. M., Fedigan, L., and Chapman, L. J. (1989c). Post-weaning resource competition and sex ratios in spider monkeys. *Oikos* 54:315–319.

Chapman, C. A., Chapman, L. J. C., and Lefebvre, L. (1990). Spider monkey alarm calls: Honest advertisement or warning kin? *Animal Behaviour* 39(1):197–198.

Charles-Dominique, P. (1977). *Ecology and Behaviour of Nocturnal Primates*. New York: Colombia University Press.

Charles-Dominique P., and Hladik, C. M. (1971). Le lepilemur du sud de Madagascar: ecologie, alimentation, et vie sociale. *Terre et la Vie* 25:3–66.

Charles-Dominique, P., Cooper, H. M., Hladik, C. M., Pages, E., Pariente, G. F., Petter-Rousseaux, A., Petter, J. J., and Schilling, A. (1980). *Nocturnal Malagasy Primates: Ecology, Physiology and Behavior*. New York: Academic Press.

Cheney, D. L., and Wrangham, R. W. (1987). Predation. In B. B. Smuts, D. L. Cheney, R. M. Seyfarth, R. W. Wrangham, and T. T. Struhsaker (eds.), *Primate Societies* (pp. 227–239). Chicago: University Chicago Press.

Cheney, D. L., Seyfarth, R. M., Andelman, S. J., and Lee, P. C. (1988). Reproductive success in vervet monkeys. In T. H. Clutton-Brock (ed.), *Reproductive Success* (pp. 384–402). Chicago: University of Chicago Press.

Chevalier-Skolnikoff, S. (1989). Spontaneous tool use and sensorimotor intelligence in *Cebus* compared with other monkeys and apes. *Behavior and Brain Sciences* 12:561–627.

Chevalier-Skolnikoff, S. (1990). Tool use by wild cebus monkeys at Santa Rosa National Park, Costa Rica. *Primates* 31(2):375–383.

Cheveraud, J. M., and Moore, A. J. (1990). Subspecific morphological variation in the saddle-back tamarin (*S. fuscicollis*). *American Journal of Primatology* 21(1):1–15.

Chiarello, A. G. (1993a). Activity pattern of the brown howler monkey *Alouatta fusca*, Geoffroy 1812, in a forest fragment of southeastern Brazil. *Primates* 34(3):289–293.

Chiarello, A. G. (1993b). Home range of the brown howler monkey, *Alouatta fusca*, in a forest fragment of Southeastern Brazil. *Folia Primatologica* 60:173–175.

Chiochon, R. L., and Chiarelli, A. B. (1980). *Evolutionary Biology of the New World Monkeys and Continental Drift*. New York: Plenum.

Chiochon, R. L., and Corruccini, R. S. (1975). Morphometric analysis of platyrrhine femora with taxonomic implications and notes on two fossil forms. *Journal of Human Evolution* 4:193–217.

Chivers, D. J., and Hladik, C. M. (1980). Morphology of the gastrointestinal tract in primates: Comparisons with other mammals in relation to diet. *Journal of Morphology* 166:337–386.

Chivers, D. J., and Langer, P. (1994). *The Digestive System in Mammals: Food, Form, and Function*. Cambridge: Cambridge University Press.

Chomsky, N. (1957). *Syntactic Structures*. The Hague: Mouton.

Christen, A. (1968). Haltung und Brutbiologie von *Cebuella*. *Folia Primatologica* 8:41–49.

Christen, A. (1974). Fortpflanzungsbiologie und Verhalten bei *Cebuella pygmaea* und Tamarin tamarin (Primates, Platyrrhini, Callithricidae). *Fortschritte der Verhaltensforschung. Zeitschrift für Tierpsychologie* Suppl. 14:1–79.

Christen, A., and Geissmann, T. (1994). A primate survey in northern Bolivia, with special reference to Goeldi's monkey, *Callimico goeldii*. *International Journal of Primatology* 15(2):239–274.

Christen, A. M., Doebeli, M., Kempken, B., Zachmann, M., and Martin, R. D. (1989). Urinary excretion of oestradiol-17beta in the female cycle of Goeldi's monkey (*Callimico goeldii*). *Folia Primatologica* 52(3–4):191–200.

Ciochan, R. L., and Chiarelli, A. B. (1980). *Evolutionary Biology of the New World Monkeys and Continental Drift*. New York: Plenum Press.

Clark, W. E. LeGros. (1949). *History of the Primates*. London: British Museum of Natural History.

Clark, W. E. LeGros. (1959). *The Antecedents of Man*. Edinburgh: Edinburgh University Press.

Clark, W. E. LeGros (1966). *History of the Primates*, 5th ed. Chicago: University of Chicago Press.

Clarke, B. C. (1979). The evolution of genetic diversity. *Proceedings of the Royal Society of London Series B* 205:453–474.

Clarke, M. R. (1981). Aspects of male behavior in the manted howlers (*Alouatta pallita* Gray) in Costa Rica. *American Journal of Primatology* 3:1–22.

Clarke, M. R. (1982). Socialization, infant mortality, and infant-nonmother interactions in howling monkeys (*Alouatta pallita* Gray) in Costa Rica. Unpublished Ph.D. thesis, University of California, Davis.

Clarke, M. R. (1983). Infant-killing and infant disappearance following male takeovers in a group of free-ranging howling monkeys (*Alouatta palliata*) in Costa Rica. *American Journal of Primatology* 5:241–247.

Clarke, M. R. (1985). Behavior of adult females toward group infants in howlers (*Alouatta palliata*). *American Journal of Primatology* 8:336.

Clarke, M. R. (1986). Interactions of adult male howling monkeys (*Alouatta palliata*) with immatures in a free-ranging social group. *American Joural of Physical Anthropology* 69(2):188.

Clarke, M. R. (1990). Behavioral development and socialization of infants in a free-ranging group of howling monkeys (*Alouatta palliata*). *Folia Primatologica* 54(1–2):1–15.

Clarke, M. R., and Glander, K. E. (1981). Adoption of infant howling monkeys (*Alouatta palliata*). *American Journal of Primatology* 1:469–472.

Clarke, M. R., and Glander, K. E. (1984). Female reproductive success in a group of free-ranging howling monkeys (*Alouatta palliata*) in Costa Rica. In M. F. Small (ed.), *Female Primates: Studies by Women Primatologists* (pp. 111–126). New York: Alan R. Liss.

Clarke, M. R., and Zucker, E. L. (1994). Survey of the howling monkey population at La Pacifica: A seven-year follow-up. *International Journal of Primatology* 15(1):61–74.

Clarke, M. R., Zucker, E. L., and Scott, N. J. (1986). Population trends of the mantled howler groups of La Pacifica, Guanacaste, Costa Rica. *American Journal of Primatology* 11(1):79–88.

Clarke, M. R., Zucker, E. L., and Harrison, R. M. (1991). Fecal estradiol, sexual swelling, and sociosexual behavior of free-ranging female howling monkeys in Costa Rica. *American Journal of Primatology* 24(2):93.

Clay, J. W. (1988). *Indigenous Peoples and Tropical Forests*. Cambridge: Cultural Survival, Inc.

Cleveland, J., and Snowdon, C. T. (1982). The complex vocal repertoire of the adult cotton-top tamarin (*Saguinus oedipus*). *Zeitschrift für Tierpsychologie* 58:231–270.

Cleveland, J., and Snowdon, C. T. (1984). Social development during the first twenty weeks in the cotton-top tamarin (*Saguinus o. oedipus*). *Animal Behaviour* 32:432–444.

Clutton-Brock, T. H., and Harvey, P. H. (1977). Primate ecology and social organization. *Journal of the Zoological Society, London* 183:1–39.

Coe, C. L., and Rosenblum, L. A. (1978). Annual reproductive strategy of the squirrel monkey (*Saimiri sciureus*). *Folia Primatologica* 29:19–42.

Coe, C. L., Smith, E. R., and Levine, S. (1985). The endocrine system of the squirrel monkey. In L. A. Rosenblum and C. L. Coe (eds.), *Handbook of Squirrel Monkey Research* (pp. 191–218). New York: Plenum.

Coe, C. L., Savage, A., and Bromley, L. J. (1992). Phylogenetic influences on hormone levels across the primate order. *American Journal of Primatology* 28(2):81–100.

Coelho, A. M., Jr., Bramblett, C., Quick, L., and Bramblett, S. (1976a). Resource availability and population density in primates: A socio-bioenergetic analysis of the energy budgets of Guatamalan howler and spider monkeys. *Primates* 17:63–80.

Coelho, A. M., Jr., Coehlo, L., Bramblett, C., Bramblett, S., and Quick, L. (1976b). Ecology, population characteristics, and sympatric association in primates: A socio-bioenergetic analysis of howler and spider monkeys in Tikal, Guatemala. *Yearbook of Physical Anthropology* 20:96–235.

Coelho, A. M., Jr., Bramblett, C., and Quick, L. (1977). Social organization and food resource availability in primates: A socio-bioenergetic analysis of diet and disease hypotheses. *American Journal of Physical Anthropology* 46:253–264.

Coimbra, C. E. A., Jr. (1995). Epidemiological factors and human adaptation in Amazonia. In L. E. Sponsel (ed.), *Indigenous Peoples and the Future of Amazonia: An Ecological Anthropology of an Endangered World* (pp. 167–181). Tucson: University of Arizona Press.

Coimbra-Filho, A. F. (1984). Situação atual dos Calitriquideos que ocorrem no Brasil. In M.T. de Mello (ed.), *A Primatologia no Brasil-1* (pp. 15–33). Brasilia: Sociedade Brasileira de Primatologia.

Coimbra-Filho, A. F., and Mittermeier, R. A. (1973). Distribution and ecology of the genus *Leontopithecus* Lesson, 1840. *Primates* 14:47–66.

Coimbra-Filho, A. F., and Mittermeier, R. A. (1977). Conservation of the Brazilian lion tamarins (*Leontopithecus rosalia*). In Prince Rainier and G. H. Bourne (eds.), *Primate Conservation* (pp. 59–94). New York: Academic Press.

Coimbra-Filho, A. F., and Mittermeier, R. A. (1981). *Ecology and Behavior of Neotropical Primates*, Vol. 1. Rio de Janeiro: Academia Brasileira de Ciências.

Coimbra-Filho, A. F., Rocha e Silva, R. da, and Pissinatti, A. (1981). Sobre a dieta de Callitrichidae em cativeiro. *Revista Biotérios* 1:83.

Coimbra-Filho, A. F., Rocha e Silva, R. da, and Alekstich, S. (1984). Gomas enriquecidas na alimentação de saguis em cativeiro. In M. T. de Mello (ed.), *A Primatologia no Brasil-1* (pp. 133–136). Brasilia: Sociedade Brasileira de Primatologia.

Coimbra-Filho, A. F., Rocha e Silva, R. da, and Pissinatti, A. (1991). Acerca da distribuição geográfica original de *Cebus apella xanthosternos* Wied, 1820 (Cebidae, Primates). In A. B. Rylands and A. T. Bernardes (eds.), *A Primatologia no Brasil-3* (pp. 215–224). Belo Horizonte: Fundação Biodiversitas.

Coimbra-Filho, A. F., Pissinatti, A., and Rylands, A. B. (1993). Experimental multiple hybridism and natural hybrids among Callithrix species from eastern Brazil. In A. B. Rylands (ed.), *Marmosets and Tamarins; Systematics, Behaviour, and Ecology* (pp. 95–120). Oxford: Oxford University Press.

Coimbra-Filho, A. F., Pissinatti, A., and Rylands, A. B. (1994). Muriquis at the Rio de Janeiro Primate Centre. *Neotropical Primates* 2(1):5–7.

Coley, P. D., Bryant, J. P., and Chapin, F. S. III (1985). Resource availability and plant antiherbivore defense. *Science* 230:895–899.

Colgrove, E. E., Sheperd, R. E., Lyons, D. M., and Mendoza, S. P. (1990). Functional significance of squirrel monkey genital displays. *American Journal of Primatology* 20(3):182.

Colillas, O., and Coppo, J. (1978). Breeding *Alouatta caraya* in Centro Argentino de Primates. In D. J. Chivers and W. L. Petter (eds.), *Recent Advances in Primatology, Vol. 2, Conservation* (pp. 201–214). London: Academic Press.

Colinvaux, P. A. (1989). The past and future Amazon. *Scientific American* 260(5):102–109.

Colinvaux, P. A., Miller, M. C., Kam-biu, L., Steinitz-Kannan, M., and Frost, I. (1985). Discovery of permanent Amazon lakes and hyrdaulic disturbances in the upper Amazon Basin. *Nature (London)* 313:42–45.

Collias, N., and Southwick, C. (1952). A field study of population density and social organization in howling monkeys. *Proceedings of the American Philosophical Society* 96:143–156.

Connolly, C. J. (1950). *External Morphology of the Primate Brain*. Springfield, IL: Charles C Thomas.

Conroy, G. C. (1976). Primate postcranial remains from the Oligocene of Eygpt. *Contributions to Primatology* 8:1–134.

Coon, C. S. (1954). *The Story of Man*. New York: Knopf.

Corillas, O. J., Ruiz, J. C., Travi, B. L. (1984). *Callithrix jacchus* reproduccionn y patologias en cautiveiro. In M.T. de Mello (ed.), *A Primatologia no Brasil-1* (pp. 115–132). Brasilia: Sociedade Brasileira de Primatologia.

Corner, B. D., and Richtsmeier, J. T. (1991). Morphometric analysis of craniofacial growth in *Cebus apella. American Journal of Physical Anthropology* 84:323–342.

Costello, M. B. (1991). Troop progressions of free-ranging howler monkeys (*Alouatta palliata*) (sociospatial organization). Unpublished Ph.D. dissertation, University of California, Riverside.

Costello, R. K., Dickinson, C., Rosenberger, A. L., Boinski, S., and Szalay, F. S. (1993). Squirrel monkey (Genus *Saimiri*) taxonomy; a multidisciplinary study of the biology of species. In W. H. Kimbel and L. B. Martin (eds.),

Species, Species Concepts, and Primate Evolution (pp. 177–210). New York: Plenum.

Crandlemire-Sacco, J. L. (1986). The ecology of the saddle-backed tamarin, *Saguinus fuscicollis,* of Southeastern Peru. Ph.D. thesis. University of Pittsburgh, Pittsburgh.

Crandlemire-Sacco, J. (1988). An ecological comparison of two sympatric primates: *Saguinus fuscicollis* and *Callicebus moloch* of Amazonian Peru. *Primates* 29(4):465–475.

Crespo, J. A. (1982). Ecología de la communidad de mamíferos del Parque Nacional Iguazu, Misiones. *Revista de Museo Argentino de Ciencias Naturales "Bernadino Rivadavia."*

Cristoffer, C. (1987). Choice of feeding trees by three species of Brazilian primates: Effects of slope and tree size. *Primate Conservation* 8:39.

Crockett, C. M. (1984). Emigration by female red howler monkeys and the case for female competition. In M. F. Small (ed.), *Female Primates: Studies by Women Primatologists* (pp. 159–173). New York: Alan R. Liss.

Crockett, C. M. (1985). Population studies of red howler monkeys (*Alouatta seniculus*). *National Geographic Research* 1:264–273.

Crockett, C. M. (1987). Diet, dimorphism, and demography: Perspectivies from howlers to hominids. In W. G. Kinzey (ed.), *The Evolution of Human Behavior: Primate Models* (pp. 115–135). Albany: SUNY Press.

Crockett, C. M. (1991). Population growth and new troop formation in Venezuelan red howler monkeys. *American Journal of Primatology* 24(2):95.

Crockett, C. M., and Eisenberg, J. F. (1987). Howlers: Variation in group size and demography. In B. B. Smuts, D. L. Cheney, R. M. Seyfarth, R. W. Wrangham, and T. T. Struhsaker (eds.), *Primate Societies* (pp. 54–68). Chicago: University of Chicago Press.

Crockett, C. M., and Pope, T. (1988). Inferring patterns of aggression from red howler monkey injuries. *American Journal of Primatology* 15:289–308.

Crockett, C. M., and Pope, T. (1993). Consequences of sex differences in dispersal for juvenile red howler monkeys. In M. E. Pereira and L. A. Fairbanks (eds.), *Juvenile Primates Life History, Development, and Behavior* (pp. 104–118). New York: Oxford University Press.

Crockett, C. M., and Rudran, R. (1987a). Red howler monkey birth data. I: Seasonal variation. *American Journal of Primatology* 13:347–368.

Crockett, C. M., and Rudran, R. (1987b). Red howler monkey birth data. II: Interannual, habitat, and sex comparisons. *American Journal of Primatology* 13:369–384.

Crockett, C. M., and Sekulic, R. (1972). Gestation length in red howler monkeys. *American Journal of Primatology* 3:291–294.

Crockett, C. M., and Sekulic, R. (1984). Infanticide in red howler monkeys (*Alouatta seniculus*). In G. Hausfater and S. B. Hrdy (eds.), *Infanticide: Comparative and Evolutionary Perspectives* (pp. 173–191). New York: Aldine.

Cronin, J. E., and Sarich, V. M. (1978). Marmoset evolution: The molecular evidence. *Primates in Medicine* 10:12–19.

Crovella, S., Montagnon, D., and Rumpler, (1992). Taxonomic position of *Cal-*

limico goeldii (Primates, Platyrrhini) on the basis of its highly repeated DNA pattern. *Abstracts of the XIVth Congress of the International Primatological Society*, Strasbourg, France, p. 15.

Cuaron, A. D. (1987). An assessment of the present status of primates in Chiapas, Mexico. *International Journal of Primatology* 8(5):520.

Cuaron, A. D. (1992). The role of remote sensing in primate conservation. Part II. A case study from Mexico and Central America. *Abstracts of the XIVth Congress of the International Primatological Society*, Strasbourg, France, pp. 90–91.

Cubicciotti, D. III, and Mason, W. A. (1975). Comparative studies of social behavior in *Callicebus* and *Saimiri*: Male-female emotional attachment. *Behavioral Biology* 16:185–197.

Cunha da, Alexia C., and Barnett, A. (1989). *Project Uakari. First Report; The preliminary Survey; Part One—Zoology*. Unpublished report to WWF-Netherlands and Royal Geographical Society, London.

Cunningham, E. P. (1994). A preliminary analysis of a *Pithecia pithecia* long call. *American Journal of Physical Anthropology* Suppl. 18:74–75.

Cusick, C. G. (1988). Anatomical organization of the superior colliculus in monkeys: Corticotectal pathways for visual and visuomotor functions. In T. P. Hicks and G. Benedek (eds.), *Progress in Brain Research*, Vol. 75 (pp. 1–15). New York: Elsevier.

Cusick, C. G., Wall, J. T., and Kaas, J. H. (1986). Representations of the face, teeth and oral cavity in areas 3b and 1 of somatosensory cortex in squirrel monkeys. *Brain Research* 370:359–364.

Cusick, C. G., Wall, J. T., Felleman, D. J., and Kaas, J. H. (1989). Somatotopic organization of the lateral sulcus of owl monkeys: Area 3b, S-II and a ventral somatosensory area. *Journal of Comparative Neurology* 282:169–190.

D'Amato, M. R., and Colombo, M. (1989). Serial learning with wild card items by monkeys (*Cebus apella*): Implications for knowledge of ordinal position. *Journal of Comparative Psychology* 103(3):252–261.

D'Amato, M. R., and Colombo, M. (1990). The symbolic distance effect in monkeys (*Cebus apella*). *Animal Learning and Behavior* 18(2):133–140.

D'Amato, M. R., and Salmon, D. P. (1984). Processing and retention of complex auditory stimuli in monkeys (*Cebus apella*). *Canadian Journal of Psychology* 38(2):237–255.

D'Amato, M. R., and Van Sant, P. (1988). The person concept in monkeys (*Cebus apella*). *Journal of Experimental Psychology—Animal Behavior Processes* 14(1):43–55.

Da Silva, E. C. (1981). A preliminary survey of brown howler monkeys (*Alouatta fusca*) at the Cantareira Reserve (São Paulo, Brazil). *Revista do Brasil Biologia* 41:897–909.

Dagosto, M. (1989). Locomotion of *Propithecus diadema* and *Varecia variegata* at Ranomafana National Park, Madagascar. *American Journal of Physical Anthropology* 78:209.

Dagosto, M. (1992). Effect of habitat structure on positional behavior and substrate use in Malagasy lemurs. *American Journal of Physical Anthropology* Suppl. 14:67.

Dagosto, M. (1994a). Seasonal variation in positional behavior of Malagasy lemurs. *American Journal of Physical Anthropology* Suppl. 18:75–76.

Dagosto, M. (1994b). Testing positional behavior of Malagasy lemurs: A randomization approach. *American Journal of Anthropology* 94:189–202.

Dagosto, M. (1995). Seasonal variation in positional behavior of Malagasy lemurs. *International Journal of Primatology* 16(5):807–834.

Dahl, J. F., and Hemingway, C. A. (1988). An unusual activity pattern for the mantled howler monkey of Belize. *American Journal of Physical Anthropology* 75:200.

Daniels, H. L. (1984). Oxygen consumption in *Lemur fulvus*: Deviation from the ideal model. *Journal of Mammalogy* 65:584–592.

Dare, R. (1974). Food-sharing in free-ranging *Ateles geoffroyi* (red spider monkeys). *Laboratory Primate Newsletter* 13:19–21.

Dare, R. (1975). The effects of fruit abundance on movement patterns of free-ranging spider monkeys, *Ateles geoffroyi. American Journal of Physical Anthropology* 42:297.

Darms, K. (1987). Interindividual distances as an indicator of social relationships in groups of common marmosets. *International Journal of Primatology* 8(5):499.

Darms, K. (1990). "Life-histories" of breeding units in common marmosets (*Callithrix jacchus*) under laboratory conditions: Dynamics of group size and composition. *Proceedings of the XIIIth Congress of the International Primatological Society*, p. 47.

Dawson, G. A. (1976). Behavioral ecology of the Panamanian tamarin, *Saguinus oedipus*. Unpublished Ph.D. Dissertation, Michigan State University, East Lansing, MI.

Dawson, G. A. (1977). Composition and stability of social groups of the tamarin, *Saguinus oedipus geoffroyi*, in Panama: Ecological and behavioral implications. In D. G. Kleiman (ed.), *The Biology and Conservation of the Callitrichidae* (pp. 23–37). Washington DC: Smithsonian Institution Press.

Dawson, G. A. (1979). The use of time and space by the Panamanian tamarin, *Saguinus oedipus. Folia Primatologica* 31:253–284.

Dawson, G. A., and Dukelow, W. R. (1976). Reproductive characteristics of free ranging Panamanian tamarins (*Saguinus oedipus geoffroyi*). *Journal of Medical Primatology* 5:266–275.

Day, M. (1986). *Guide to Fossil Man*. Chicago: World Publishing Co.

De Bois, H. (1994). Golden-headed lion tamarin (*Leontopithecus chrysomelas*). *1993 International Studbook Number Six*. Antwerp: Royal Zoological Society of Antwerp.

de Carvalho, C. T., Albernaz, A. L. K. M., and de Lucca, C. A. T. (1989). Aspectos da bionomia do mico-leão preto (*Leontopithecus chrysopygus* Mikan) (Mammalia, Callithricidae). *Revista do Instituto Florestal São Paulo* 1(1):67–83.

de Carvalho, C. T., and de Carvalho, C. F. (1989). A organização social dos sauís-pretos, (*Leontopithecus chrysopygus* Mikan), na reserva em Teodoro Sampaio, São Paulo (Primates Callithricidae). *Revista Brasileira de Zoologia* 6(4):707–717.

de Faria, D. S. (1984a). Aspectos gerais do comportamento de *Callithrix jacchus penicillata* em mata cilliar do cerrado. In M. T. de Mello (ed.), *A Primatologia no Brasil-1* (pp. 56–65). Brasilia: Sociedade Brasileira de Primatologia.

de Faria, D. S. (1984b). Uso de árvores gomíferas do cerrado por *Callithrix jacchus penicillata*. In M. T. de Mello (ed.), *A Primatologia no Brasil-1* (pp. 83–96). Brasilia: Sociedade Brasileira de Primatologia.

de Faria, D. S. (1986). Tamanho, composição de um grupos social e area de vivencia (home-range) do sagui *Callithrix jacchus penicillata* na mata ciliar do corrego caatinga, Brasilia, DF. In M. T. de Mello (ed.), *A Primatologia no Brasil-2* (pp. 87–105). Campinas: Sociedade Brasileira de Primatologia.

de Faria, D. S. (1987). Grouping and related activities in the marmoset *Callithrix penicillata*. *International Journal of Primatology* 8(5):529.

de Faria, D. S. (1989). O estudo de campo com o mico-estrela no planalto central brasileiro. In C. Ades (ed.), *Etologia de Animais e de Homens* (pp. 109–121). São Paulo: EDICON/EDUSP.

de Ruiter, J. R. (1986). The influence of group size on predator scanning and foraging behavior of wedgecapped capuchin monkeys (*Cebus olivaceus*). *Behaviour* 98(1–4):240–258.

De Valois, R. L., and De Valois, K. K. (1988). *Spatial Vision.* New York: Oxford University Press.

De Valois, R. L., and Jacobs, G. H. (1968). Primate color vision. *Science* 162:533–540.

De Valois, R. L., and Morgan, H. C. (1974). Psychophysical studies of monkey vision. II. Squirrel monkey wavelength and saturation discrimination. *Vision Research* 14:69–73.

De Valois, R. L., Morgan, H. C., Polson, M. C., Mead, W. R., and Hull, E. M. (1974). Psychophysical studies of monkey vision. I. Macaque luminosity and color vision tests. *Vision Research* 14:53–67.

de Waal, F. B. M. (1986). The integration of dominance and social bonding in primates. *Quarterly Review of Biology* 61:459–479.

DeBruyn, E. J., and Casagrande, V. A. (1981). Demonstration of ocular dominance columns in a New World primate by means of monocular deprivation. *Brain Research* 207:453–458.

Defler, T. R. (1981). The density of *Alouatta seniculus* in the eastern llanos of Colombia. *Primates* 22:564–569.

Defler, T. R. (1982). A comparison of intergroup behavior in *Cebus albifrons* and *C. apella*. *Primates* 23(3):385–392.

Defler, T. R. (1983). Some population characteristics of *Callicebus torquatus lugens* (Humboldt, 1812) (Primates: Cebidae) in eastern Colombia. *Lozania (Acta Zoologica Colombiana)* 38:1–9.

Defler, T. R. (1987). Ranging and the use of space in a group of woolly monkeys (*Lagothrix lagotricha*) in the NW Amazon of Colombia. *International Journal of Primatology* 8(5):420.

Defler, T. R. (1989). Ranging and the use of space in a group of woolly monkeys (*Lagothrix lagotricha*) in Mono Lanudo, Churuco, Colombian Amazon. [in Spanish] *Trianea (Acta Cientifica Teen., Inderena)* 3:183–205.

Defler, T. R. (1994). Callicebus torquatus is not a white-sand specialist. *American Journal of Primatology* 33(2):149–154.

Defler, T. R. (1995). Genus *Lagothrix*, E. Geoffroy St. Hilaire (Atelinae, Cebidae, Platyrrhini): *Lagothrix lagotricha. Illustrated Monographs of Living Primates*. The Netherlands: Istitut voor Ontwikkelingsopdrachten. In press.

Dehay, C., and Kennedy, H. (1993). Control mechanisms of primate corticogenesis. In B. Gulyas, D. Ottoson, and P. E. Roland (eds.), *Functional Organization of the Human Visual Cortex* (pp. 13–27). New York: Pergamon Press.

Delson, E. (1988). Classification of the Primates. In I. Tattersall, E. Delson, and J. V. Couvering (eds.), *Encyclopedia of Human Evolution and Prehistory*. New York: Garland.

Delson, E., and Rosenberger, A. L. (1980). Phyletic perspectives on platyrrhine origins and anthropoid relationships. In R. L. Ciohon and A. B. Chiarelli (eds.), *Evolutionary Biology of the New World Monkeys and Continental Drift* (pp. 445–458). New York: Plenum.

Delson, E., and Rosenberger, A. L. (1984). Are there any anthropoid primate living fossils? In N. Eldridge and S. M. Stanley (eds.), *Living Fossils* (pp. 50–61). New York: Springer-Verlag.

Demes, B., Jungers, W. L., Fleagle, J. G., and Wunderlich, R. G. (1995). Leaping kinematics of indriid primates. *American Journal of Physical Anthropology* 20:83.

Denevan, W. M. (1992). *The Native Population of the Americas in 1492*, 2nd ed. Madison: University of Wisconsin Press.

Denslow, J. S. (1987). Tropical rainforest gaps and tree species diversity. *Annual Review of Ecology and Systematics* 18:431–451.

Denslow, J. S., and Padoch, C. (eds.) (1988). *People of the Tropical Rain Forest*. Berkeley: University of California Press.

Deputte, B. L. (1982). Duetting in male and female songs of the white-cheeked gibbon (*Hylobates concolor leucogenys*). In C. T. Snowdon, C. H. Brown, and M. R. Petersen (eds.), *Primate Communication* (pp. 67–93). New York: Cambridge University Press.

Deshmukh, I. (1986). *Ecology and Tropical Biology*. Boston: Blackwell Scientific Publications.

Dew, L., and Wright, P. (1995). Conservation implication of seed dispersal by primates in a Malagasy rainforest (Ranomafana National Park). *Biotropica*. In press.

Dewar, R. E. (1984). Extinctions in Madagascar: The loss of the subfossil fauna. In P. S. Martin and R. G. Klein (eds.), *Quaternary Extinctions: A Prehistoric Revolution* (pp. 574–593). Tucson: University of Arizona Press.

Diamond, I. T., Conley, M., Itoh, K., and Fitzpatrick, D. (1985). Laminar organization of geniculocortical projections in *Galago senegalensis* and *Aotus trivirgatus. Journal of Comparative Neurology* 242:584–610.

Diamond, J. (1988). Factors controlling species diversity: Overview and synthesis. *Annals of the Missouri Botanical Garden* 75:117–129.

Dickinson, C., Harrison, H., and Miller, K. (1990). Phylogenetic relationships among species of the genus *Cebus*. *American Journal of Physical Anthropology* 81(2):215–216.

Dietz, J. M. (1993a). Conservation, ecology and behavior of golden-headed lion tamarins (*Leontopithecus chrysomelas*) in Una Reserve, Brazil. Unpublished Progress Re₁ ₁rt, World Wildlife Fund, February, 1993, Washington, DC.

Dietz, J. M., ana Baker, A. J. (1992). Correlates of polygyny in golden lion tamarins. *Abstracts of the XIVth Congress of the International Primatological Society*, Strasbourg, France, p. 59.

Dietz, J. M., and Baker, A. J. (1993a). Polygyny and female reproductive success in golden lion tamarins (*Leontopithecus rosalia*). *Animal Behaviour* 46(6):1067–1078.

Dietz, J. M., and Baker, A. J. (1993b). Polygyny and female reproductive success in golden lion tamarins, *Leontopithecus rosalia*. *Animal Behaviour* 46(6):1067–1078.

Dietz, J. M., and Kleiman, D. G. (1987). Sexual dimorphism and alternative reproductive tactics in the golden lion tamarin. *International Journal of Primatology* 8(5):506.

Dietz, J. M., Kleiman, D. G., and Beck, B. B. (1992). A decade of research on native golden lion tamarins. *Abstracts of the XIVth Congress of the International Primatological Society*, Strasbourg, France, p. 95.

Dietz, J. M., Baker, A. J., and Miglioretti, (1995). The effect of seasonality on reproduction, juvenile weight gain and adult weight in golden lion tamarins. In preparation.

Digby, L. (1990). An experimental test of dispersal choice in *Callithrix jacchus*. *American Journal of Primatology* 20(3):185.

Digby, L. J. (1994). Social organization and reproductive strategies in a wild population of common marmosets (*Callithrix jacchus*). Unpublished Ph.D. dissertation, University of California, Davis.

Digby, L. J., and Barreto, C. E. (1993). Social organization in a wild population of *Callithrix jacchus*. I. Group composition & dynamics. *Folia Primatologica* 61(3):123–134.

Digby, L. J., and Ferrari, S. F. (1994). Multiple breeding females in free-ranging groups of *Callithrix jacchus*. *International Journal of Primatology* 15(3):389–398.

Diong, C. H. (1973). Studies of the Malayan wild pig in Perak and Johore. *Malay Nature Journal* 26:130–150.

Dixson, A. F. (1983). The owl monkey (*Aotus trivirgatus*). In J. P. Hearn (ed.), *Reproduction in New World Primates: New Models in Medical Science* (pp. 69–114). Boston: MTP Press.

Dixson, A. F. (1986). Plasma testosterone concentrations during post-natal development in the male common marmoset. *Folia Primatologica* 47:166–170.

Dixson, A. F. (1987). Baculum length and copulatory behavior in primates. *American Journal of Primatology* 13:51–60.

Dixson, A. F. (1992). Observations on postpartum changes in hormones and sexual behavior in callitrichid primates: Do females exhibit postpartum "estrus"? In N. Itoigawa, Y. Sugiyama, G. P. Sackett, and R. K. Thompson (eds.), *Topics in Primatology, Vol 2, Behavior, Ecology, and Conservation* (pp. 141–149). Tokyo: University of Tokyo Press.

Dixson, A. F. (1993). Callitrichid mating systems: Laboratory and field approaches to studies of monogamy and polyandry. In A. B. Rylands (ed.), *Marmosets and Tamarins; Systematics, Behaviour, and Ecology* (pp. 164–175). Oxford: Oxford University Press.

Dixson, A. F., and Fleming, D. (1981). Parental behaviour and infant development in owl monkeys (*Aotus trivirgatus griseimembra*). *Journal of Zoology (London)* 194:25–39.

Dixson, A. F., Anzenberger, G., Monteiro Da Cruz, M. A. O., Patel I., and Jeffreys, A. J. (1992). DNA fingerprinting of free-ranging groups of common marmosets (*Callithrix jacchus jacchus*) in NE Brazil. In R. D. Martin, A. F. Dixson, and E. J. Wickings (eds.), *Paternity in Primates: Genetic Tests and Theories* (pp. 192–202). Basel: Karger.

Dollman, G. (1937). Exhibition and remarks upon a series of skins of marmosets and tamarins. *Proceedings of the Zoological Society of London, Series C* 107:64–65.

Dudley, R., and Milton, K. (1990). Parasite deterrence and the energetic costs of slapping in howler monkeys, *Alouatta palliata*. *Journal of Mammalogy* 71(3):463–465.

DuFour, D. L. (1995). A closer look at the implications of bitter cassava use. In *Indigenous Peoples and the Future of Amazonia: An Ecological Anthropology of an Endangered World* (pp. 149–165).Tucson: University of Arizona Press.

Dugmore, S. J. (1986). Behavioral observations on a pair of captive white-faced saki monkeys (*Pithecia pithecia*). *Folia Primatologica* 46(2):83–90.

Dukelow, W. R., and Dukelow, K. B. (1989). Reproductive and endocrinological measures of stress and nonstress in nonhuman primates. *American Journal of Primatology* Suppl. 1:17–24.

DuMond, F. V., and Hutchinson, T. C. (1967). Squirrel monkey reproduction: The "fatted" male phenomenon and seasonal spermatogenesis. *Science* 158:1067–1070.

DuMond, F. V., Hoover, B. L., and Norconk, M. A. (1979). Hand-feeding parent reared golden lion tamarins, *Leontopithecus r. rosalia*, at Monkey Jungle. *International Zoo Yearbook* 19:155–158.

Dunbar, R. I. M. (1984). *Reproductive Decisions*. Princeton: Princeton University Press.

Dunbar, R. I. M. (1993). Coevolution of neocortical size, group size and language in humans. *Behavioral and Brain Sciences* 16(4):681–735.

Durham, N. M. (1975). Some ecological, distributional and group behavioral patterns of Atelinae in southern Peru: With comments on interspecific relations. In R. H. Tuttle (ed.), *Socioecology and Psychology of Primates* (pp. 87–101). The Hague: Mouton.

Dutrillaux, B. (1988a). Chromosome evolution in primates. *Folia Primatologica* 50(1–2):134–135.

Dutrillaux, B. (1988b). New interpretation of the presumed common ancestral karyotype of platyrrhine monkeys. *Folia Primatologica* 50(3–4):226–229.

Dutrillaux, B., Lombard, M., Carroll, J. B., and Martin, R. D. (1988). Chromosomal affinities of *Callimico goeldii* (Platyrrhini) and characterization of a Y-autosome translocation in the male. *Folia Primatologica* 50(3–4):230–236.

Dykyj, D. (1982). Allometry of trunk and limbs in New World primates. Ph.D. thesis, City University of New York.

Easley, S. P. (1982). Ecology and behavior of Callicebus torquatus, Cebidae, Primates. Unpublished Ph.D. dissertation, Washington University, St. Louis.

Easley, S. P., and Kinzey, W. G. (1986). Territorial Shift in the yellow-handed titi monkey. *American Journal of Primatology* 11(4):307–318.

Eason, P. (1989). Harpy eagle attempts predation on adult howler monkey. *Condor* 91:469–470.

Egler, S. G. (1983). Current status of the pied bare-faced tamarin in Brazilian Amazonia. *Primate Conservation* 3:20.

Egler, S. G. (1986). Estudos bion micos de *Saguinus bicolor* (Spix, 1823) (Callitrihidae,: Primates), em mata tropical alterada, Manaus (AM). M.Sc. thesis, Universidade Estadual de Campinas, Campinas, Brasil.

Egler, S. G. (1992). Feeding ecology of *Saguinus bicolor bicolor* (Callitrichidae: Primates) in a relict forest in Manaus, Brazilian Amazonia. *Folia Primatologica* 59(2):61–76.

Eisenberg, J. F. (1976). Communication mechanisms and social integration in the black spider monkey, *Ateles fusciceps robustus*, and related species. *Smithsonian Contributions to Zoology* 213:1–108.

Eisenberg, J. F. (1977). Comparative ecology and reproduction of New World monkeys. In D. G. Kleiman (ed.), *The Biology and Conservation of the Callitrichidae* (pp. 13–22). Washington DC: Smithsonian Institution Press.

Eisenberg, J. F. (1979). Habitat, economy, and society: Some correlations and hypotheses for the neotropical primates. In I. S. Bernstein and E. O. Smith (eds.), *Primate Ecology and Human Origins: Ecological Influences on Social Organization* (pp. 215–262). New York: Garland.

Eisenberg, J. F. (1981). *The Mammalian Radiations*. Chicago: The University of Chicago Press.

Eisenberg, J. F. (1989). *Mammals of the Neotropics. Volume 1. The Northern Neotropics*. Chicago: Chicago University Press.

Eisenberg, J. F. (1991). Mammalian social organization and the case of *Alouatta*. In J. G. Robinson and L. Tiger (eds.), *Man and Beast Revisited* (pp. 127–138). Washington, DC: Smithsonian Institution Press.

Ellefson, J. (1968). Territorial behavior in the common white-handed gibbon, *Hylobates lar*. In P. Jay (ed.), *Primates: Studies in Adaptation and Variability* (pp. 180–199). New York: Holt, Rinehart & Winston.

Elliot, D. G. (1912). A Review of the Primates, Monograph 1, 3 Vols. New York: American Museum of Natural History.

Elowson, A. M., Sweet, C. S., and Snowdon, C. T. (1992). Ontogeny of trill and J-call vocalizations in the pygmy marmoset (*Cebuella pygmaea*). *Animal Behavior* 42:703–715.

Emmons, L. E. (1987). Comparative feeding ecology of felids in a Neotropical rain forest. *Behavioral Ecology and Sociobiology* 20:271–283.

Emmons, L. H. (1995). Mammals of rain forest canopies. In M. D. Lowman and N. M. Nadkarni (eds.), *Forest Canopies* (pp. 199–223). New York: Academic Press.

Epple, G. (1967). Vergleichende untersuchungen über sexual und sozialverhalten de Krallenaffen (Hapalidae). *Folia Primatologica* 7:37–65.

Epple, G. (1968). Comparative studies on vocalization in marmoset monkeys (Hapalidae). *Folia Primatologica* 8:1–40.

Epple, G. (1975). Paternal behavior in *Saguinus fuscicollis* ssp. (Callitrichidae). *Folia Primatologica* 24:221–238.

Epple, G. (1978). Reproductive and social behavior of marmosets with special reference to captive breeding. *Primate Medicine* 10:50–62.

Epple, G., Belcher, A. M., Küderling, I., Zeller, U., Scolnick, L., Greenfield, K. L., and Smith, A. B. III. (1993). Making sense out of scents: Species differences in scent glands, scent-marking behaviour, and scent-mark composition in the Callitrichidae. In A. B. Rylands (ed.), *Marmosets and Tamarins; Systematics, Behaviour, and Ecology* (pp. 123–151). Oxford: Oxford University Press.

Erikson, G. E. (1957). The hands of New World primates, with comparative functional observations on the hands of other primates. *American Journal of Physical Anthropology* 15:446.

Erkert, H. G. (1989). Lighting requirements of nocturnal primates in captivity: A chronobiological approach. *Zoo Biology* 8(2):179–191.

Erkert, H. G. (1991). Influence of ambient temperature on circadian rhythms in Colombian owl monkeys, *Aotus lemurinus griseimembra*. In A. Ehara, T. Kimura, O. Takenaka, and M. Iwamoto (eds.), *Primatology Today*, (pp. 435–438). Amsterdam: Elsevier Scientific Publishers.

Escobar-Paramo, P. (1989a). The development of black-capped capuchin (*Cebus apella*) in La Macarena, Colombia. *Field Studies of New World Monkeys, La Macarena, Colombia* 2:45–56.

Escobar-Paramo, P. (1989b). Social relations between infants and other group members in the wild black-capped capuchin (*Cebus apella*). *Field Studies of New World Monkeys, La Macarena, Colombia* 2:57–63.

Estrada, A. (1982). Survey and census of howler monkeys (*Alouatta palliata*) in the rain forest of "Los Tuxtlas," Veracruz, Mexico. *American Journal of Primatology* 2:363–372.

Estrada, A. (1983). Primate studies at the Biological Reserve "Los Tuxtlas," Veracruz, Mexico. *Primate Conservation* 3:21.

Estrada, A. (1984). Resource use by howler monkeys (*Alouatta palliata*) in the rain forest of Los Tuxtlas, Vera Cruz, Mexico. *International Journal of Primatology* 5:105–131.

Estrada, A., and Coates-Estrada, R. (1983). Rain forest in Mexico: Research and conservation at Los Tuxtlas. *Oryx* 17:201–204.

Estrada, A., and Coates-Estrada, R. (1984a). Fruit-eating and seed dispersal by howling monkeys (*Alouatta palliata*) in the tropical rain forest of Los Tuxtlas, Mexico. *American Journal of Primatology* 6:77–91.

Estrada, A., and Coates-Estrada, R. (1984b). Some observations on the present distribution and conservation of *Alouatta* and *Ateles* in southern Mexico. *American Journal of Primatology* 7:133–137.

Estrada, A., and Coates-Estrada, R. (1985). A preliminary study of resource overlap between howling monkeys (*Alouatta palliata*) and other arboreal mammals in the tropical rain forest of Los Tuxtlas, Mexico. *American Journal of Primatology* 9:27–37.

Estrada, A., and Coates-Estrada, R. (1986a). Frugivory by howling monkeys

(*Alouatta palliata*) at Los Tuxtlas, Mexico: Dispersal and fate of seeds. In A. Estrada and T. H. Fleming (eds.), *Frugivores and Seed Dispersal* (pp. 93–104). Dordrecht: Dr W. Junk.

Estrada, A., and Coates-Estrada, R. (1986b). Use of leaf resources by howling monkeys (*Alouatta palliata*) and leaf-cutting ants (*Atta cephalotes*) in the tropical rain forest of Los Tuxtlas, Mexico. *American Journal of Primatology* 10:51–66.

Estrada, A., and Trejo, W. (1978). Dieta y selectividad en el mono aullador (*Alouatta villosa*), en le selva alta perennifolia de Estación de Biología Tropical "Los Tuxtlas" en Veracruz. *Memorias. Il Congreso Nacional de Zoología, Facultad de Ciencias Biologicas, Universidad Autonomo de Nuevo Leon, Mexico*, pp. 493–517.

Estrada, A., Coates-Estrada, R., and Vazquez-Yanes, C. (1984). Observation on fruiting and dispersers of *Cecropia obtusifolia* at Los Tuxtlas, Mexico. *Biotropica* 16(4):315–318.

Evans, R. M. (1974). *The Perception of Color*. New York: John Wiley.

Evans, S. (1983). The pair-bond of the common marmoset, *Callithrix jacchus jacchus*: An experimental investigation. *Animal Behavior* 31:651–658.

Evans, S., and Hodges J. K. (1984). Reproductive status of adult daughters in family groups of common marmosets (*Callithrix jacchus jacchus*). *Folia Primatologica* 42:127–133.

Falk, D. (1981). Comparative study of endocranial casts of New and Old World monkeys. In R. L. Ciochon and B. Chiarelli (eds.), *Evolutionary Biology of the New World Monkeys and Continental Drift* (pp. 275–292). New York: Plenum.

Fanning, T. G., Forman, L., and Seuánez, H. N. (1987). A recent DNA amplification event in Goeldi's marmoset (*Callimico goeldii*). *International Journal of Primatology* 8(5):504.

Fanning, T. G., Seuánez, H. N., and Forman, L. (1989). Satellite DNA sequences in the neotropical marmoset *Callimico goeldii* (Primates, Platyrrhini). *Chromosoma* 98(6):396–401.

Fedigan, L. M. (1986). Demographic trends in the *Alouatta palliata* and *Cebus* capuchins population of Santa Rosa National Park, Costa Rica. In J. G. Else and P.C. Lee (eds.), *Primate Ecology and Conservation* (pp. 285–293). New York: Cambridge University Press.

Fedigan, L. M. (1990). Vertebrate predation in *Cebus capucinus*: Meat eating in a neotropical monkey. *Folia Primatologica* 54:196–205.

Fedigan, L. M., and Baxter, M. F. (1984). Sex differences and social organization of free-ranging spider monkeys (*Ateles geoffroyi*). *Primates* 25:279–294.

Fedigan, L. M., Fedigan, L., and Chapman, C. (1985). A census of *Alouatta palliata* and *Cebus capucinus* monkeys in Santa Rosa National Park, Costa Rica. *Brenesia* 23:309–322.

Fedigan, L. M., Fedigan, L., Chapman, C., and Glander, K. E. (1988). Spider monkey home ranges: A comparison of radio telemetry and direct observation. *American Journal of Primatology* 16(1):19–29.

Felleman, D. J., Nelson, R. J., Sur, M., and Kaas, J. H. (1983). Representations of the body surface in areas 3b and 1 of postcentral parietal cortex of *Cebus* monkeys. *Brain Research* 268:15–26.

Fernandes, M. E. B. (1987). Preliminary observations of *Chiropotes satanas utahicki* in captivity. *International Journal of Primatology* 8(5):539.

Fernandes, M. E. B. (1991). Comunicação social dos cuxiús (*Chiropotes satanas utahicki*, Cebidae, Primates) em cativeiro. In A. B. Rylands and A. T. Bernardes (eds.), *A Primatologia no Brasil-3* (pp. 297–305). Belo Horizonte: Fundação Biodiversitas.

Ferrari, S. F. (1987a). Food transfer in wild marmoset group. *Folia Primatologica* 48:203–206.

Ferrari, S. F. (1987b). Social organization and behavior of the marmoset *Callithrix flaviceps*. *International Journal of Primatology* 8(5):492.

Ferrari, S. F. (1988). The behaviour and ecology of the buffy-headed marmoset, *Callithrix flaviceps* (O. Thomas, 1903). Unpublished Ph.D. Thesis, University College, London.

Ferrari, S. F. (1991a). Preliminary report on a field study of *Callithrix flaviceps*. In A. B. Rylands and A. T. Bernardes (eds.), *A Primatologia no Brasil-3* (pp. 159–171). Belo Horizonte: Fundação Bidiversitas.

Ferrari, S. F. (1991b). Diet for a small primate. *Natural History* 100(1):38–42.

Ferrari, S. F. (1992). The care of infants in a wild marmoset (*Callithrix flaviceps*) group. *American Journal of Primatology* 26(2):109–118.

Ferrari, S. F. (1993). Ecological differentiation in the Callitrichidae. In A. B. Rylands (ed.), *Marmosets and Tamarins; Systematics, Behaviour, and Ecology* (pp. 314–328). Oxford: Oxford University Press.

Ferrari, S. F. (1994). The distribution of the black-headed marmoset, *Callithrix nigriceps*: A correction. *Neotropical Primates* 2(1):11–12.

Ferrari, S. F., and de Souza, A. P. (1994). More untufted capuchins in southeastern Amazonia? *Neotropical Primates* 2(1):9–10.

Ferrari, S. F., and Lopes Ferrari, M. A. (1989). A re-evaluation of the social organisation of the Callitrichidae, with reference to the ecological differences between genera. *Folia Primatologica* 52:132–147.

Ferrari, S. F., and Lopes Ferrari, M. A. (1990). Predator avoidance behaviour in the buffy-headed marmoset (*Callithrix flaviceps*). *Primates* 31(3):323–338.

Ferrari, S. F., and Lopes Ferrari, M. A. (1992). A new species of marmoset, genus *Callithrix*, Erxleben 1777 (Callitrichidae, Primates) from western Brazilian Amazonia. *Goeldiana, Zoologia* 12:1–13.

Ferrari, S. F., and Martins, E. L. (1995). Gummivory and gut morphology in two sympatric callitrichids (*Callithrix emiliae* and *Saguinus fuscicollis weddelli*) from western Brazilian Amazonia. *American Journal of Physical Anthropology*. In press.

Ferrari, S. F., and Strier, K. B. (1992). Explotation of *Mabea fistulifera* nectar by marmosets (*Callithrix flaviceps*) and muriquis (*Brachyteles arachnoides*) in south-east Brazil. *Journal of Tropical Ecology* 8:225–239.

Ficken, M. S., and Ficken, R. W. (1967). Singing behavior of blue-winged and golden-winged warblers and their hybrids. *Behaviour* 28:149–181.

Figueroa, R. (1989). Social interactions of a fourth month adopted infant in a wild group of *Alouatta seniculus*. *Field Studies of New World Monkeys, La Macarena, Colombia* 2:37–39.

Finlay, B. L., Wikler, K. C., and Senegelaub, D. R. (1987). Regressive events in brain development and scenarios for vertebrate brain evolution. *Brain Behavior and Evolution* 30:102–117.

Fishkind, A. S., and Sussman, R. W. (1987). Preliminary survey of the primates of the Zona Protectora and La Selva Biological Station, northeast Costa Rica. *Primate Conservation* 8:63–66.

Fittkau, E. J., and Klinge, H. (1973). On biomass and trophic structure of the central Amazonian rain forest ecosystem. *Biotropica* 5(1):2–14.

Fitzpatrick, J. W., and Woolfenden, G. E. (1988). Components of lifetime reproductive success in the Florida scrub jay. In T. H. Clutton-Brock (ed.), *Reproductive Success* (pp. 305–324). Chicago: University of Chicago Press.

Fleagle, J. G. (1983). Locomotor adaptations of Oligocene and Miocene hominoids and their phyletic implications. In R. L. Ciochon and R. S. Corruccini (eds.), *New Interpretations of Ape and Human Ancestry* (pp. 301–324). New York: Plenum.

Fleagle, J. G. (1986). The fossil record of early catarrhine evolution. In B. Wood, L. Martin, and P. Andrews (eds.), *Major Topics in Primate and Human Evolution* (pp. 130–149). Cambridge: Cambridge University Press.

Fleagle, J. G. (1988). *Primate Adaptation and Evolution*. New York: Academic Press.

Fleagle, J. G. (1990). New fossil platyrrhines from the Pinturas Formation, southern Argentina. *Journal of Human Evolution* 19:61–85.

Fleagle, J. G., and Bown, T. (1983). New primate fossils from late Oligocene (Colhuehuapian) localities of Chubet Province, Argentina. *Folia Primatologica* 41:240–266.

Fleagle, J. G., and Jungers, W. L. (1982). Fifty years of higher primate phylogeny. In F. Spencer (ed.), *A History of American Physical Anthropology 1930–1980* (pp. 187–230). New York: Academic Press.

Fleagle, J. G., and Kay, R. F. (1983). New interpretations of the phyletic position of Oligocene hominiods. In R. L. Ciochon and R. S. Corruccini (eds.), *New Interpretations of Ape and Human Ancestry* (pp. 181–210). New York: Plenum.

Fleagle, J. G., and Kay, R. F. (1985). The paleobiology of catarrhines. In E. Delson (ed.), *Ancestors: The Hard Evidence* (pp. 23–36). New York: Alan R Liss.

Fleagle, J. G., and Kay, R. F. (1988). The phyletic position of the Parapithecidae. *Journal of Human Evolution* 16:483–531.

Fleagle, J. G., and Kay, R. F. (1989). The dental morphology of *Dolichocebus gaimanensis*, a fossil monkey from Argentina. *American Journal of Physical Anthropology* 78:221.

Fleagle, J. G., and Kay, R. F. (1994). Anthropoid origins, past, present, and future. In J. G. Fleagle and R. F. Kay (eds.), *Anthropoid Origins* (pp. 675–708). New York: Plenum.

Fleagle, J. G., and Meldrum, D. J. (1988). Locomotor behavior and skeletal morphology of two sympatric pithecine monkeys, *Pithecia pithecia* and *Chiropotes satanas*. *American Journal of Primatology* 16(3):227–249.

Fleagle, J. G., and Mittermeier, R. A. (1980). Locomotor behavior, body size, and comparative ecology of seven Surinam monkeys. *American Journal of Physical Anthropology* 52(3):301–314.

Fleagle, J. G., and Rosenberger, A. L. (1983). Cranial morphology of the earliest anthropoids. In M. Sakka (ed.), *Morphologie Evolutive, Morphogenese du Crane, et Origine de l'Homme* (pp. 141–153). Paris: CNRS.

Fleagle, J. G., Simons, E. L., and Conroy, G. C. (1975). Ape limb bones from the Oligocene of Egypt. *Science* 189:135–137.

Fleagle, J. G., Mittermeier, R. A., and Skopec, A. L. (1981). Differential habitat use of *Cebus apella* and *Saimiri sciureus* in central Surinam. *Primates* 22(3):361–367.

Fleagle, J. G., Bown, T. M., Orbradovich, J. O., and Simons, E. L. (1986a). Age of the earliest African anthropoids. *Science* 234:1247–1249.

Fleagle, J. G., Bown, T. M., Obradovich, J. O., and Simons, E. L. (1986b). How old are the Fayum primates? In J. G. Else and P. C. Lee, (eds.), *Primate Evolution* (pp. 133–142). New York: Cambridge University Press.

Fleagle, J. G., Bown, T. M., and Swisher, C. (1995). Age of the Pinturas and Santa Cruz formations. *VI Congreso Argentino de Paleontología y Bioestratigrafía, Actas* (pp. 211–212). Buenos Aires: Paleontological Society of Argentina.

Fleagle, J. G., Kay, R. F., and Anthony, M. T. (1996). Fossil New World monkeys. In R. F. Kay, R. Madden, R. Cifelli, and J. J. Flynn (eds.), *The History of Neotropical Fauna: Vertebrate Paleobiology of the Miocene of Tropical South America*. Washington, DC: Smithsonian Institution Press.

Fleming, T. H., Williams, C. F., Bonaccorso, F. J., and Herbst, L. H. (1985). Phenology, seed dispersal, and colonization in *Muntingia calabura*, a neotropical pioneer tree. *American Journal of Botany* 72:383–391.

Flesness, N. R. (1986). Captive studies and genetic considerations. In K. Bernishke (ed.), *Primates: The Road to Self-sustaining Populations* (pp. 845–856). New York: Springer-Verlag.

Flinn, M. V. (1988). Mate-guarding in a Caribbean village. *Ethology and Sociobiology* 9:1–28.

Florence, S. L., Conley, M., and Casagrande, V. A. (1986). Ocular dominance columns and retinal projections in New World spider monkeys (*Ateles ater*). *Journal of Comparative Neurology* 243:234–248.

Flynn, J. J., Wyss, A. R., Charrier, R., and Swisher, C. C. (1995). An early Miocene anthropoid skull from the Chilean Andes. *Nature (London)* 373:603–607.

Foerg, R. (1982). Reproductive behaviour in *Varecia variegata*. *Folia Primatologica* 38(1–2):108–121.

Fonseca, G. A. B. (1985). Observations on the ecology of the muriqui (*Brachyteles arachnoides* E. Geoffroy 1806): Implications for its conservation. *Primate Conservation* 5:48–52.

Fonseca, G. A. B. (1986). Observações sobre a ecologia do mono carvoeiro ou muriqui (*Brachyteles arachnoides*) e sugestões para a sua conservação. In M. T. de Mello (ed.), *A Primatologia no Brasil-2* (pp. 177–183). Brasilia: Instituto de Ciências Biológicas.

Fonseca, G. A. B. da, and Lacher, T. E., Jr. (1984). Exudate-feeding by *Callithrix jacchus penicillata* in semideciduous woodland (cerradão) in central Brazil. *Primates* 25:441–450.

Fonseca G. A. B., Robinson, J., and Mittermeier, R. A. (1987). Conservation of the Atelinae. *International Journal of Primatology* 8(5):420.

Fontaine, R. (1979). Survey of the red uakari (*Cacajao calvus rubicundus*) in Eastern Peru. Unpublished report to New York Zoological Society.

Fontaine, R. (1981). The uakaris, genus *Cacajao*. In A. F. Coimbra-Filho and R. A. Mittermeier (eds.), *Ecology and Behavior of Neotropical Primates* (pp. 443–493). Rio de Janeiro: Academia Brasileira de Cincias.

Fontaine, R., and DuMond, F. V. (1977). The red uakari in a seminatural environment: Potentials for propagation and study. In Prince Rainer of Monaco and G. H. Bourne (eds.), *Primate Conservation* (pp. 167–236). New York: Academic Press.

Fontaine, R., and Hench, M. (1982). Breeding new world monkeys at Miami's monkey jungle. *International Zoo Yearbook* 22:77–83.

Fooden, J. (1963). A revision of the woolly monkeys (genus *Lagothrix*). *Journal of Mammalogy* 44(2):213–247.

Ford, H. A. (1971). *Ecological Genetics*, 3rd ed. London: Chapman and Hall.

Ford, S. M. (1980a). A systematic revision of the Platyrrhini based on features of the postcranium. Unpublished Ph.D. dissertation, University of Pittsburgh.

Ford, S. M. (1980b). Callitrichids as phyletic dwarfs, and the place of the Callitrichidae in Platyrrhini. *Primates* 21:31–43.

Ford, S. M. (1986a). Comment on the evolution of claw-like nails in callitrichids (marmosets/tamarins). *American Journal of Physical Anthropology* 70(1):25–26.

Ford, S. M. (1986b). Systematics of the New World monkeys. In D. Swindler and J. Erwin (eds.), *Comparative Primate Biology Vol. 1: Systematics, Evolution, and Anatomy* (pp. 73–135). New York: Alan R. Liss.

Ford, S. M. (1990a). Locomotor adaptations of fossil platyrrhines. *Journal of Human Evolution* 19:141–173.

Ford, S. M. (1990b). Platyrrhine evolution in the West Indies. *Journal of Human Evolution* 19:237–254.

Ford, S. M., and Davis, L. C. (1992). Systematics and body size: Implications for feeding adaptations in New World monkeys. *American Journal of Physical Anthropology* 88(4):415–468.

Forman, L., Kleiman, D. G., Bush. R. M., Dietz, J. M., Ballou, J. D., Phillips, L. G., Coimbra-Filho, A. F., and O'Brien, S. J. (1986). Genetic variation within and among lion tamarins. *American Journal of Physical Anthropology* 71:1–11.

Forman, M. F., and French, J. A. (1992). Scent marking and food availability in golden-lion tamarins (*Leontopithecus rosalia*) and Wied's black tufted-ear marmoset (*Callithrix kuhli*). *American Journal of Primatology* 27(1):28.

Fowler, C. S. (1976). Ethnoecology. In D. L. Hardesty (ed.), *Ecological Anthropology* (pp. 215–243). New York: John Wiley.

Fragaszy, D. M. (1978). Contrasts in feeding behaviour in squirrel and titi monkeys. In D. J. Chivers and J. Herbert (eds.), *Recent Advances in Primatology, Vol. 1, Behavior* (pp. 363–367). London: Academic Press.

Fragaszy, D. M. (1986). Comparative studies of squirrel monkeys (*Saimiri sciureus*) and titi monkeys (*Callicebus moloch*): Performance on choice tasks in near space. *Journal of Comparative Psychology* 100(4):392–400.

Fragaszy, D. M. (1990). Early behavioral development in capuchins (*Cebus*). *Folia Primatologica* 54:119–128.

Fragaszy, D. M., and Adams-Curtis, L. E. (1994). Growth and reproduction in tufted capuchins: A ten-year retrospective of a captive colony. *American Journal of Primatology* 33(3):208.

Fragaszy, D. M., and Visalberghi, E. (1990). Social processes affecting the appearance of innovative behavior in capuchin monkeys. *Folia Primatologica* 54:155–165.

Fragaszy, D. M., Visalberghi, E., and Robinson, J. G. (1990). Variability and adaptability in the genus *Cebus*. *Folia Primatologica* 54:114–118.

Frazão, E. (1991). Insectivory in free-ranging bearded saki (*Chiropotes satanas chiropotes*). *Primates* 32(2):243–245.

Frazão, E. (1992). Dieta e Estratégia de Forregear de *Chiropotes satanas chiropotes* (Cebidae:Primates) na Amazônia Central Brasileira. Instituto Nacional de Pesquisas da Amazônia (INPA), Manaus, Brazil.

Freese, C. H. (1975). Final report: A census of non-human primates in Peru. PAHO Project AMR 0719 (pp. 1–49). Washington, DC: Pan American Health Organization.

Freese, C. H. (1976). Censusing *Alouatta palliata*, *Ateles geoffroyi* and *Cebus capucinus* in the Costa Rican dry forest. In R. W. Thorington, Jr. and P. G. Heltne (eds.), *Neotropical Primates: Field Studies and Conservation* (pp. 4–9). Washington, DC: National Academy of Sciences.

Freese, C. H., and Oppenheimer, J. R. (1981). The Capuchin Monkeys, genus *Cebus*. In A. F. Coimbra-Filho and R. A. Mittermeier (eds.), *Ecology and Behavior of Neotropical Primates* (pp. 331–390). Rio de Janeiro: Academia Brasileira de Ciencias.

Freese, C. H., Heltne, P. G., Castro, N., and Whitesides, G. (1982). Patterns and determinants of monkey densities in Peru and Bolivia, with notes on distributions. *International Journal of Primatology* 3(1):53–90.

French, J. A. (1992a). The behavioral endocrinology of captive breeding in *Leontopithecus* and *Callithrix*. *Abstracts of the XIVth Congress of the International Primatological Society*, Strasbourg, France, p. 96.

French, J. A. (1992b). Intruder studies in captive callitrichid primates: Are they models of intergroup encounters or immigration encounters? *Abstracts of the XIVth Congress of the International Primatological Society*, Strasbourg, France, p. 292.

French, J. A., and Inglett, B. J. (1989). Female-female aggression and male indifference in response to unfamiliar intruders in lion tamarins. *Animal Behaviour* 37:487–497.

French, J. A., and Snowdon, C. T. (1981). Sexual dimorphism in responses to unfamiliar intruders in the tamarin, *Saguinus oedipus*. *Animal Behaviour* 29:822–829.

Froehlich, J. W., and Froehlich, P. H. (1986). Dermatoglyphics and subspecific systematics of mantled howler monkeys (*Alouatta palliata*). In D. M. Taub and F. A. King (eds.), *Current Perspectives in Primate Biology* (pp. 107–121). New York: Van Nostrand Reinhold.

Froehlich, J. W., and Froehlich, P. H. (1987). The status of Panama's endemic howling monkeys. *Primate Conservation* 8:58–62.

Froehlich, J. W., and Thorington, R. W., Jr. (1982). The genetic structure and socio-ecology of howler monkeys (*Alouatta palliata*) on Barro Colorado Island. In E. G. Leigh, A. S. Rand, and D. M. Windsor (eds.), *The Ecology of a Tropical Forest* (pp. 291–305). Washington DC: Smithsonian Institution Press.

Froehlich, J. W., Supriatna, J., and Froehlich, P. H. (1991). Morphometric analyses of *Ateles*: Systematic and biogeographic implications. *American Journal of Primatology* 25:1–22.

Galbreath, G. J. (1983). Karyotypic evolution in *Aotus*. *American Journal of Primatology* 4:245–251.

Galetti, M., and Pedroni, F. (1994). Seasonal diet of capuchin monkeys (*Cebus apella*) in a semideciduous forest in south-east Brazil. *Journal of Tropical Ecology* 10:27–39.

Galetti, M., Laps, R., and Pedroni, F. (1987). Feeding behaviour of the brown howler monkey (*Alouatta fusca clamitans*) in a forest fragment in state of Sao Paulo, Brazil. *International Journal of Primatology* 8(5):542 (abstract).

Galetti, M., Pedroni, F., and Morellato, L. P. C. (1994). Diet of the brown howler monkey (*Alouatta fusca*) in a forest fragment in Brazil. *Mammalia* 58:111–118.

Galliari, C. A. (1985). Dental eruption in captive-born *Cebus apella*: From birth to 30 months old. *Primates* 26(4):506–510.

Ganzhorn, J. U. (1988). Food partitioning among Malagasy primates. *Oecologia* 7:436–450.

Ganzhorn, J. U. (1989). Niche separation of seven lemur species in the eastern rainforest of Madagascar. *Oecologia* 79:279–286.

Ganzhorn, J. U., Abraham, J. P., and Razanahoera-Rakotomalala, (1985). Some aspects of the natural history and food selection of *Avahi laniger*. *Primates* 26:452–463.

Garber, P. A. (1980a). Locomotor behavior and feeding ecology of the Panamanian tamarin (*Saguinus oedipus geoffroyi*, Callitrichidae, Primates). Unpublished Ph.D. Dissertation, Washington University, St, Louis, MO.

Garber, P. A. (1980b). Locomotor behavior and feeding ecology of the Panamanian tamarin (*Saguinus oedipus geoffroyi*, Callitricidae, Primates). *International Journal of Primatology* 1:185–201.

Garber, P. A. (1984a). Use of habitat and positional behavior in a neotropical primate, *Saguinus oedipus*. In P. Rodman and J. Cant (eds.), *Adaptations for Foraging in Nonhuman Primates* (pp. 112–133). New York: Columbia University Press.

Garber, P. A. (1984b). Proposed nutritional importance of plant exudates in the diet of the Panamanian tamarin, *Saguinus oedipus geoffroyi*. *International Journal of Primatology* 5:1–15.

Garber, P. A. (1986). The ecology of seed dispersal in two species of callitrichid primate (*Saguinus mystax* and *Saguinus fuscicollis*). *American Journal of Primatology* 10:155–170.

Garber, P. A. (1988a). Diet, foraging patterns, and resources defense in a mixed species troop of *Saguinus mystax* and *Saguinus fuscicollis* in Amazonian Peru. *Behaviour* 105(1–2):18–34.

Garber, P. A. (1988b). Foraging decisions during nectar feeding by tamarin monkeys (*Saguinus mystax* and *Saguinus fuscicollis*, Callitrichidae, Primates) in Amazonian Peru. *Biotropica* 20(2):100–106.

Garber, P. A. (1991). A comparative study of positional behavior in three species of tamarin monkeys. *Primates* 32:219–230.

Garber, P. A. (1992). Vertical clinging, small body size, and the evolution of feeding adaptations in the Callitrichinae. *American Journal of Physical Anthropology* 88(4):469–482.

Garber, P. A. (1993a). Feeding ecology and behaviour of the genus *Saguinus*. In A. B. Rylands (ed.), *Marmosets and Tamarins: Systematics, Behaviour and Ecology* (pp. 273–295). Oxford: Oxford University Press.

Garber, P. A. (1993b). Seasonal patterns of diet and ranging in two species of tamarin monkeys: Stability versus variability. *International Journal of Primatology* 14(1):145–166.

Garber, P. A. (1994a). A phylogenetic approach to the study of tamarin and marmoset social systems. *American Journal of Primatology* 34(2):111–114.

Garber, P. (1994b). Social and reproductive patterns in neotropical primates: Relation to ecology, body size, and infant care. *American Journal of Primatology* 34(2):111–114.

Garber, P. A., and Dolins, F. L. (1994). Use of spatial information and perceptual cues in primate foraging behavior. *American Journal of Physical Anthropology* Suppl. 18:92.

Garber, P. A., and Pruetz, J. D. (1995). Positional behavior in moustached tamarin monkeys: Effects of habitat on locomotor variability and locomotor stability. *Journal of Human Evolution* (in press).

Garber, P. A., and Teaford, M. F. (1986). Body weights in mixed species troops of *Saguinus mystax mystax* and *Saguinus fuscicollis nigrifrons* in Amazonian Peru. *American Journal of Physical Anthropology* 71:331–336.

Garber, P. A., Encarnación, F., Moya, L., and Pruetz, J. D. (1993). Demographic and reproductive patterns in moustached tamarin monkeys (*Saguinus mystax*): Implications for reconstructing platyrrhine mating systems. *American Journal of Primatology* 29:235–254.

Garber, P. A., Moya, L., Prueta, J. D., and Ique, C. (1995). Social and seasonal influences on reproductive biology in male moustached tamarins (*Saguinus mystax*). *American Journal of Primatology* (in press).

Garcia, F., Rodriguez, E., Canales, D., and Hermida, J. (1987). Situation of *Ateles* and *Alouatta* in Mexico. *International Journal of Primatology* 8(5):549.

Garcia Yuste, J. E. (1988). Patrones Etologicos y Ecologicos del Mono Nocturno, *Aotus azarae boliviensis*. Unpublished Ph.D. dissertation, University of Valencia, Spain.

Garcia, J. E., and Braza, F. (1987). Activity rhythms and use of space of a group of *Aotus azarae* in Bolivia during the rainy season. *Primates* 28(3):337–342.

Garcia, J. E., and Braza, F. (1989). Densities comparisons using different analytic methods in *Aotus azarae*. *Primate Report* 25:45–52.

Garraghty, P. E., Florence, S. L., Tenhula, W. N., and Kaas, J. H. (1991). Parallel thalamic activation of the 1st and 2nd somatosensory areas in prosimian primates and tree shrews. *Journal of Comparative Neurology* 311:289–299.

Gaulin, S. J. C. (1977). The ecology of *Alouatta seniculus* in Andean cloud forest. Unpublished Ph.D. thesis, Harvard University, Cambridge, MA.

Gaulin, S. J. C. (1979). A Jarmon/Bell model of primate feeding niches. *Human Ecology* 7(1):1–17.

Gaulin, S. J. C., and Gaulin, C. K. (1982). Behavioral ecology of *Alouatta seniculus* in Andean cloud forest. *International Journal of Primatology* 3(1):1–52.

Gebo, D. L. (1985). The nature of the primate grasping foot. *American Journal of Physical Anthropology* 67:269–278.

Gebo, D. L. (1987). Locomotor diversity in prosimian primates. *American Journal of Primatology* 13:271–282.

Gebo, D. L. (1989). Locomotor and phylogenetic considerations in anthropoid evolution. *Journal of Human Evolution* 18(3):201–233.

Gebo, D. L. (1992). Locomotor and postural behavior in Aloutta *palliata* and *Cebus capucinus*. *American Journal of Primatology* 26(4):277–290.

Gebo, D. L., Dagosto, M., Rosenberger, A. L., and Setoguchi, T. (1990). New platyrrhine tali from La Venta, Colombia. *Journal of Human Evolution* 19(6–7):737–746.

Gentry, A. H. (1988). Tree species richness of upper Amazonian forests. *Proceedings of the National Academy of Science U.S.A.* 85:156–159.

Gentry, A. H. (1993). Diversity and floristic composition of lowland tropical forest in Africa and South America. In P. Goldblatt (ed.), *Biological Relationships Between Africa and South America* (pp. 500–547). New Haven: Yale University Press.

Gibson, K. R. (1990). Tool use, imitation, and deception in a captive cebus monkey. In S. T. Parker and K. R. Gibson (eds.), *"Language" and Intelligence in Monkeys and Apes, Comparative Developmental Perspectives* (pp. 205–218). New York: Cambridge University Press.

Gilbert, K. A., and Stouffer, P. C. (1989). Use of a ground water source by mantled howler monkeys (*Alouatta palliata*). *Biotropica* 21(4):380.

Gilbert, K. A., Stouffer, P. C., and Styles, E. W. (1991). Dry season diet of *Cebus capucinus*. *American Journal of Primatology* 20(3):103.

Gingerich, P. D. (1984). Primate evolution: Evidence from the fossil record, comparative morphology, and molecular biology. *Yearbook of Physical Anthropology* 27:57–72.

Gingerich, P. D. (1985). South American mammals in the Paleocene of North America. In F. G. Stehli and S. D. Webb (eds.), *The Great American Biotic Interchange* (pp. 123–139). New York: Plenum.

Gingerich, P. D. (1993). Oligocene age of the Gebel Qatrani formation, Fayum, Egypt. *Journal of Human Evolution* 24(3):207–218.

Glander, K. E. (1975a). Habitat and resource utilization: An ecological view of social organization in mantled howling monkeys. Unpublished Ph.D. thesis, University of Chicago.

Glander, K. E. (1975b). Habitat description and resource utilization: A preliminary report on mantled howling monkey ecology. In R. Tuttle (ed.), *Socioecology and Psychology of Primates* (pp. 37–57). The Hague: Mouton.

Glander, K. E. (1977). Poison in a monkey's Garden of Eden. *Natural History* 86(3):34–41.

Glander, K. E. (1978a). Howling monkey feeding behavior and plant secondary compounds: A study of strategies. In G. G. Montgomery (ed.), *The Ecology of Arboreal Folivores* (pp. 561–573). Washington DC: Smithsonian Institution Press.

Glander, K. E. (1978b). Drinking from arboreal water sources by mantled howling monkeys (*Alouatta palliata* Gray). *Folia Primatologica* 29:206–217.

Glander, K. E. (1979). Feeding associations between howling monkeys and basilisk lizards. *Biotropica* 11:235–236.

Glander, K. E. (1980). Reproduction and population growth in free-ranging mantled howler monkeys. *American Journal of Physical Anthropology* 53:25–36.

Glander, K. E. (1981). Feeding patterns in mantled howling monkeys. In A. C. Kamil and T. D. Sargent (eds.), *Foraging Behavior, Ecological, Ethological, and Psychological Approaches* (pp. 231–257). New York: Garland.

Glander, K. E. (1992). Dispersal patterns in Costa Rican mantled howling monkeys. *International Journal of Primatology* 13(4):415–436.

Glander, K. E. (1994). Morphometrics and growth in captive aye-ayes (*Daubentonia madagascariensis*). *Folia Primatologica* 62:108–114.

Glander, K. E. (1994). Nonhuman primate self-medication with wild plant foods. In N. Etkin (ed.), *Eating on the Wild Side: The Pharmacologic, Ecologic, and Social Implications of Using Noncultigens* (pp. 227–239). Tucson: University of Arizona Press.

Glander, K. E., Wright, P. C., Seigler, D. S., Randrianasolo, V., and Randrianasolo, B. (1989). Consumption of cyanogenic bamboo by a newly discovered species of bamboo lemur. *American Journal of Primatology* 19:119–124.

Glander, K. E., Wright, P. C., Merenlender, A., and Daniels, P. S. (1992). Morphometrics and testicle size of prosimian primates in southeastern Madagascar. *Journal of Human Evolution* 22:1–17.

Godfrey, L. R. (1988). Adaptive diversification of Malagasy strepsirrhines. *Journal of Human Evolution* 17:93–134.

Godfrey, L. R., Lyon, S. K., and Sutherland, M. R. (1993). Sexual dimorphism in large-bodied primates: The case of the subfossil lemurs. *American Journal of Physical Anthropology* 90:315–334.

Godinot, M. (1994). Early North African primates and their significance for the origin of Simiiformes (=Anthropoidea). In J. G. Fleagle and R. F. Kay (eds.), *Anthropoid Origins* (pp. 235–295). New York: Plenum.

Goldizen, A. W. (1987a). Tamarins and marmosets: Communal care of offspring. In B. B. Smuts, D. L. Cheney, R. M. Seyfarth, R. W. Wrangham, and T. T. Struhsaker (eds.), *Primate Societies* (pp. 34–43). Chicago: University of Chicago Press.

Goldizen, A. W. (1987b). Facultative polyandry and the role of infant-carrying in wild saddle-back tamarins (*Saguinus fuscicollis*). *Behavioral Ecology and Sociobiology* 20:99–109.

Goldizen, A. W. (1988). Tamarin and marmoset mating systems: Unusual flexibility. *Trends in Ecology and Evolution* 3:36–40.

Goldizen, A. W. (1990). A comparative perspective on the evolution of tamarin and marmoset social systems. *International Journal of Primatology* 11(1):63–83.

Goldizen, A. W., and Terborgh, J. (1986). Cooperative polyandry and helping behavior in saddle-backed tamarins (*Saguinus fuscicollis*). In J. Else and P. Lee (eds.), *Primate Ecology and Conservation* (pp. 191–198). New York: Cambridge University Press.

Goldizen, A. W., Terborgh, J., Cornejo, F., Pooras, D. T., and Evans, R. (1988). Seasonal food shortage, weight loss, and the timing of births in saddle-back tamarins (*Saguinus fuscicollis*). *Journal of Animal Ecology* 57:893–901.

Goldsmith, M. L. (1987). Different feeding methods alter the social behavior of captive squirrel monkeys (*Saimiri sciureus* and *S. boliviensis*). *American Journal of Primatology* 12(3):345.

Good, K. R. (1986). Limiting factors in Amazonian ecology. In M. Harris and E. B. Ross (eds.), *Food and Evolution: Toward a Theory of Human Food Habits* (pp. 407–425). Philadelphia: Temple University Press.

Good, K. R. (1995). The Yanomami; keep on trekking. *Natural History* 104(4):56–65.

Goodall, J. (1986). *The Chimpanzees of Gombe*. Cambridge: Harvard University Press.

Goodman, S. M. (1991). The enigma of antipredator behavior in lemurs: Evidence of a large extinct eagle on Madagascar. *International Journal of Primatology* 15:129–134.

Goodman, S. M., O'Connor, S., and Langrand, O. (1993). A review of predation on lemurs: Implications for the evolution of social behavior in small nocturnal primates. In P. M. Kappeler and J. U. Ganzhorn (eds.), *Lemur Social Systems and Their Ecological Basis* (pp. 51–66). New York: Plenum.

Goosen, C. (1987). Social grooming in primates. In G. Mitchell and J. Erwin (eds.), *Comparative Primate Biology, Vol. 2B: Behavior, Cognition, and Motivation* (pp. 107–131). New York: Alan R. Liss.

Gould, L. (1992). Alloparental care in free-ranging *Lemur catta* at Berenty Reserve, Madagascar. *Folia Primatologica* 53:72.

Gould, S. J. (1977). *Ontogeny and Phylogeny*. Cambridge, MA: Harvard University Press.

Gozalo, A., and Montoya, E. (1990). Reproduction of the owl monkey (*Aotus nancymai*) in captivity. *American Journal of Primatology* 21:61–68.

Grand, T. I. (1972). A mechanical interpretation of terminal branch feeding. *Journal of Mammalogy* 53:198–201.

Gray, J. P. (1985). *Primate Sociobiology*. New Haven: HRAF.

Green, G. M., and Sussman, R. W. (1990). Deforestation history of the eastern rain forests of Madagascar from satellite images. *Science* 248:212–215.

Green, K. M. (1978). Primate censusing in northern Colombia: A comparison of two techniques. *Primates* 19:537–550.

Green, K. M. (1979). Vocalizations, behavior, and ontogeny of the golden lion tamarin, *Leontopithecus rosalia rosalia*. Unpublished Ph.D. dissertation, Johns Hopkins University, Baltimore, MD.

Green, K. M. (1980). An assessment of the Poço das Antas Reserve, Brazil, and prospects for survival of the golden lion tamarin, *Leontopithecus rosalia rosalia*. Report to World Wildlife Fund-US, Washington, DC.

Green, S. (1975). Variation of vocal pattern with social situation in the Japanese monkey (*Macaca fuscata*): A field study. In L. A. Rosenblum (ed.), *Primate Behavior*, Vol. 4 (pp. 1–102). New York: Academic Press.

Greene, J. C. (1959). *The Death of Adam: Evolution and Its Impact on Western Thought*. Ames: Iowa State University Press.

Gregory, W. K. (1922). *The Origin and Evolution of the Human Dentition*. Baltimore: Williams & Wilkins.

Gregory, W. K. (1955). *Evolution Emerging*. New York: MacMillan.

Grether, W. F. (1939). Color vision and color blindness in monkeys. *Comparative Psychological Monographs* 15:1–38.

Gross, D. R. (1975). Protein capture and cultural development in the Amazon basin. *American Anthropologist* 77(3):526–549.

Gross, D. R. (1982). Proteina y cultura en la Amazonia: Una segunda revision. *Amazonica Peruana* 3(6):85–97.

Gunter, R., Feigenson, L., and Blakeslee, P. (1965). Color vision in the cebus monkey. *Journal of Comparative Physiology and Psychology* 60:107–113.

Gursky, S. L. (1994). Infant care in the spectral Tarsier (*Tarsius spectrum*) Sulawesi, Indonesia. *International Journal of Primatology* 15(6):843–853.

Hailman, J. P., Ficken, M. S., and Ficken, R. W. (1985). The "chick-a-dee" call of *Parus atricapillus*: A recombinant system of animal communication compared with written English. *Semiotica* 56:191–224.

Hall, K. R. L. (1965). Social organization of the old-world monkeys and apes. *Symposium of the Zoological Society of London* 14:265–289.

Hallsworth, E. G. (1982). The human ecology of tropical forest. In E. G. Hallsworth (ed.), *Socio-economic Effects and Constraints in Tropical Forest Management* (pp. 1–8). New York: John Wiley.

Hamburger, V. (1934). The effects of wing-bud extirpation on the development of central nervous system in chick embryos. *Journal of Experimental Zoology* 68:449–494.

Hamburger, V., and Yip, J. W. (1984). Reduction of experimentally induced neuronal death in spinal ganglia of the chick embryo by nerve growth factor. *Journal of Neuroscience* 4:767–774.

Hames, R. B. (1979a). Relatedness and interaction among the Ye'kwana: A preliminary analysis. In N. A. Chagnon and W. Irons (eds.), *Evolutionary Biology and Human Social Behavior* (pp. 239–249). N. Scituate, MA: Duxbury.

Hames, R. B. (1979b). A comparison of the efficiencies of the shotgun and the bow in neotropical forest hunting. *Human Ecology* 7(3):219–252.

Hames, R. B., and Vickers, W. T. (1982). Optimal diet breadth theory as a model to explain variability in Amazonian hunting. *American Ethnologist* 9(2):358–378.

Hamilton, W. D. (1964). The genetic evolution of social behavior. *Journal of Theoretical Biology* 7: 1–52.

Hampton, J. K., Hampton, S. H., and Landwehr, B. T. (1966). Observations on a successful breeding colony of the marmoset, *Oedipomidas oedipus*. *Folia Primatologica* 4:265–287.

Hanihara, T., and Natori, M. (1989). Evolutionary trends of the hairy-facey *Saguinus* in terms of dental and cranial morphology. *Primates* 30(4):531–541.

Happel, R. E. (1982). Ecology of *Pithecia hirsuta* in Peru. *Journal of Human Evolution* 11(7):581–590.

Haq, B. U., Hardenbol, J., and Vail, P. R. (1987). Chronology of fluctuating sea levels since the Triassic. *Science* 235:1156–1167.

Harad, M. L., Schneider, H., Schneider, M. P. C., Sampaio, I., Czelusniak, J., and Goodman, M. (1995). DNA evidence on the phylogenetic systematics of New World monkeys: Support for the sister grouping of *Cebus* and *Saimiri* from two unlinked nuclear genes. *Molecular Phylogenetics and Evolution* (in press).

Haraway, D. (1989). *Primate Visions; Gender, Race, and Nature in the World of Modern Science.* New York: Routledge.

Harcourt, C. S. (1987). Brief trap/retrap study of the brown mouse lemur (*Microcebus rufus*). *Folia Primatologica* 49:209–211.

Harcourt, C. S. (1991). Diet and behaviour of a nocturnal lemur, *Avahi laniger*, in the wild. *Journal Zoology, London* 223:667–674.

Harcourt, C. S., and Thornback, J. (1990). *Lemurs of Madagascar and the Comoros.* Gland, Switzerland and Cambridge, UK: IUCN.

Harding, R. S. O., and Miller, L. E. (1991). Regulation of intergroup spacing through vocalization in *Cebus olivaceus*. *American Journal of Physical Anthropology* Suppl. 12:87.

Haring, D. M., Wright, P. C., and Simons, E. L. (1987). Conservation of Madagascar's Sifakas (*Propithecus*) in captivity and in the wild. In B. L. Dresser, R. W. Reece, and E. J. Maruska (eds.), *5th World Conference. Breeding Endangered Species in Captivity.* Cincinnati: Cincinnati Zoo.

Harris, J. C., and Newman, J. D. (1988). Combined opiate/adrenergic receptor blockade enhances squirrel monkey vocalization. *Pharmacology, Biochemistry and Behavior* 31(1):223–226.

Harrison, M. L., and Tardif, S. D. (1989). Gum-feeding, gum-foraging and social behavior in marmosets and tamarins. *American Journal of Primatology* 72(2):269.

Harrison, T. (1986). The phylogenetic relationships of the early catarrhine primates: A review of the current evidence. *Journal of Human Evolution* 16:41–80.

Hartwig, W. E. (1994). Patterns, puzzles and perspectives on Platyrrhine origins. In R. S. Corruccini and R. L. Ciochon (eds.), *Integrative Paths to the Past* (pp. 69–94). Englewood Cliffs, NJ: Prentice-Hall.

Hartwig, W. E. (1995). A giant New World monkey from the Pleistocene of Brazil. *Journal of Human Evolution* 28:189–195.

Hartwig, D. M., and Cortelle, C. (1995). New large platyrrhine primates from the pleistocene of Bahia, Brasil. *American Journal of Vertebrate Paleontology* 33a.

Harvey, P. H., Martin, R. D., and Clutton-Brock, T. H. (1987). Life histories in comparative perspective. In B. Smuts, D. Cheney, R. Seyfarth, R. Wrangham, and T. Struhsaker (eds.), *Primate Societies* (pp. 181–196). Chicago: University of Chicago Press.

Hasegawa, M. (1990). Phylogeny and molecular evolution in primates. *Japanese Journal of Genetics* 65:243–266.

Hayes, S. L., and Snowdon, C. T. (1990). Responses to predators in cotton-top tamarins. *American Journal of Primatology* 20:283–291.

Headland, T. N. (1987). The wild yam question: How well could independent hunter-gatherers live in a tropical rainforest ecosystem? *Human Ecology* 15:465–493.

Headland, T. N., and Bailey, R. C. (1991). Introduction: Have hunter-gatherers ever lived in tropical rain forest independently of agriculture? *Human Ecology* 19(2):115–122.

Headland, T. N., and Reid, L. A. (1989). Hunter-gatherers and their neighbors from prehistory to the present. *Current Anthropology* 30(1):43–66.

Heistermann, M., Kleis, E., Proeve, E., and Wolters, H. J. (1989). Fertility status, dominance, and scent marking behavior of family-housed female cotton-top tamarins (*Saguinus oedipus*) in absence of their mothers. *American Journal of Primatology* 18(3):177–189.

Heltne, P. G. (1977). Census of *Aotus* in the north of Colombia. Report on PAHO project AMRO-3171. Washington, DC: Pan American Health Organization.

Heltne, P. G., Turner, D. C., and Wolhandler, J. (1973). Maternal and paternal periods in the development of infant *Callimico goeldii*. *American Journal of Physical Anthropology* 38:555–560.

Heltne, P. G., Wojcik, J. F., and Pook, A. G. (1981). Goeldi's monkey, Genus *Callimico*. In A. F. Coimbra-Filho and R. A. Mittermeier (eds.), *Ecology and Behavior of Neotropical Primates*, Vol. 1 (pp. 169–210). Rio de Janeiro: Academia Brasileira de Ciências.

Hemmer, H. (1971). Beitrag zur Erfassung der Progressiven Cephalisation bei primaten. In H. Hemmer, H. Biegert, and W. Leutenegger (eds.), *Proceedings of the 3rd International Congress of Primatology* 1 (pp. 99–107). Karger.

Hendrickson, A. E., and Wilson, J. R. (1979). A difference in (^{14}C) deoxyglucose autoradiographic patterns in striate cortex between *Macaca* and *Saimiri* monkeys following monocular stimulation. *Brain Research* 170:353–358.

Hendrickson, A. E., Wilson, J. R., and Ogren, M. E. (1978). The neuroanatomical organization of pathways between the dorsal lateral geniculate nucleus and visual cortex in Old World and New World primates. *Journal of Comparative Neurology* 182:123–136.

Hennessy, M. B. (1986a). Multiple, brief maternal separations in the squirrel monkey: Changes in hormonal and behavioral responsiveness. *Physiology and Behavior* 36(2):245–250.

Hennessy, M. B. (1986b). Maternal separation alters later consumption of novel liquids in the squirrel monkey. *Behavioral and Neural Biology* 45(2):254–260.

Hernández-Comacho, J., and Cooper, R. W. (1976). The non-human primates of Colombia. In R. W. Thorington, Jr. and P. W. Heltne (eds.), *Neotropical Primates* (pp. 35–69). Washington, DC: National Academy of Sciences.

Hershkovitz, P. (1949). Mammals of Northern Colombia. Preliminary Report No. 4: Monkeys (Primates), with taxonomic revisions of some forms. *Proceedings of the United States National Museum* 98:323–427.

Hershkovitz, P. (1963). A systematic and zoogeographic account of the monkeys of the genus *Callicebus* (Cebidae) of the Amazonas and Orinoco River basins. *Mammalia* 27:1–79.

Hershkovitz, P. (1966). Taxonomic notes on tamarins, genus *Saguinus* (Callitrichidae, Primates) with descriptions of four new forms. *Folia Primatologica* 4(5):381–395.

Hershkovitz, P. (1968). Metachromism or the principle of evolutionary change in mammalian tegumentary colors. *Evolution* 22:556–575.

Hershkovitz, P. (1972). The recent mammals of the neotropical region: A zoogeographic and ecological review. In A. Keast, F. C. Erk, and B. Glass (eds.), *Evolution, Mammals, and Southern Continents* (pp. 311–430). Albany: State University New York Press.

Hershkovitz, P. (1974). A new genus of late Oligocene monkey (Cebidae, Platyrrhini) with notes on postorbital closure and platyrrhine evolution. *Folia Primatologica* 21:1–35.

Hershkovitz, P. (1977). *Living New World Monkeys (Platyrrhini) with an Introduction to Primates*, Vol. 1. Chicago: University of Chicago Press.

Hershkovitz, P. (1979). The species of sakis, genus *Pithecia* (Cebidae, Primates), with notes on sexual dichromatism. *Folia Primatologica* 31:1–22.

Hershkovitz, P. (1983). Two new species of night monkeys, genus *Aotus* (Cebidae, Platyrrhini): A preliminary report on *Aotus* taxonomy. *American Journal of Primatology* 4(3):209–244.

Hershkovitz, P. (1984). Taxonomy of the squirrel monkey, genus *Saimiri* (Cebidae, Platyrrhini): A preliminary report with description of a hitherto unnamed form. *American Journal of Primatology* 6(4):257–312; 7(2):155–210.

Hershkovitz, P. (1985). A preliminary taxonomic review of the South American bearded saki monkeys, genus *Chiropotes* (Cebidae, Platyrrhini), with the description of a new subspecies. *Fieldiana, Zoology*, 27(1363):1–46.

Hershkovitz, P. (1987a). Uacaries, New World monkeys of the genus *Cacajao* (Cebidae, Plaltyrrhini): A preliminary taxonomic review with the description of a new subspecies. *American Journal of Primatology* 12:1–53.

Hershkovitz, P. (1987b). The taxonomy of South American sakis, genus *Pithecia* (Cebidae, Platyrrhini): A preliminary report and critical review of the description of a new species and a new subspecies. *American Journal of Primatology* 12(4):387–468.

Hershkovitz, P. (1988). Origin, speciation, and distribution of South American titi monkeys, genus *Callicebus* (Family Cebidae, Platyrrhini). *Proceedings of the Academy of Natural Science Philadelphia* 140(1):240–272.

Hershkovitz, P. (1990). Titis, New World monkeys of the genus *Callicebus* (Cebidae, Platyrrhini): A preliminary taxonomic review. *Fieldiana, Zoology* 55:1–109.

Herzog, M., and Hopf, S. (1984). Behavioral responses to species-specific warning calls in infant squirrel monkeys reared on social isolation. *American Journal of Primatology* 7:99–106.

Herzog, M., and Meyer, S. (1987). Alarm behavior in free-ranging squirrel monkeys (*Saimiri oerstedii oerstedii* and *S. o. citrinellus*). *International Journal of Primatology* 8(5):537.

Hess, D. T., and Edwards, M. A. (1988). Anatomical demonstration of ocular segregation in the retinogeniculocortical pathway of the New World capuchin monkey (*Cebus appella*). *Journal of Comparative Neurology* 264:409–420.

Heymann, E. W. (1987). A field observation of predation on a moustached tamarin (*Saguinus mystax*) by an anaconda. *International Journal of Primatology* 8:193–195.

Heymann, E. W., and Bartecki, U. (1990). A young saki monkey, *Pithecia hirsuta*, feeding on ants, *Cephalotes atratus*. *Folia Primatologica* 55(3–4):181–1845.

Hick, U. (1968). The collection of saki monkeys at Cologne Zoo. *International Zoo Yearbook* 8:192–194.

Hill, W. C. O. (1953a). Observations on the genitalia of the woolly monkey (*Lagothrix*). *Proceedings of the Zoological Society of London* 122:973–984.

Hill, W. C. O. (1953b). *Primates: Comparative Anatomy and Taxonomy, I. Strepsirhini*. Edinburgh: Edinburgh University Press.

Hill, W. C. O. (1957). *Primates: Comparative Anatomy and Taxonomy, III. Hapalidae*. Edinburgh: Edinburgh University Press.

Hill, W. C. O. (1959). The anatomy of *Callimico goeldii* (Thomas). *Transactions of the Philosophical Society* 49(5):1–116.

Hill, W. C. O. (1960). *Primates, Comparative Anatomy and Taxonomy, IV. Cebidae, Part A*. Edinburgh: Edinburgh University Press.

Hill, W. C. O. (1962). *Primates, Comparative Anatomy and Taxonomy, V. Cebidae, Part B*. Edinburgh: Edinburgh University Press.

Hinde, R. A. (1977). On assessing the basis of partner preferences. *Behavior* 62:1–9.

Hirabuki, Y., and Izawa, K. (1990). Chemical properties of soils eaten by wild red howler monkeys (*Alouatta seniculus*): A preliminary study. *Field Studies of New World Monkeys, La Macarena, Colombia* 3:25–28.

Hladik, A. (1980). The dry forest on the west coast of Madagascar: Climate, phenology and food available for prosimians. In P. Charles-Dominique, M. M. Cooper, A. Hladik, C. M. Hladik, E. Payes, G. F. Pariente, A. Petter Rousseaux, and A. Schilling (eds.), *Nocturnal Malagasy Primates* (pp. 3–40). New York: Academic Press.

Hladik, C. M. (1967). Surface relative du tractus digestif de quelques primates, morphologie des villosities intestinales et correlations avec le regime alimentaire. *Mammalia* 31:120–147.

Hladik, A., and Hladik, C. M. (1969). Rapports trophiques entre vegetation et primates dans la foret de Barro-Colorado (Panama). *Terre et la Vie* 1:25–117.

Hladik, C. M., Hladik, A., Bousset, J., Valdebouze, P., Viroben, G., and Delort-Laval, J. (1971). Le regime alimentaire des primates de l'ile de Barro Colorado (Panama). *Folia Primatologica* 16:85–122.

Hoage, R. J. (1977). Parental care in *Leontopithecus rosalia rosalia*: Sex and age differences in carrying behavior and the role of prior experience. In D. G. Kleiman (ed.), *The Biology and Conservation of the Callitrichidae* (pp. 293–305). Washington, DC: Smithsonian Institution Press.

Hoage, R. J. (1982). Social and physical maturation in captive lion tamarins, *Leontopithecus rosalia rosalia* (Primates: Callitrichidae). *Smithsonian Contributions to Zoology* 354:1–56.

Hodun, A., Snowdon, C. T., and Soini, P. (1981). Subspecific variation in the long calls of the tamarin (*Saguinus fuscicollis*). *Zeitschrift für Tierpsychologie* 57:97–110.

Hoffman, K. A., Mendoza, S. P., Hennessy, M. B., and Mason, W. A. (1994). Pituitary-adrenal responses to separation in infant titi monkeys (*Callicebus moloch*): Evidence for paternal attachment. *American Journal of Primatology* 33(3):214–215.

Hoffstetter, R. (1972). Relationships, origins, and history of the ceboid monkeys and caviomorph rodents: A modern reinterpretation. In T. Dobzhansky, M. K. Hecht, and W. C. Steere (eds.), *Evolutionary Biology* (pp. 323–347). New York: Appeleton-Century-Crofts.

Hoffstetter, R. (1980). Origin and deployment of New World monkeys emphasizing the southern continents route. In R. Ciochon and A. B. Chiarelli (eds.), *Evolutionary Biology of New World Monkeys and Continental Drift* (pp. 103–122). New York: Plenum.

Holloway, R. L., Jr., and Post, D. G. (1982). The relativity of relative brain measures and hominid mosaic evolution. In E. Armstrong and D. Falk (eds.), *Primate Brain Evolution: Methods and Concepts* (pp. 57–76). New York: Plenum.

Homburg, I. (1987). Social behavior of family groups of white-faced sakis (*Pithecia pithecia*) under captive conditions. *International Journal of Primatology* 8(5):506.

Homburg, I. (1989). Soziale Strukturmerkmale von Weissgesicht-Sakis (*Pithecia pithecia*). Bielefeld: Diplomarbeit, Fakultt fr Biologie, Universitt Bielefeld.

Hopf, S., and Ploog, D. (1991). Insect-catching behavior in socially inexperienced infant squirrel monkeys. *Primates* 32(1):111–114.

Horwich, R. H. (1983a). Breeding behaviors of the black howler monkey, *Alouatta pigra*, of Belize. *Primates* 24:222–230.

Horwich, R. H. (1983b). Species status of the black howler monkey, *Alouatta pigra*, of Belize. *Primates* 24:288–289.

Horwich, R. H., and Gebhard, K. (1983). Roaring rhythms in black howler howler monkeys, (*Alouatta pigra*) of Belize. *Primates* 24:290–296.

Horwich, R. H., and Johnson, E. D. (1986). Geographical distribution of the black howler (*Alouatta pigra*) in Central America. *Primates* 27:53–62.

Hoshino, J., Mori, A., Kudo, K., and Kawai, M. (1984). Preliminary report on the grouping of mandrills (*Mandrillus sphinx*) in Cameroon. *Primates* 25:295–307.

Howe, H. F.(1983). Annual variation in a neotropical seed-dispersal system. In S. L. Sutton, T. C. Whitmore, and A. C. Chadwick (eds.), *Tropical Rain Forest: Ecology and Management* (pp. 211–227). London: Blackwell.

Hrdy, S. B. (1988). Daughters or sons. *Natural History* 97(4):63–83.

Hubel, D. H., and Wiesel, T. N. (1968). Receptive fields and functional architecture of monkey striate cortex. *Journal of Physiology (London)* 195:215–243.

Hubel, D. H., Wiesel, T. N., and LeVay, S. (1976). Functional architecture of area 17 in normal and monocularly deprived macaque monkeys. *Cold Spring Harbor Symposium on Quantitative Biology* 40:581–599.

Hubrecht, R. C. (1984). Studies on the common marmosets (*Callithrix jacchus jacchus*) in captivity and in the wild: The social and reproductive status of daughters and non-breeding females. *Primate Eye* 26:23.

Hubrecht, R. C. (1985). Home range size and use and territorial behavior in the common marmosets *Callithrix jacchus jacchus* at the Tapacura Field Station, Recife, Brazil. *International Journal of Primatology* 6(5):531–548.

Hubrecht, R. C. (1986). Operation Raleigh Primate Census in the Maya Mountains, Belize. *Primate Conservation* 7:15–17.

Hunter, J., Martin, R. D., Dixson, A. F., and Rudder, B. C. C. (1979). Gestation and interbirth intervals in the owl monkey (*Aotus tirvirgatus griseimembra*). *Folia Primatologica* 31:165–175.

Hutterer, K. L. (1988). The prehistory of the Asian rain forests. In J. S. Denslow and C. Padoch (eds.), *People of the Tropical Rain Forest* (pp. 63–72). Berkeley: University California Press.

Inglett, B. J., and French, J. A. (1987a). The dynamics of intra-family aggression and social integration in lion tamarins. *International Journal of Primatology* 8(5):511.

Inglett, B. J., and French, J. A. (1987b). Social preferences of mated golden lion tamarins. *International Journal of Primatology* 8(5):508.

Inglett, B. J., and French, J. A. (1987c). Sexually dimorphic aggressive responses to intruders in lion tamarins (*Leontopithecus rosalia*). *American Journal of Primatology* 12(3):350.

Inglett, B. J., French, J. A., Simmons, L. G., and Vires, K. W. (1989). Dynamics of intrafamily aggression and social reintegration in lion tamarins. *Zoo Biology* 8(1):67–78.

Irons, W. (1981). Why lineage exogamy? In R. D. Alexander and D. W. Tinkle (eds.), *Natural Selection and Social Behavior* (pp. 276–289). New York: Chiron.

Isaac, G. (1977). The food sharing behavior of proto human Hominids. *Scientific American* 238(4):90–109.

Iwano, T., and Iwakawa, C. (1988). Feeding behaviour of the aye-aye (*Daubentonia madagascariensis*) on nuts of Ramy (*Canarium madagascariensis*). *Folia Primatologica* 50:136–142.

Izawa, K. (1975). Foods and feeding behavior of monkeys in the Upper Amazon Basin. *Primates* 16(3):295–316.

Izawa, K. (1976). Group sizes and compositions of monkeys in the Upper Amazon basin. *Primates* 17:367–399.

Izawa, K. (1978a). Frog-eating behavior of wild black-capped capuchin (*Cebus apella*). *Primates* 19(4):633–642.

Izawa, K. (1978b). A field study of the ecology and behavior of the black-mantled tamarin (*Saguinus nigricollis*). *Primates* 19:241–274.

Izawa, K. (1979). Studies on peculiar distribution pattern of *Callimico*. *Kyoto University Overseas Research Reports of New World Monkeys* 1:1–19.

Izawa, K. (1980). Social behavior of the wild black-capped capuchin (*Cebus apella*). *Primates* 21:443–467.

Izawa, K. (1988). Preliminary report on social changes of black-capped capuchins (*Cebus apella*). *Field Studies of New World Monkeys, La Macarena, Colombia* 1:13–18.

Izawa, K. (1989). The adoption of an infant observed in a wild group of red howler monkeys (*Alouatta seniculus*). *Field Studies of New World Monkeys, La Macarena, Colombia* 2:33–36.

Izawa, K. (1990). Social changes within a group of wild black-capped capuchins (*Cebus apella*) in Colombia (II). *Field Studies of New World Monkeys, La Macarena, Colombia* 3:1–5.

Izawa, K., and Bejarano, G. (1981). Distribution ranges and patterns of nonhuman primates in western Pando, Bolivia. *Kyoto University Overseas Research Reports of New World Monkeys* 2:1–11.

Izawa, K., and Lozano, H. (1989). Social changes within a group and reproduction of wild howler monkeys (*Alouatta seniculus*) in Colombia. *Field Studies of New World Monkeys, La Macarena, Colombia* 2:1–6.

Izawa, K., and Lozano, H. M. (1990a). Frequency of soil-eating by a group of wild howler monkeys (*Alouatta seniculus*) in La Macarena, Colombia. *Field Studies of New World Monkeys, La Macarena, Colombia* 4:47–56.

Izawa, K., and Lozano, H. M. (1990b). River crossing by a wild howler monkey (*Alouatta seniculus*). *Field Studies of New World Monkeys, La Macarena, Colombia* 3:29–33.

Izawa, K., and Nishimura, A. (1988). Primate fauna at the study site. *Field Studies of New World Monkeys, La Macarena, Colombia* 1:5–11.

Izawa, K., and Yoneda, M. (1981). Habitat utilization of nonhuman primates in a forest of the western Pando, Bolivia. *Kyoto University Overseas Research Reports of New World Monkeys* 2:13–22.

Jacobs, G. H. (1963). Spectral sensitivity and color vision of the squirrel monkey. *Journal of Comparative Physiology and Psychology* 56:616–621.

Jacobs, G. H. (1977a). Visual capacities of the owl monkey (*Aotus trivirgatus*). I. Spectral sensitivity and color vision. *Vision Research* 17:811–820.

Jacobs, G. H. (1977b). Visual capacities of the owl monkey (*Aotus trivirgatus*). II. Spatial contrast sensitivity. *Vision Research* 17:821–825.

Jacobs, G. H. (1981). *Comparative Color Vision*. New York: Academic Press.

Jacobs, G. H. (1983a). Differences in spectral response properties of LGN cells in male and female squirrel monkeys. *Vision Research* 23:461–468.

Jacobs, G. H. (1983b). Within-species variations in visual capacity among squirrel monkeys (*Saimiri sciureus*): Sensitivity differences. *Vision Research* 23:239–248.

Jacobs, G. H. (1984). Within-species variations in visual capacity among squirrel monkeys (*Saimiri sciureus*): Color vision. *Vision Research* 24:1267–1277.

Jacobs, G. H. (1990a). Evolution of mechanisms for color vision. In M. H. Brill (ed.), *Perceiving, Measuring and Using Color. SPIE Proceedings* 1250:287–292.

Jacobs, G. H. (1990b). Discrimination of luminance and chromaticity differences by dichromatic and trichromatic monkeys. *Vision Research* 30:387–397.

Jacobs, G. H. (1991). Variations in colour vision in non-human primates. In D. H. Foster (ed.), *Inherited and Acquired Colour Vision Deficiencies* (pp. 199–214). London: Macmillan.

Jacobs, G. H. (1993). The distribution and nature of colour vision among the mammals. *Biological Review* 68:413–471.

Jacobs, G. H., and Blakeslee, B. (1984). Individual variations in color vision among squirrel monkeys (*Saimiri sciureus*) of different geographical origin. *Journal of Comparative Psychology* 98:347–357.

Jacobs, G. H., and Deegan, J. F. II (1994). Photopigment polymorphism in cotton-top tamarins (*Saguinus oedipus*). *American Journal of Primatology* 33:217.

Jacobs, G. H., and Harwerth, R. S. (1989). Color vision variations in Old and New World primates. *American Journal of Primatology* 18:35–44.

Jacobs, G. H., and Neitz, J. (1985a). Spectral positioning of mammalian cone pigments. *Journal of the Optical Society of America* A2:P23.

Jacobs, G. H., and Neitz, J. (1985b). Color vision in squirrel monkeys: Sex-related differences suggest the mode of inheritance. *Vision Research* 25:141–144.

Jacobs, G. H., and Neitz, J. (1987a). Inheritance of color vision in a New World monkey (*Saimiri sciureus*). *Proceedings of the National Academy of Science U.S.A.* 84:2545–2549.

Jacobs, G. H., and Neitz, J. (1987b). Polymorphism of the middle wavelength cone in two species of South American monkey: *Cebus apella* and *Callicebus moloch*. *Vision Research* 27:1263–1268.

Jacobs, G. H., Bowmaker, J. K., and Mollon, J. D. (1981). Behavioural and microspectrophotometric measurements of colour vision in monkeys. *Nature (London)* 292:541–543.

Jacobs, G. H., Neitz, J., and Crognale, M. (1987). Color vision polymorphism and its photopigment basis in a callitrichid monkey (*Saguinus fuscicollis*). *Vision Research* 27:2089–2100.

Jacobs, G. H., Neitz, J., Crognale, M., and Brammer, G. T. (1991). Spectral sensitivity of vervet monkeys (*Cercopithecus aethiops sabeus*) and the issue of catarrhine trichromacy. *American Journal of Primatology* 23:185–195.

Jacobs, G. H., Deegan, J. F. II, Neitz, J., Crognale, M. A., and Neitz, M. (1993a). Photopigments and color vision in the nocturnal monkey, *Aotus*. *Vision Research* 33(13):1773–1783.

Jacobs, G. H., Neitz, J., and Neitz, M. (1993b). Genetic basis of polymorphism in the color vision of platyrrhie monkeys. *Vision Research* 33:269–274.

Jaimez, Salgado E., Gutierrez, Calvache D., MacPhee, R. D. E., and Gould, G. C. (1992). The monkey caves of Cuba. *Transactions of the British Cave Research Association* 19:25–30.

Jameson, D., and Hurvich, L. M. (1989). Essay concerning color constancy. *Annual Review of Psychology* 40:1–22.

Janson, C. H. (1975). Ecology and population densities of primates in a Peruvian rainforest. Unpublished undergraduate thesis, Princeton University.

Janson, C. H. (1983). Adaptation of fruit morphology to dispersal agents in a neotropical rainforest. *Science* 219:187–189.

Janson, C. H. (1984a). Female choice and mating system of the brown capuchin monkey *Cebus apella* (Primates: Cebidae). *Zeitschrift für Tierpsychologie* 65:177–200.

Janson, C. H. (1984b). Capuchin-like monkeys. In D. Macdonald (ed.), *The Encyclopedia of Mammals* (pp. 352–357). New York: Facts on File Publications.

Janson, C. H. (1985). Aggressive competition and individual food consumption in wild brown capuchin monkeys (*Cebus apella*). *Behavioral Ecology and Sociobiology* 18:125–138.

Janson, C. H. (1986a). Capuchin counterpoint. *Natural History* 95(2):44–53.

Janson, C. H. (1986b). Ecological and social consequences of food competition in brown capuchin monkeys. Unpublished Ph.D. Dissertation, University of Washington, Seattle.

Janson, C. H. (1986c). The mating system as a determinant of social evolution in capuchin monkeys (*Cebus*). In J. G. Else and P. C. Lee (eds.), *Primate Ecology and Conservation* (pp. 169–179). New York: Cambridge University Press.

Janson, C. H. (1988a). Intra-specific food competition and primate social structure: A synthesis. *Behaviour* 105(1–2):1–17.

Janson, C. H. (1988b). Food competition in brown capuchin monkeys (*Cebus apella*): Quantitative effects of group size and tree productivity. *Behaviour* 105(1–2):53–76.

Janson, C. H. (1990a). Social correlates of individual spatial choice in foraging groups of brown capuchin monkeys, *Cebus apella*. *Animal Behaviour* 40(5):910–921.

Janson, C. H. (1990b). Ecological consequences of individual spatial choices in foraging groups of brown capuchin monkeys, *Cebus apella*. *Animal Behaviour* 40(5):922–934.

Janson, C. H. (1993). Primate group size, brains and communication: A New World perspective. *Behavioral and Brain Sciences* 16(4):711–712.

Janson, C. H. (1994). Comparison of mating system across two populations of brown capuchin monkeys. *American Journal of Primatology* 33(3):217.

Janson, C. H., and Boinski, S. (1992). Morphological and behavioral adaptations for foraging in generalist primates: The case of the Cebines. *American Journal of Physical Anthropology* 88(4):483–498.

Janson, C. H., and van Schaik, C. P. (1993). Ecological risk aversion in juvenile primates: Slow and steady wins the race. In M. E. Pereira and L. A. Fairbanks (eds.), *Juvenile Primates; Life History, Development, and Behavior* (pp. 57–76). New York: Oxford University Press.

Janson, C. H., Terborgh, J., and Emmons, L. H. (1981). Non-flying mammals as pollinating agents in the Amazonian forest. *Biotropica, Reproductive Botany* Supplement:1–6.

Janzen, D. H. (1986). The future of tropical ecology. *Annual Review of Ecology and Systematics* 17:305–324.

Jenkins, P. D., and Albrecht, G. H. (1991). Sexual dimorphism and sex ratios in Madagascan prosimians. *American Journal of Primatology* 24:1–14.

Jerison, H. J. (1973). *Evolution of the Brain and Intelligence*. New York: Academic Press.

Jernvall, J., Wright, P. C., and Rakotonirina, G. (1995). Ontogeny of toothwear of *Propithecus diadema edwardsi* in relation to seed predation. *XVth Congress of the International Primatology Society, Bali, Indonesia: Handbook and Abstracts*, p. 364.

Johns, A. D. (1985). First field observations of *Pithecia albicans*. *Primate Eye* 26:17–18.

Johns, A. D. (1986a). Current issues in Amazonian primate conservation. *Primate Eye* 29(Suppl.):42–45.

Johns, A. D. (1986b). Notes on the ecology and current status of the buffy saki, *Pithecia albicans*. *Primate Conservation* 7:26–29.

Johns, A. D. (1986c). Effects of habitat disturbance on rainforest wildlife in Brazilian Amazonia. Final report of WWF U.S. Project US-302. Washington, DC: World Wildlife Fund.

Johns, A. D., and Ayres, J. M. (1987). Southern bearded sakis beyond the brink. *Oryx* 21(3):164–167.

Johnson, B., and Buol, S. (in prep.). Comparisons of soil in Ranomafana National Park. Ranomafana Symposium.

Johnson, L. D., Petto, A. J., Boy, D. S., Sehgal, P. K., and Beland, M. E. (1986). The effect of perinatal and juvenile mortality on colony-born production at the New England Regional Primate Research Center. In K. Benirschke (ed.), *Primates: The Road to Self-Sustaining Populations* (pp. 771–779). New York: Springer-Verlag.

Jolly, A. (1966a). *Lemur Behavior—A Madagascar Field Study*. Chicago: University Chicago Press.

Jolly, A. (1966b). Lemur social behavior and primate intelligence. *Science* 153:501–506.

Jolly, A. (1972). Troop continuity and troop spacing in *Propithecus verreauxi* and *Lemur catta* at Berenty (Madagascar). *Folia Primatologica* 17:335–362.

Jolly, A. (1984). The puzzle of female feeding priority. In M. F. Small (ed.), *Female Primates: Studies by Women Primatologists* (pp. 197–215). New York: Alan R. Liss.

Jolly, A. (1985). *The Evolution of Primate Behavior*, 2nd ed. New York: Macmillan.

Jones, C. B. (1978). Aspects of reproductive behavior in the mantled howler monkey, *Alouatta palliata* Gray. Unpublished Ph.D. thesis, Cornell University, Ithaca, NY.

Jones, C. B. (1979). Grooming in the mantled howler monkey, *Alouatta palliata* Gray. *Primates* 20:289–292.

Jones, C. B. (1980a). The functions of status in the mantled howler monkey, *Alouatta palliata* Gray: Intraspecific competition for group membership in a folivorous neotropical primate. *Primates* 21:389–405.

Jones, C. B. (1980b). Seasonal parturition mortality and dispersal in the mantled howler monkey *Alouatta palliata* Gray. *Brenesia* 17:1–10.

Jones, C. B. (1982). A field manipulation of spatial relations among male mantled howler monkeys. *Primates* 23:130–134.

Jones, C. B. (1983a). Social organization of captive black howler monkeys (*Alouatta caraya*): "Social competition" and the use of non-damaging behavior. *Primates* 24:25–39.

Jones, C. B. (1983b). Do howler monkeys feed upon legume flowers preferentially at flower opening time? *Brenesia* 21:41–46.

Jones, C. B. (1985). Reproductive patterns in mantled howler monkeys: Estrus, mate choice and copulation. *Primates* 26:130–142.

Jones, M. L. (1986). Successes and failures of captive breeding. In K. Benirschke (ed.), *Primates: The Road to Self-Sustaining Populations* (pp. 251–260). New York: Springer-Verlag.

Jorgensen, M., Novak, M., and Suomi, S. (1991). Social factors and individual differences affecting performance on a foraging, spatial memory task in cebus monkeys. *American Journal of Primatology* 24(2):111.

Jungers, W. L. (1977). Hindlimbs and pelvic adaptations to vertical climbing and clinging in Megaladapis, a giant subfossil prosimian from Madagascar. *Yearbook of Physical Anthropology* 20:508–524.

Jungers, W. L. (1980). Adaptive diversity in subfossil Malagasy prosimians. *Zeitschrift für Morphologie Anthropologie* 71(2):177–186.

Jungers, W. L. (1985). *Size and Scaling in Primate Biology*. New York: Plenum.

Jungers, W. L., Godfrey, L. R., Simons, E. L., Chatrath, P. S., and Rakoto-samimanana, G. (1991). Phylogenetic and functional affinities of *Babakotia* (Primates), a fossil lemur from northern Madagascar. *Proceedings of the National Academy of Sciences U.S.A.* 88:9082–9086.

Jungers, W. L., Godfrey, L. R., Simons, E. L., and Chatrath, P. S. (1995). Subfossil *Indri indri* from the Ankarana Massif of northern Madagascar. *American Journal of Physical Anthropology* 97:357–366.

Kaas, J. (1983). What, if anything, is S-1? Organization of first somatosensory area of cortex. *Physiological Review* 63:206–231.

Kaas, J. H. (1993). The organization of visual cortex in primates: Problems, conclusions and the use of comparative studies in understanding the human brain. In B. Gulyas, D. Ottoson, and P. E. Roland (eds.), *Functional Organization of the Human Visual Cortex* (pp. 1–11). New York: Pergamon Press.

Kaas, J. H., Huerta, M. F., Weber, J. T., and Harting, J. K. (1978). Patterns of retinal termination and laminar organization of the lateral geniculate nucleus of primates. *Journal of Comparative Neurology* 182:517–554.

Kaemmerer, K., Mitchell, L., Roberts, M., and Winslow, M. (1987). Parturition in captive spider monkeys. *International Journal of Primatology* 8(5):427.

Kappeler, P. M. (1987). The acquisition process of a novel behavior pattern in a group of ring-tailed lemurs (*Lemura catta*). *Primates* 28:225–228.

Kappeler, P. M. (1990). The evolution of sexual size dimorphism in prosimian primates. *American Journal of Primatology* 21:201–214.

Kappeler, P. M. (1993). Sexual selection and lemur social systems. In P. M. Kappeler and J. U. Ganzhorn (eds.), *Lemur Social Systems and Their Ecological Basis* (pp. 223–240). New York: Plenum.

Kaufman, E. M. (1987). A morphometric study of platyrrhine phylogeny, using measurements of the crania and postcrania. *American Journal of Physical Anthropology* 72(2):218.

Kavanagh, M. (1983). *A Complete Guide to Monkeys, Apes and Other Primates*. New York: Viking Press.

Kavanagh, M., and Dresdale, L. (1975). Observations on the woolly monkey (*Lagothrix lagotricha*) in northern Colombia. *Primates* 16:285–294.

Kay, R. F. (1980). Platyrrhine origins: A reappraisal of the dental evidence. In R. L. Ciochon and A. B. Chiarelli (eds.), *Evolutionary Biology of the New World Monkeys and Continental Drift* (pp. 159–188). New York: Plenum.

Kay, R. F. (1984). On the use of anatomical features to infer foraging behavior in extinct primates. In P. S. Rodman and J. G. H. Cant (eds.), *Adaptations for Foraging in Nonhuman Primates: Contributions to an Organismal Biology of Prosimians, Monkeys and Apes* (pp. 21–53). New York: Columbia University Press.

Kay, R. F. (1990). The phyletic relationships of extant and fossil Pithecinae (Platyrrhini, Anthropoidea). *Journal of Human Evolution* 19(1/2):175–208.

Kay, R. F. (1994). "Giant" Tamarin from the Miocene of Colombia. *American Journal of Physical Anthropology* 95:333–353.

Kay, R. F., and Frailey, C. D. (1993). Large fossil platyrrhines from the Rio Acre local fauna, late Miocene, western Amazonia. *Journal of Human Evolution* 25:319–327.

Kay, R. F., and Madden, R. H. (1995). Paleogeography and paleoecology. In R. F. Kay, R. H. Madden, R. L. Cifelli, and J. J. Flynn (eds.), *Vertebrate Paleontology in the Neotropics: The Miocene Fauna of La Venta, Colombia* (pp. 444–480). Washington, DC: Smithsonian Institution Press.

Kay, R. F., and Meldrum, D. J. (1996). A new small platyrrhine from the Miocene of Colombia and the phyletic position of Callitrichidae. In R. F. Kay, R. H. Madden, R. L. Cifelli, and J. J. Flynn (eds.), *Vertebrate Paleontology in the Neotropics: The Miocene Fauna of La Venta, Colombia*. Washington, DC: Smithsonian Institution Press.

Kay, R. F., and Simons, E. L. (1980). The ecology of Oligocene African Anthropoidea. *International Journal of Primatology* 1:21–37.

Kay, R. F., and Simons, E. L. (1983). Dental formulae and dental eruption patterns in the Parapithecidae (Primates, Anthropoidea). *American Journal of Physical Anthropology* 62:363–375.

Kay, R. F., and Simons, E. L. (1986). *Parapithecus grangeri* of the African Oligocene: An archaic catarrhine without lower incisors. *Journal of Human Evolution* 15:205–213.

Kay, R. F., and Williams, B. A. (1994). Dental evidence for anthropoid origins. In J. G. Fleagle and R. F. Kay (eds.), *Anthropoid Origins* (pp. 361–445). New York: Plenum.

Kay, R. F., and Williams, B. A. (1995). Recent finds of monkeys from the Oligocene/Miocene of Salla, Bolivia. *American Journal of Physical Anthropology* 20:203.

Kay, R. F., Madden, R. H., Plavcan, J. M., Cifelli, R. L., and Guerrero-D, J. (1987). *Stirtonia victoriae*, a new species of Miocene Colombian primate. *Journal of Human Evolution* 16:173–196.

Kay, R. F., Plavcan, J. M., Glander, K. E., and Wright, P. C. (1988). Behavioral and size correlates of canine dimorphism in platyrrhine primates. *American Journal of Physical Anthropology* 77(3):385–397.

Kellogg, R., and Goldman, E. A. (1944). Review of the spider monkeys. *Proceedings of the United States Nature Museum* 96:1–45.

King, J. E., Hsiao, S., and Leeming, M. N. (1986). Licking patterns for sucrose solutions by young and aged squirrel monkeys. *Physiology and Behavior* 37(5):765–771.

King, J. E., Landau, VI, Scott, A. G., and Berning, A. L. (1987). Hand preference during capture of live fish by squirrel monkeys. *International Journal of Primatology* 8(5):540.

Kingston, W. R. (1969). Marmosets and tamarins. *Laboratory Animal Handbook* 4:243–250.

Kingston, W. R. (1975). The breeding of endangered species of marmosets and tamarins. In R. D. Martin (ed.), *Breeding Endangered Species in Captivity* (pp. 213–222). London: Academic Press.

Kingston, W. R. (1986). Establishment of primates rescued from the Tucurui Dam flooding in the Brazilian National Primate Center. *Primate Eye* 28:18–20.

Kingston, W. R. (1987). Captive breeding of *Alouatta belzebul* and *Chiropotes satanas utahicki*. *Laboratory Primate Newsletter* 26(3):8.

Kinzey, W. G. (1971). Male reproductive system and spermatogenesis. In E. S. E. Hafez (ed.), *Comparative Reproduction of Nonhuman Primates* (pp. 85–114). Springfield, IL: Charles C Thomas.

Kinzey, W. G. (1972). Canine teeth of the monkey, *Callicebus moloch*: Lack of sexual dimorphism. *Primates* 13(4):365–369.

Kinzey, W. G. (1977a). Diet and feeding behaviour of *Callicebus torquatus*. In T. H. Clutton-Brock (ed.), *Primate Ecology: Studies of Feeding and Ranging Behaviour in Lemurs, Monkeys and Apes* (pp. 127–151). London: Academic Press.

Kinzey, W. G. (1977b). Positional behavior and ecology in *Callicebus torquatus*. *Yearbook of Physical Anthropology* 20:468–480.

Kinzey, W. G. (1978). Feeding behaviour and molar features in two species of titi monkey. In D. J. Chivers and J. Herbert (eds.), *Recent Advances in Primatology, Vol. 1, Behaviour* (pp. 373–385). London: Academic Press.

Kinzey, W. G. (1981). The titi monkeys, genus *Callicebus*. In A. F. Coimbra-Filho and R. A. Mittermeier (eds.), *Ecology and Behavior of Neotropical Primates*, Vol. 1 (pp. 241–276). Rio de Janeiro: Academia Brasileira de Ciências.

Kinzey, W. G. (1982). Distribution of primates and forest refuges. In G. T. Prance (ed.), *Biological Diversification in the Tropics* (pp. 455–482). New York: Columbia University Press.

Kinzey, W. G. (1983). Is daily path length determined by group size? *American Journal of Physical Anthropology* 60(2):214.

Kinzey, W. G. (1986a). New World primate field studies: What's in it for anthropology? *Annual Review of Anthropology* 15:121–148.

Kinzey, W. G. (1986b). Feeding, travel distance and group size in *Callicebus torquatus*. *Primate Report* 14:11.

Kinzey, W. G. (1987a). *The Evolution of Human Behavior: Primate Models*. Albany: SUNY Press.

Kinzey, W. G. (1987b). A primate model for human mating systems. In W. G. Kinzey (ed.), *Primate Models for the Evolution of Human Behavior* (pp. 105–114). Albany: State University of New York Press.

Kinzey, W. G. (1987c). Comparative functional morphology of the dentition of bearded saki and spider monkeys. *The Anatomical Record* 218(1):72A.

Kinzey, W. G. (1992). Dietary and dental adaptations in the Pitheciinae. *American Journal of Physical Anthropology* 88(2):499–514.

Kinzey, W. G. (in press). History of New World primate field studies. In F. Spencer (ed.), *History of Physical Anthropology: An Encyclopedia*. New York: Garland.

Kinzey, W. G., and Becker, M. (1983). Activity pattern of the masked titi monkey, *Callicebus personatus*. *Primates* 24(3):337–343.

Kinzey, W. G., and Cunningham, E. P. (1994). Variability in platyrrhine social organization. *American Journal of Primatology* 34:185–198.

Kinzey, W. G., and Gentry, A. H. (1979). Habitat utilization in two species of *Callicebus*. In R. W. Sussman (ed.), *Primate Ecology: Problem-Oriented Field Studies* (pp. 89–100). New York: John Wiley.

Kinzey, W. G., and Norconk, M. A. (1990). Hardness as a basis of fruit choice in two sympatric primates. *American Journal of Physical Anthropology* 81(1):5–15.

Kinzey, W. G., and Norconk, M. A. (1992). Seed predation by bearded saki monkeys. *American Journal of Primatology* 27(1):38–39 (abstract).

Kinzey, W. G., and Norconk, M. A. (1993). Physical and chemical properties of fruit and seeds eaten by *Pithecia* and *Chiropotes* in Surinam and Venezuela. *International Journal of Primatology* 14(2):207–227.

Kinzey, W. G., and Redhead, C. S. (1982). Male titi monkey fails to preserve territory without mate. *American Journal of Physical Anthropology* 57(2):202.

Kinzey, W. G., and Robinson, J. G. (1983). Intergroup loud calls, range size and spacing in Callicebus torquatus. *American Journal of Physical Anthropology* 60:539–544.

Kinzey, W. G., and Wright, P. C. (1982). Grooming behavior in the titi monkey, *Callicebus torquatus*. *American Journal of Primatology* 3(1–4):267–275.

Kinzey, W. G., Rosenberger, A. L., and Ramirez, M. (1975). Vertical clinging and leaping in a neotropical anthropoid. *Nature (London)* 255:327–328.

Kinzey, W. G., Rosenberger, A. L., Heisler, P. S., Prowse, D.L., and Trilling, J. S. (1977). A preliminary field investigation of the yellow handed titi monkey, *Callicebus torquatus torquatus*, in northern Peru. *Primates* 18(1):159–181.

Kinzey, W. G., Norconk, M. A., and Alvarez-Cordero, E. (1988). Primate survey of eastern Bolívar, Venezuela. *Primate Conservation* 9:66–70.

Kinzey, W. G., Norconk, M. A., and Leighton, M. (1990). Preliminary data on physical and chemical properties of fruit eaten by *Pithecia pithecia*. *American Journal of Primatology* 20(3):204–205.

Kleiman, D. G. (1977a). Monogamy in mammals. *Quarterly Review of Biology* 52:39–69.

Kleiman, D. G. (1977b). Characteristics of reproduction and sociosexual interactions in pairs of lion tamarins (*Leontopithecus rosalia*) during the reproductive cycle. In D. G. Kleiman (ed.), *The Biology and Conservation of the Callitrichidae* (pp. 181–190). Washington, DC: Smithsonian Institution Press.

Kleiman, D. G. (1980). The sociology of captive propagation. In M. E. Soulé and B. A. Wilcox (eds.), *Conservation Biology* (pp. 243–261). Sunderland, MA: Sinauer Associates.

Kleiman, D. G. (1981). *Leontopithecus rosalia*. *Mammalian Species* (148)1–7.

Kleiman, D. G. (1985). Paternal care in New World primates. *American Zoologist* 25:857–859.

Kleiman, D. G., Gracey, D. W., and Hodgen, G. D. (1978). Urinary chorionic gonadotropin levels in pregnant golden lion tamarins: Preliminary observations. *Journal of Medical Primatology* 7:333–338.

Kleiman, D. G., Beck, B. B., Dietz, J. M., Dietz, L. A., Ballou, J. D., and Coimbra-Filho, A. F. (1986). Captive research and management, ecological studies, educational strategies, and reintroduction. In K. Benirschke (ed.), *Primates: The Road to Self-Sustaining Populations* (pp. 959–979). New York: Springer-Verlag.

Kleiman, D. G., Hoage, R. J., and Green, K. M. (1988). The lion tamarins, genus *Leontopithecus*. In R. A. Mittermeier, A. B. Rylands, A. F. Coimbra-Filho, and G. A. B. da Fonseca (eds.), *Ecology and Behavior of Neotropical Primates*, Vol. 2. (pp. 299–347). Washington, DC: World Wildlife Fund.

Kleiman, D. G., Beck, B. B., Dietz, J. M., and Dietz, L. A. (1991). Costs of reintroduction and criteria for success: Accounting and accountability in the Golden Lion Tamarin Conservation Program. In J. H. W. Gipps (ed.), *Beyond Captive Breeding. Symposia of the Zoological Society of London* 62:125–142.

Klein, L. L., and Klein, D. J. (1971). Aspects of social behavior in a colony of spider monkeys, *Ateles geoffroyi*, at the San Francisco Zoo. *International Zoo Yearbook* 11:175–181.

Klein, L. L., and Klein, D. (1973). Observations on two types of neotropical primate intertaxa associations. *American Journal of Physical Anthropology* 3:649–654.

Klein, L. L., and Klein, D. J. (1975). Social and ecological contrasts between four taxa of neotropical primates. In R. H. Tuttle (ed.), *Socioecology and Psychology of Primates* (pp. 59–85). The Hague: Mouton.

Klein, L. L., and Klein, D. J. (1976). Neotropical primates: Aspects of habitat usage, population density, and regional distribution in La Macarena, Colombia. In R. W. Thorington, Jr. and P. G. Heltne (eds.), *Neotropical Primates: Field Studies and Conservation* (pp. 70–78). Washington, DC: National Academy of Sciences.

Klein, L. L., and Klein, D. J. (1977). Feeding behavior of the Colombian spider monkey, *Ateles belzebuth*. In T. H. Clutton-Brock (ed.), *Primate Ecology: Studies of Feeding and Ranging Behavior in Lemurs, Monkeys and Apes* (pp. 153–181). London: Academic Press.

Koenig, A., and Rothe, H. (1991). Social relationships and individual contribution to cooperative behaviour on common marmosets. *Primates* 32(2):183–195.

Konstant, W. R. (1986). Considering subspecies in the captive management of *Ateles*. In K. Benirschke (ed.), *Primates: The Road to Self-Sustaining Populations* (pp. 911–920). New York: Springer-Verlag.

Konstant, W. R., Mittermeier, R. A., and Nash, S. D. (1985). Spider monkeys in captivity and in the wild. *Primate Conservation* 5:82–109.

Krause, D. W., Hartman, J. H., and Wells, N. A. (1996). Late Cretaceous vertebrates from Madagascar: Implications for biotic change in deep time. In B. D. Patterson and S. M. Goodman (eds.), *Environmental Change in Madagascar*. Washington, DC: Smithsonian Institute. In press.

Kreig, H. (1930). Die affen der Gran Chaco und seiner grenzbegeite. *Zeitschrift für Morphologie und Oekologie der Tiere* 4:760–785.

Kroodsma, D. E. (1981). Geographic variation and functions of song types in warblers (Parulidae). *Auk* 98:743–751.

Krubitzer, L. A., and Kaas, J. H. (1990). Cortical connections of MT in four species of primates: Areal, modular, and retinotopic patterns. *Vision Neuroscience* 5:165–204.

Küderling, I., Evans, C. S., and Abbott, D. H. (1992). Differential excretion of urinary estradiol in alpha-females and adult daughters of red bellied tamarins (*Saguinus labiatus*, Callitrichidae). *Abstracts of the XIVth Congress of the International Primatological Society*, Strasbourg, France, pp. 211–212.

Kuhlmann, M. (1975). Adenda alimentar dos bugios. *Silvicultura, São Paulo* 9:57–62.

Kummer, H. (1968). *Social Organization of Hamadryas Baboons*. Chicago: University of Chicago Press.

Kunzle, H. (1976). Thalamic projections from the precentral motor cortex in Macaca fascicularis. *Brain Research* 105:253–267.

Kurland, J. A. (1977). *Kin Selection in the Japanese Monkey*. New York: S. Karger.

Kurten, B. (1972). *Not from the Apes*. New York: Vantage Books.

Kuypers, H. G. J. M. (1983). A new look at the organization of the motor system. *Progress in Brain Research* 57:381–403.

Lacher, T. E., Jr., Fonseca, G. A. B. da, Alves, C., Jr., and Magalhães-Castro, B. (1984). Parasitism of trees by marmosets in a central Brazilian gallery forest. *Biotropica* 16:202–209.

Lahm, S. A. (1986). Diet and habitat preference of *Mandrillus sphinx* in Gabon: Implications of foraging strategy. *American Journal of Primatology* 11:9–26.

Lamas, G. (1982). A preliminary zoogeographical division of Peru, based on butterfly distributions (Lepidoptera, Papilionoida). In G. T. Prance (ed.), *Biological Diversification in the Tropics* (pp. 336–357). New York: Columbia University Press.

Lancaster, J. B. (1991). A feminist and evolutionary biologist looks at women. *Yearbook of Physical Anthropology* 34:1–11.

Landau, VI. (1987). The adaptation of New World monkeys to new environmental situations: Food acquisition and food processing behaviors. *Dissertation Abstracts International* B48(3):901–902.

Langguth, A., Martins, Teixeira D. L., Mittermeier, R. A., and Bonvicino, C. R. (1987). The red-handed howler monkey in northeastern Brazil. *Primate Conservation* 8:36–39.

Langrand, O. (1990). *Guide to the birds of Madagascar*. New Haven: Yale University Press.

Lehman, S. M., and Robertson, K. L. (1994). Preliminary survey of *Cacajao melanocephalus melanocephalus* in southern Venezuela. *International Journal of Primatology* 33(3):223.

Leigh, S, (1991). The ontogeny of body size dimorphism in anthropoid primates. *American Journal of Physical Anthropology* Suppl. 12:113.

Lein, M. N. (1978). Song variation in a population of chestnut-sided warblers (*Dendroica pennsylvanica*): Its nature and suggested significance. *Canadian Journal of Zoology* 56:1266–1283.

LeMaho, G. M., Rochas, M., Felbabel, H., and Chatonnet, J. (1981). Thermoregulation in the nocturnal simian: The night monkey *Aotus trivirgatus*. *Journal of Physiology* 240:R156–R165.

Lemelin, P. (1989). Functional myology of the platyrrhine prehensile tail. *American Journal of Physical Anthropology* 78(2):260.

Lemos de Sá, R. M. (1987). Group structure, foraging behavior and activity budget in *Brachyteles arachnoides*: A contribution to conservation and management perspectives. *International Journal of Primatology* 8(5):422.

Lemos de Sá, R. M., and Glander, K. (1993). Capture techniques and morphometrics for the woolly spider monkey, or muriqui (*Brachyteles arachnoides* E. Geoffroy 1806). *American Journal of Primatology* 29(2):145–153.

Lemos de Sá, R. M., and Strier, K. B. (1992). A preliminary comparison of forest structure and use by two isolated groups of woolly spider monkeys, *Brachyteles arachnoides*. *Biotropica* 24(3):455–459.

Lemos de Sá, R. M., Pope, T. R., Struhsaker, T. T., and Glander, K. E. (1995). Sexual dimorphism in canine length of woolly spider monkeys (*Brachyteles arachnoides* E. Geoffroy 1806). *International Journal of Primatology* (in press).

Leo Luna, M. (1980). First field study of the yellow-tailed woolly monkey. *Oryx* 15:386–389.

Leo Luna, M. (1982). Conservation of the yellow-tailed woolly monkey, *Lagothrix flavicauda*. *International Zoo Yearbook* 22:47–52.

Leo Luna, M. (1987). Primate conservation in Peru: A case study of the yellow-tailed woolly monkey. *Primate Conservation* 8:122–123.

Leo Luna, M., and Ortiz, E. (1981). Evaluación preliminar de la distribución y situación del mono choro amarilla (*Lagothrix flavicauda*). *Informe Tecnico, Dirección de Conservatión, PAHO, Ministerio de Agricultura, Lima*.

Leutenegger, W. (1973). Maternal-fetal weight relationships in primates. *Folia Primatologica* 20:280–293.

Leutenegger, W. (1980). Monogamy in callitrichids: A consequence of phyletic dwarfism? *International Journal of Primatology* 1:95–98.

Leutenegger, W., and Larson, S. (1985). Sexual dimorphism in the postcranial skeleton of New World primates. *Folia Primatologica* 44:82–95.

Levene, H. (1953). Genetic equilibrium when more than one niche is available. *American Naturalist* 104:331–334.

Liberman, A. M. (1982). On finding that speech is special. *American Psychologists* 37:148–167.

Liberman, A. M., Cooper, F. S., Shankweiler, D. P., and Studdert-Kennedy, M. (1967). Perception of the speech code. *Psychological Review* 24:431–461.

Lieblich, A. K., Symmes, D., Newman, J. D., and Shapiro, M. (1980). Development of the isolation peep in laboratory bred squirrel monkeys. *Animal Behavior* 28:1–9.

Lillehei, R. A., and Snowdon, C. T. (1978). Individual and situational differences in the vocalizations of young stumptail macaques. *Behaviour* 65:270–281.

Linares, O. (1976). Garden hunting in the American tropics. *Human Ecology* 4(4):331–350.

Lindsay, N. B. D. (1980). A report on a field study of Geoffroy's tamarin, *Saguinus geoffroyi*. *Dodo, Journal of the Jersey Wildlife Preservation Trust* 17:27–51.

Linn, G. S., Mase, D., Lafrancois, D., O'Keefe, R. T., and Lifshitz, K. (1991). Social behavior of group-housed *Cebus apella* over the menstrual cycle. *American Journal of Primatology* 24(2):116.

Lizot, J. (1977). Population, resources, and warfare among the Yanomama. *Man* 12:449–517.

Lopes, M. A., and Ferrari, S. F. (1994). Foraging behavior of a tamarin group

(*Saguinus fuscicollis weddelli*) and interactions with marmosets (*Callithrix emiliae*). *International Journal of Primatology* 15(3):373–388.

Lorenz, R., and Mason, W. A. (1971). Establishment of a colony of titi monkeys. *International Zoo Yearbook* 11:168–175.

Lorini, M. L., and Persson, V. G. (1990). Nova espécie de *Leontopithecus* Lesson, 1840, do sul do Brasil (Primates, Callitrichidae). *Boletim do Museu Nacional de Rio de Janeiro—Zoologia* 338:1–14.

Lovejoy, T. E., Bierregaard, R. O., Jr., Rylands, A. B., Malcolm, J. R., Quintela, C. E., Harper, L. H., Brown, K. S., Jr., Powell, A. H., Powell, G. V. N., Schubart, H. O. R., and Hays, M. B. (1986). Edge and other effects of isolation on Amazon forest fragments. In M. E. Soule (ed.), *Conservation Biology: The Science of Scarcity and Diversity* (pp. 257–285). Sunderland, MA: Sinauer Associates.

Lucas, P. W., Corlett, R. T., and Luke, D. A. (1986). Sexual dimorphism of tooth size in anthropoids. *Human Evolution* 1(1):23–39.

Luchterhand, K., Kay, R. F., and Madden, R. H. (1986). *Mohanamico hershkovitzi*, gen et sp. nov., un primate du Miocène moyen d'Amérique du Sud. *Comptes Rendus Académies des Sciences*, tome 311, Série II 303:1753–1758.

Luckett, W. P. (1975). Placentation in haplorrhines and strepsirrhines. In W. P. Luckett and F. S. Szalay (eds.), *Phylogeny of the Primates: A Multidisciplinary Approach* (pp. 300–313). New York: Plenum.

Lund, P. W. (1839). *Annales des Sciences Naturelles* (2) 11:214–234.

Lund, P. W. (1841). Blik paa Brasiliens dyreverden for sidste Jordomsaelting. Overs. K. Danske Vidensk. Selskabs natur. *Math Afhandlinger* 8:219–272 (or 29–144).

Lyon, M. F. (1962). Sex chromatin and gene action in the mammalian X-chromosome. *American Journal of Human Genetics* 14:135–148.

Ma, N. S.-F. (1981a). Chromosome evolution in the owl monkey, Aotus. *American Journal of Physical Anthropology* 54(3):293–303.

Ma, N. S.-F. (1981b). Errata: Chromosome evolution in the owl monkey, *Aotus. American Journal of Physical Anthropology* 56(3):326.

Ma, N. S-F., Aquino, R., and Collins, W. E. (1985). Two new karyotypes in the Peruvian owl monkey (*Aotus trivirgatus*). *American Journal of Primatology* 9(4):333–341.

MacDonald, D. (1984). *The Encyclopedia of Mammals.* New York: Facts on File.

Mace, G. M. (1986). Captive breeding and conservation. *Primate Eye* 29(Suppl.): 53–58.

MacFadden, B. J. (1990). Chronology of Cenozoic primate localities in South America. *Journal of Human Evolution* 19:7–21.

Mack, D. S., and Kafka, H. (1978). Breeding and rearing of woolly monkeys at the National Zoological Park, Washington. *International Zoo Yearbook* 18:117–122.

Mack, D. S., and Kleiman, D. G. (1978). Distribution of scent marks in different contexts in captive lion tamarins, *Leontopithecus rosalia* (Primates). In H. Rothe, H. J. Wolters, and J. P. Hearn (eds.), *Biology and Behavior of Marmosets* (pp. 181–188). Göttingen: Eigenverlag H. Rothe.

MacLean, P. D. (1964). Mirror display in the squirrel monkey. *Science* 146:950–952.

MacPhee, R. D. E. (1993). From Cuba: A mandible of *Paralouatta*. *Evolutionary Anthropology* 2(2):42.

MacPhee, R. D. E., and Fleagle, J. G. (1991). Postcranial remains on *Xenothrix macgregori* (Primates, Xenotrichidae) and other Late Quaternary mammals from Long Mile Cave, Jamaica. *Bulletin of the American Museum of Natural History* 206:287–321.

MacPhee, R. D. E., and Iturralde-Vinent, M. A. (1995). Earliest monkey from Greater Antilles. *Journal of Human Evolution* 28:197–200.

MacPhee, R. D. E., and Woods, C. A. (1982). A new fossil cebine from Hispaniola. *American Journal of Physical Anthropology* 58:419–436.

MacPhee, R. D. E., Burney, D. A., and Wells, N. A. (1985). Early holocene chronology and environment of Ampasambazimba, a Malagasy subfossil site. *International Journal of Primatology* 6:463–489.

MacPhee, R. D. F., and Raholimavo, E. M. (1988). Modified subfossil aye-aye incisors from southwestern Madagascar: Species allocation and paleoecological significance. *Folia Primatologica* 51:126–142.

MacPhee, R. D. E., Horowitz, I., Arredondo, O., and Vasquez, O. J. (1995). A new genus for the extinct Hispaniolan monkey *Saimiri bernensis* Rimoli, 1977, with notes on its systematic position. *American Museum Novitates* 3134. New York: American Museum of Natural History.

Madden, R. H., and Albuja, L. (1987). Conservation status of *Ateles fusciceps fusciceps* in northwestern Ecuador. *International Journal of Primatology* 8(5):513.

Maier, W., Alonson, C., and Langguth, A. (1982). Field observations on *Callithrix jacchus jacchus* L. *Z. Säugetierk.* 47:334–346.

Malaga, C. (1985). Non standard mating systems for *Saguinus mystax*. *American Journal of Physical Anthropology* 66:201.

Malmo, R. B., and Grether, W. F. (1947). Further evidence of red blindness (protanopia) in *Cebus* monkeys. *Journal of Comparative Physiology and Psychology* 40:143–147.

Marks, S. (1976). *Large Mammals and a Brave People: Subsistence Hunters in Zambia.* Seattle: University of Washington Press.

Marks, S. (1977). Hunting behavior and strategies of the Valley Bisa in Zambia. *Human Ecology* 5(1):1–36.

Marler, P. (1970). Bird song and human speech: Could there be parallels? *American Scientist* 58:669–673.

Marler, P. (1977). The structure of animal communication sounds. In T. H. Bullock (ed.), *Dahlem Workshop on the Recognition of Complex Acoustic Signals* (pp. 17–35). Berlin: Dahlem Konferenzen.

Marler, P., and Peters, S. (1982). Subsong and plastic song: Their role in the vocal learning process. In D. E. Kroodsma and E. H. Miller (eds.), *Acoustic Communication in Birds: Vol. 2: Song Learning and Its Consequences* (pp. 25–50). New York: Academic Press.

Marler, P., and Tenaza, R. (1977). Signaling behavior of apes with special reference to vocalization. In T. A. Sebeok (ed.), *How Animals Communicate* (pp. 965–1033). Bloomington: Indiana University Press.

Marsh, C. W., and Mittermeier, R. A. (1987). *Primate Conservation in the Tropical Rainforest.* New York: Alan R. Liss.

Marsh, C. W., Johns, A. D., and Ayres, J. M. (1987). Effects of habitat disturbance on rain forest primates. In C. W. Marsh and R. A. Mittermeier (eds.), *Primate Conservation in the Tropical Rain Forest* (pp. 83–107). New York: Alan R. Liss.

Martau, P. A., Caine, N. G., and Candlen, D. J. (1985). Reliability of the emotions profile index, primate form with *Papio hamadryas, Macaca fuscata* and two *Saimiri* species. *Primates* 26:501–505.

Martin, L. B., Kinzey, W. G., and Maas, M. C. (1994). Enamel thickness in pitheciine primates. *American Journal of Physical Anthropology* Suppl. 18:138.

Martin, P. S., and Klein, R. G., (eds.), (1984). *Quaternary Extinctions: A Prehistoric Revolution.* Tucson: University of Arizona Press.

Martin, R. D. (1973). A review of the behavior and ecology of the lesser mouse lemur (*Microcebus murinus* J. F. Miller, 1777). *Z. Comp. Ethol.* Suppl. 9:43–89.

Martin, R. D. (1981). Relative brain size and basal metabolic rate in terrestrial vertebrates. *Nature (London)* 293:57–60.

Martin, R. D. (1983). Human brain evolution in an ecological context. 52nd James Arthur Lecture. New York: American Museum of Natural History.

Martin, R. D. (1990). *Primate Origins and Evolution.* Princeton: Princeton University Press.

Martin, R. D. (1992). Goeldi and the dwarfs: The evolutionary biology of the small New World monkeys. *Journal of Human Evolution* 22:367–393.

Martin, R. L., Wood, C., Baehr, W., and Applebury, M. L. (1986). Visual pigment homologies revealed by DNA hybridization. *Science* 232:1266–1269.

Martuscelli, P., Petroni, L. M., and Olmos, F. (1994). Fourteen new localities for the muriqui *Brachyteles arachnoides. Neotropical Primates* 2(2):12–15.

Masataka, N. (1981). A field study of the social behavior of Goeldi's monkeys (*Callimico goeldii*) in North Bolivia. I. Group composition, breeding cycle, and infant development. II. Grouping pattern and intragroup relationship. *Kyoto University Overseas Research Reports of New World Monkeys* 2:23–41.

Masataka, N. (1982). A field study on the vocalizations of Goeldi's monkeys (*Callimico goeldii*). *Primates* 23(2):206–219.

Masataka, N. (1983). Categorical responses to natural and synthesized alarm calls in Goeldi's monkeys (*Callimico goeldii*). *Primates* 24:40–51.

Masataka, N. (1986). Interspecific and intraspecific responses to some species-specific vocalizations in marmosets, tamarins, and Goeldi's monkeys. In D. M. Taub and F. A. King (eds.), *Current Perspectives in Primate Social Dynamics* (pp. 368–377). New York: Van Nostrand Reinhold.

Masataka, N. (1988). The response of red-chested mustached tamarins to long calls from their natal and alien populations. *Animal Behavior* 36(1):55–61.

Masataka, N. (1990). Handedness of capuchin monkeys. *Folia Primatologica* 55:189–192.

Masataka, N., and Kohda, M. (1988). Primate play vocalizations and their functional significance. *Folia Primatologica* 50:152–156.

Masataka, N., and Symmes, D. (1986). Effect of separation distance on isolation call structure in squirrel monkeys (*Saimiri sciureus*). *American Journal of Primatology* 10:271–278.

Mason, W. A. (1966). Social organization of the South American monkey, *Callicebus moloch*, a preliminary report. *Tulane Studies in Zoology* 13:23–28.

Mason, W. A. (1968). Use of space by *Callicebus* groups. In P. Jay (ed.), *Primates, Studies in Adaptation and Variability* (pp. 200–216). New York: Holt.

Mason, W. A. (1971). Field and laboratory studies of social organization in *Saimiri* and *Callicebus*. *Primate Behavior* 2:107–138.

Mason, W. A. (1978). Ontogeny of social systems. In D. J. Chivers and J. Herbert (eds.), *Recent Advances in Primatology, Vol. 1, Behavior* (pp. 5–14). London: Academic Press.

Mason, W. A., Valeggia, C. R., Garcia, P. N., and Mendoza, S. P. (1994). What limits group size in the monogamous titi monkey (*Callicebus*)? *American Journal of Primatology* 33(3):227.

Massey, A. (1987). A population survey of *Alouatta palliata*, *Cebus capucinus*, and *Ateles geoffroyi* at Palo Verde, Costa Rica. *Revista de Biologia Tropical* 35(2):345–347.

Maurus, M., Kuehlmorgen, B., Wiesner, E., Barclay, D., and Streit, K. M. (1985). 'Dialogues' between squirrel monkeys. *Language and Communication* 5:185–191.

May, R. M., and Seger, J. (1986). Ideas in ecology. *American Scientist* 74:256–267.

McConnell, P. B., and Snowdon, C. T. (1986). Vocal inte ractions among unfamiliar groups of captive cotton-top tamarins. *Behaviour* 97:273–296.

McDade, L. A., Bawa, K. S., Hespenheide, H. A., and Hartshorn, G. S. (1994). *La Selva: Ecology and Natural History of a Neotropical Rain Forest*. Chicago: University Press.

McDougal, D. B., Jr. (1981). Defects in carbohydrate metabolism. In G. J. Siegel, R. W. Albers, B. W. Agranoff, and R. Katzman (eds.), *Basic Neurochemistry* (pp. 601–613). Boston: Little, Brown.

McFarland, M. J. (1986). Ecological determinants of fission-fusion sociality in *Ateles* and *Pan*. In J. G. Else and P. C. Lee (eds.), *Primate Ecology and Conservation* (pp. 181–190). Cambridge: Cambridge University Press.

McGrew, W. C. (1989). Why is ape tool use so confusing? In V. Standen and R. Foley (eds.), *Comparative Socioecology* (pp. 457–472). London: Blackwell.

McGrew, W. C., and McLuckie, E. C. (1986). Philopatry and dispersion in the cotton-top tamarin, *Saguinus (o.) oedipus*: An attempted laboratory simulation. *International Journal of Primatology* 7:399–420.

McKenna, J. J. (1982). Primate field studies: The evolution of behavior and its sociobiology. In J. L. Fobes and J. E. King (eds.), *Primate Behavior* (pp. 53–83). New York: Academic Press.

McNab, B., and Wright, P. C. (1987). Temperature regulation and oxygen consumption in the Philippine tarsier (*Tarsius syrichta*). *Physiological Zoology* 60(5):596–600.

McNeely, J. A., and Pitt, D. (1985). *Culture and Conservation: The Human Dimension in Environmental Planning*. Dover: Croom Helm.

McNeely, J. A., and Wachtel, P. S. (1988). *Soul of the Tiger: Searching for Nature's Answers in Exotic Southeast Asia*. New York: Doubleday.

Meggers, B. J. (1985). Aboriginal adaptation in Amazonia. In G. T. Prance and T. E. Lovejoy (eds.), *Amazonia* (pp. 307–327). New York: Pergamon Press.

Meggers, B. J. (1995). *Amazonia: Man and Culture in a Counterfeit Paradise*. Washington, DC: Smithsonian Institution Press.

Meier, B., and Albignac, R. (1991). Rediscovery of *Allocebus trichotis* Gunther 1875 (Primates) in Northeast Madagascar. *Folia Primatologica* 56:57–63.

Meier, B., Albignac, R., Peyrieras, A., Rumpler, Y., and Wright, P. (1987). A new species of Hapalemur (Primates) from South East Madagascar. *Folia Primatologica* 48:211–215.

Meldrum, D. J. (1993). Postcranial adaptations and positional behavior in fossil platyrrhines. In D. L. Gebo (ed.), *Postcranial Adaptation in Nonhuman Primates* (pp. 235–251). Carbondale, IL: Northern Illinois University Press.

Meldrum, D. J., and Kay, R. F. (1995). Postcranial skeleton and laventan platyrrhines. In R. F. Kay, R. H. Madden, R. L. Cifelli, and J. J. Flynn (eds.), *Vertebrate Paleontology in the Neotropics: The Miocene Fauna of La Venta, Colombia*. Washington, DC: Smithsonian Institution Press.

Meldrum, D. J., and Lemelin, P. (1991). Axial skeleton of *Cebupithecia sarmientoi* (Pithecinae, Platyrrhini) from the middle Miocene of La Venta, Colombia. *American Journal of Primatology* 25(2):69–89.

Mendes, F. D. C. (1987). Social behavior in the muriqui: The problem of dominance relationships. *International Journal of Primatology* 8(5):422.

Mendes, F. D. C. (1990). Affiliacao e Hierarquia no muriqui: o grupo Matao de Caratinga. Masters thesis, Universidade de Sao Paulo.

Mendes, S. L. (1985). Uso do espaço, padrões de adividades diárias e organização social de *Alouatta fusca* (Primates, Cebidae) em Caratinga-MG. Unpublished Master's thesis, University of Brasília, Brasília, DF.

Mendoza, S. P., and Mason, W. A. (1986a). Parental division of labour and differentiation of attachments in a monogamous primate (*Callicebus moloch*). *Animal Behavior* 34(5):1336–1347.

Mendoza, S. P., and Mason, W. A. (1986b). Contrasting responses to intruders and to involuntary separations by monogamous and polygynous New World monkeys. *Physiology and Behavior* 38(6):795–801.

Mendoza, S. P., and Mason, W. A. (1989). Behavioral and endocrine consequences of heterosexual pair formation in squirrel monkeys. *Physiology and Behavior* 46(4):597–603.

Mendoza, S. P., and Mason, W. A. (1991). Breeding readiness in squirrel monkeys: Female-primed females are triggered by males. *Physiology and Behavior* 49(3):471–479.

Mendoza, S. P., Lowe, E. L., Davidson, J. M., and Levine, S. (1978). Annual cyclicity in the squirrel monkey (*Saimiri sciureus*): The relationship between testosterone, fatting, and sexual behavior. *Hormones and Behavior* 11:295–303.

Menzel, C. R. (1986a). Structural aspects of arboreality in titi monkeys (*Callicebus moloch*). *American Journal of Physical Anthropology* 70(2):167–176.

Menzel, C. R. (1986b). An experimental study of territory maintenance in captive titi monkeys (*Callicebus moloch*). In J. Else and P. Lee (eds.), *Primate Ecology and Conservation* (pp. 133–143). New York: Cambridge University Press.

Mesulem, M.-M. (1981). A cortical network for directed attention and unilateral neglect. *Annals of Neurology* 10:309–325.

Meyers, D. M., and Wright, P. C. (1993). Resource tracking: Food availability and *Propithecus* seasonal reproduction. In P. M. Kappeler and J. U. Ganzhorn (eds.), *Lemur Social systems and Their Ecological Basis* (pp. 181–194). New York: Plenum.

Miles, R. C., (1958a). Color vision in the squirrel monkey. *Journal of Comparative Physiology and Psychology* 51:328–331.

Miles, R. C. (1958b). Color vision in the marmoset. *Journal of Comparative Physiology and Psychology* 51:152–154.

Millar, J. S. (1977). Adaptive features of mammalian reproduction. *Evolution* 31:370–386.

Miller, K. D., Keller, J. B., and Stryker, M. P. (1989). Ocular dominance column development: Analysis and simulation. *Science* 245:605–615.

Miller, L. H. (1991). The influence of resource dispersion on group size among wedge-capped capuchins (*Cebus olivaceus*). *American Journal of Primatology* 24(2):123.

Milton, K. (1978). Behavioral adaptations to leaf-eating by the mantled howler monkey (*Alouatta palliata*). In G. G. Montgomery (ed.), *The Ecology of Arboreal Folivores* (pp. 535–549). Washington, DC: Smithsonian Institution Press.

Milton, K. (1980). *The Foraging Strategy of Howler Monkeys: A Study in Primate Economics*. New York: Columbia University Press.

Milton, K. (1981a). Distributional patterns of tropical plant foods as an evolutionary stimulus to primate mental development. *American Anthropologist* 83:534–548.

Milton, K. (1981b). Food choice and digestive strategies of two sympatric primate species. *The American Naturalist* 117(4):496–505.

Milton, K. (1981c). Estimates of reproductive parameters for free-ranging *Ateles geoffroyi*. *Primates* 22:574–579.

Milton, K. (1982). Dietary quality and population regulation in a howler monkey population. In E. G. Leigh, A. S. Rand, and D. M. Windsor (eds.), *The Ecology of a Tropical Forest* (pp. 273–289). Washington DC: Smithsonian Institution Press.

Milton, K. (1984a). Habitat, diet and activity patterns of freeranging woolly spider monkeys (*Brachyteles arachnoides* E. Geoffrey 1806). *International Journal of Primatology* 5(5):491–514.

Milton, K. (1984b). Diet and social structure of free-ranging woolly spider monkeys. *American Journal of Physical Anthropology* 63(2):195.

Milton, K. (1984c). Urine washing behavior in the woolly spider monkey. *Zeitschrift für Tierpsychologie* 67:154–160.

Milton, K. (1984d). Protein and carbohydrate resources of the Maku Indians of northwestern Amazonia. *American Anthropologist* 86(1):7–27.

Milton, K. (1984e). The role of food-processing factors in primate food choice. In P. S. Rodman and J. G. H. Cant (eds.), *Adaptations for Foraging in Nonhuman Primates: Contributions to an Organismal Biology of Prosimians, Monkeys and Apes* (pp. 249–279). New York: Columbia University Press.

Milton, K. (1985a). Multi-male mating and absence of canine dimorphism in woolly spider monkeys (*Brachyteles arachnoides*). *American Journal of Physical Anthropology* 68(4):519–523.

Milton, K. (1985b). Mating patterns of woolly spider monkeys, *Brachyteles arachnoides*: Implications for female choice. *Behavioral Ecology and Sociobiology* 17:53–59.

Milton, K. (1986). Ecological background and conservation priorities for woolly spider monkeys (*Brachyteles arachnoides*). In K. Benirschke (ed.), *Primates: The Road to Self-Sustaining Populations* (pp. 241–250). New York: Springer-Verlag.

Milton, K. (1987a). Behavior and ecology of the woolly spider monkey, *Brachyteles arachnoides*. *International Journal of Primatology* 8(5):422.

Milton, K. (1987b). Mating behaviors in woolly spider monkeys (*Brachyteles arachnoides*. *International Journal of Primatology* 8(5):460.

Milton, K. (1987c). Physiological characteristics of the genus *Alouatta*. *International Journal of Primatology* 8(5):428.

Milton, K., and Nessimian, J. L. (1984). Evidence for insectivory in two primate species (*Callicebus torquatus lugens* and *Lagothrix lagothricha lagothricha*) from northwestern Amazonia. *American Journal of Primatology* 6(4):367–371.

Milton, K., Casey, T. M., and Casey, K. K. (1979). The basal metabolism of mantled howler monkeys (*Alouatta palliata*). *Journal of Mammology* 60:373–376.

Minezawa, M., Jordan, O. C., and Valdivia, C. J. (1989). Karyotypic study of titi monkeys, *Callicebus moloch brunneus*. *Primates* 30(1):81–88.

Mishkin, M. (1972). Cortical visual areas and their interactions. In R. Russell (ed.), *Frontiers in Physiological Psychology* (pp. 93–119). New York: Academic Press.

Mitchell, C. (1987). Ecological correlates of between-troop variation of *Saimiri boliviensis* association with *Cebus* spp. in southeastern Peru. *International Journal of Primatology* 8(5):455.

Mitchell, C. (1989). The behavioral ecology of *Saimiri sciureus* in Manu National Park. Ph.D. dissertation, Princeton University, Princeton, NJ.

Mitchell, C. L. (1990). The ecological basis for female social dominance: A behavioral study of the squirrel monkey (*Saimiri sciureus*) in the wild. Ph.D. dissertation, Princeton University, Princeton, NJ.

Mitchell, C. L., Boinski, S., and van Schaik, C. P. (1991). Competitive regimes and female bonding in two species of squirrel monkeys (*Saimiri oerstedii* and *S. sciureus*). *Behavioral Ecology and Sociobiology* 28(1):55–60.

Mittermeier, R. A. (1977). Distribution, synecology and conservation of Suriname monkeys. Unpublished Ph.D. dissertation, Harvard University.

Mittermeier, R. A. (1978). Locomotion and posture in *Ateles geoffroyi* and *Ateles paniscus*. *Folia Primatologica* 30:161–193.

Mittermeier, R. A. (1986a). Primate conservation priorities in the Neotropical region. In K. Benirschke (ed.), *Primates: The Road to Self-Sustaining Populations* (pp. 221–240). New York: Springer-Verlag.

Mittermeier, R. A. (1986b). Strategies for the conservation of the highly endangered primates. In K. Bernishke (ed.), *Primates: The Road to Self-sustaining Populations* (pp. 1013–1022). New York: Springer-Verlag.

Mittermeier, R. A. (1987). Effects of hunting on rain forest primates. In C. W. Marsh and R. A. Mittermeier (eds.), *Primate Conservation in the Tropical Rain Forest* (pp. 109–146). New York: Alan R. Liss.

Mittermeier, R. A., and Cheney, D. L. (1987). Conservation of primates and their habitats. In B. B. Smuts, D. L. Cheney, R. M. Seyfarth, R. W. Wrangham, and T. T. Struhsaker (eds.), *Primate Societies* (pp. 477–490). Chicago: University of Chicago Press.

Mittermeier, R. A., and Coimbra-Filho, A. F. (1977). Primate conservation in Brazilian Amazonia. In Prince Rainer of Monaco and G. H. Bourne (eds.), *Primate Conservation* (pp. 109–146). New York: Academic Press.

Mittermeier, R. A., and Coimbra-Filho, A. F. (1981). Systematics: Species and subspecies. In A. F. Coimbra-Filho and R. A. Mittermeier (eds.), *Ecology and Behavior of Neotropical Primates*, Vol. 1 (pp. 29–109). Rio de Janeiro: Academia Brasileira de Ciências.

Mittermeier, R. A., and Fleagle, J. G. (1976). The locomotor and postural repertoires of *Ateles geoffroyi* and *Colobus guereza*, and a reevaluation of the locomotor category semibrachiation. *American Journal of Physical Anthropology* 45:235–255.

Mittermeier, R. A., and van Roosmalen, M. G. M. (1981). Preliminary observations on habitat utilization and diet in eight Suriname monkeys. *Folia Primatologica* 36:1–39.

Mittermeier, R. A., Macedo-Ruiz, H. de, Luscombe, B. A., and Cassidy, J. (1977). Rediscovery and conservation of the Peruvian yellow-tailed woolly monkey (*Lagothrix flavicauda*). In: Prince Rainer of Monaco and G. H. Bourne, (eds.), *Primate Conservation* (pp. 95–115). London: Academic Press.

Mittermeier, R. A., Coimbra-Filho, A. F., Constable, D. I., Rylands, A. B., and Valle, C. M. C. (1982). Conservation of primates in the Atlantic Forest Region of Eastern Brazil. *International Zoo Yearbook* 22:2–17.

Mittermeier, R. A., Konstant, W. R., Ginsberg, H., van Roosmalen, M. G. M., and Da Silva, E. C., Jr.(1983). Further evidence of insect consumption in the bearded saki monkey, *Chiropotes satanas chiropotes*. *Primates* 24(4):602–605.

Mittermeier, R. A., Macedo-Ruiz, H. de, Leo Luna, M., Young, A., Constable, I. D., Ponce del Prado, C., and Luscombe, B. A. (1984). Conservation education campaign for the Peruvian yellow-tailed woolly monkey to be launched in Peru. *Primate Conservation* 4:19–22.

Mittermeier, R. A., Valle, C. M. C., Alves, M. C., Santos, I. B., Machado Pinto, C. A., Strier, K. B., Young, A. L., Veado, E. M., Constable, I. D., Paccagnella, S. G., and Lemos de Sa, R. M. (1987). Current distribution of the muriqui in the Atlantic Coastal Forest Region of Eastern Brazil. *Primate Conservation* 8:143–149.

Mittermeier, R. A., Rylands, A. B., Coimbra-Filho, A. F., and Fonseca, G. A. B.

(eds.) (1988a). *Ecology and Behavior of Neotropical Primates*, Vol. 2. Washington, DC: World Wildlife Fund.

Mittermeier, R. A., Rylands, A. B., and Coimbra-Filho, A. F. (1988b). Systematics: Species and subspecies—an update. In R. A. Mittermeier, A. B., Rylands, A. Coimbra-Filho, and G. A. B. Fonseca (eds.), *Ecology and Behavior of Neotropical Primates*, Vol. 2. (pp. 13–75). Washington, DC: World Wildlife Fund.

Mittermeier, R. A., Kinzey, W. G., and Mast, R. B. (1989). Neotropical primate conservation. *Journal of Human Evolutionution* 18(7):597–610.

Mittermeier, R. A., Schwarz, M., and Ayres, J. M. (1992). A new species of marmoset, genus *Callithrix* Erxleben, 1777 (Callitrichidae, Primates), from the Rio Maués region, state of Amazonas, central Brazilian Amazonia. *Goeldiana, Zoologia* 14:1–17.

Mittermeier, R. A., Tattersall, I., Konstant, W. R., Meyers, D. M., and Mast, R. B. (1995). *Lemurs of Madagascar*. Washington, DC: Conservation International.

Miyamoto, M., and Goodman, M. (1990). DNA systematics and evolution of primates. *Annual Review of Ecology and Systematics* 21:197–220.

Mollon, J. D. (1989). "Tho she kneel'd in that place where they grew" . . .The uses and origins of primate colour vision. *Journal of Experimental Biology* 146:21–38.

Mollon, J. D. (1991). Uses and evolutionary origins of primate colour vision. In R. Gregory and J. Cronly-Dilon (eds.), *Evolution of the Eye and Visual System* (pp. 306–319). Boca Raton, FL: CRC Press.

Mollon, J. D., Bowmaker, J. K., and Jacobs, G. H. (1984). Variations in colour vision in a New World primate can be explained by polymorphism of retinal photopigments. *Proceedings of the Royal Society of London Series B* 222:373–399.

Montagnon, D., Crovella, S., and Rumpler, Y. (1993). Confirmation of the taxonomic position of *Callimico goeldi* (Primates, Platyrrhini) on the basis of its highly repeated DNA patterns. *Comptes Rendus Academie des Sciences Paris* 316(3):219–223.

Moore, J. (1984). Female transfer in primates. *International Journal of Primatology* 5:537–590.

Moran, E. F. (1993). *Through Amazonian Eyes: The Human Ecology of Amazonian Populations*. Iowa City: University of Iowa Press.

Moreno, L. I., Salas, I. C., and Glander, K. E. (1991). Breech delivery and birth-related behaviors in wild mantled howling monkeys. *American Journal of Primatology* 23:197–199.

Morgan, M. J., Mollon, J. D., and Adam, A. (1989). Dichromats break colour-camouflage of textural boundaries. *Investigation of Ophthalmology and Visual Science Supplement* 30:220.

Morland, H. S. (1990). Parental behavior and infant development in ruffed lemurs (*Varecia variegata*) in a northeast Madagascar rain forest. *American Journal of Primatology* 20:253–265.

Morland, H. S. (1991). Preliminary report on the social organization of ruffed lemurs (*Varecia variegata variegata*) in a northeast Madagascar forest. *Folia Primatologica* 56:157–161.

Morrison, P., and Middleton, E. H. (1967). Body temperature and metabolism in the pygmy marmoset. *Folia Primatologica* 6:70–82.

Morse, D. H. (1967). The context of songs in black-throated green and blackburnian warblers. *Wilson Bulletin* 79:64–74.

Mouncastle, V. B., Lynch, J. C., and Georgopoulos, A. (1975). Posterior parietal association cortex of the monkey: command functions for operations within extrapersonal space. *Journal of Neurophysiology* 38:871–908.

Moynihan, M. (1964). Some behavior patterns of platyrrhine monkeys. I. The night monkey (*Aotus trivirgatus*). *Smithsonian Miscellaneous Collection* 146(5):1–84.

Moynihan, M. (1966). Communication in *Callicebus*. *Journal of the Zoological Society, London* 150:77–127.

Moynihan, M. (1970a). The control, suppression, decay, disappearance and replacement of displays. *Journal of Theoretical Biology* 29:85–112.

Moynihan, M. (1970b). Some behavior patterns of platyrrhine monkeys. II. *Saguinus geoffroyi* and some other tamarins. *Smithsonian Contributions to Zoology* 28:1–77.

Moynihan, M. (1976a). Notes on the ecology and behavior of the pygmy marmoset (*Cebuella pygmaea*) in Amazonian Colombia. In R. W. Thorington, Jr. and P. G. Heltne (eds.), *Neotropical Primates, Field Studies and Conservation* (pp. 79–84). Washington, DC: National Academy of Science.

Moynihan, M. (1976b). *The New World Primates*. Princeton: Princeton University Press.

Müller, P. (1973). The dispersal centres of terrestrial vertebrates in the neotropical realm: A study in the evolution of the neotropical biota and its native landscapes. *Biogeographica*, Vol. 2. The Hague: W. Junk.

Muller, E. F., and Jaksche, H. (1980). Thermoregulation, oxygen consumption, heart rate and evaporative water loss in the thick-tailed bushbaby (*Galago crassicaudatus* Geoffroy, 1812). *Zeitschrift Saugetier* 45:269–278.

Murdock, A. (1991). Sexual bias in troop progression leading in the mantled howler monkeys: Male or female? *American Journal of Primatology* 24(2):124.

Murra, J. V. (1944). The historic tribes of Ecuador. In J. H. Steward (ed.), *Handbook of South American Indians*. Washington, DC: Bulletin of the Bureau of American Ethnology 143(2):785–821.

Muskin, A. (1984a). Preliminary field observations of *Callithrix aurita* (Callitrichinae, Cebidae). In M. T. de Mello (ed.), *A Primatologia no Brasil-1* (pp. 79–82). Brasilia: Sociedade Brasileira de Primatologia.

Muskin, A. (1984b). Field notes and geographic distribution of *Callithrix aurita* in eastern Brazil. *American Journal of Primatology* 7:377–380.

Muskin, A., and Fischgrund, A. J. (1981). Seed dispersal of *Stemmadenia* (Apocynaceae) and sexually dimorphic feeding strategies by *Ateles* in Tikal, Guatemala. *Biotropica, Supplement Reproductive Botany* 13:78–80.

Myers, N. (1992). *The Primary Source: Tropical Forests and Our Future*. New York: W. W. Norton.

Napier, J. R., and Napier, P. H. (1967). *A Handbook of Living Primates*. London: Academic Press.

Napier, J. R., and Napier, P. H. (1985). *The Natural History of the Primates.* Cambridge, MA: MIT Press.

Napier, J. R., and Walker, A. C. (1967). Vertical clinging and leaping & newly recognized category of locomotor behaviour in primates. *Folia Primatol.* 6:204–219.

Nash, L. T. (1976). Troop fission in free-ranging baboons in the Gombe Stream National Park. *American Journal of Physical Anthropology* 44:63–77.

Nash, L. T. (1986). Dietary, behavioral, and morphological aspects of gummivory in primates. *Yearbook of Physical Anthropology* 29:113–137.

Natale, F., Antinucci, F., Poti, P., and Spinozzi, G. (1988). Object manipulation in capuchin monkeys (*Cebus apella*). In A. Tartabini and M. L. Genta (eds.), *Perspectives in the Study of Primates: An Italian Contribution to International Primatology* (pp. 25–37). Cosenza, Italy: De Rose.

Nathans, J., Thomas, D., and Hogness, D. S. (1986). Molecular genetics of human color vision: The genes encoding blue, green and red pigments. *Science* 232:193–202.

Natori, M. (1986a). Interspecific relationships of *Callithrix* based on the dental characters. *Primates* 27(3):321–336.

Natori, M. (1986b). Phylogenetic relationships of marmosets (Callitrichidae). *Jinruigaku Zasshi/Journal of the Anthropological Society of Nippon* 94(2):247.

Natori, M. (1988). An analysis of cladistic relationships of *Leontopithecus* based on dental and cranial characters. *Jinruigaku Zasshi/Journal of the Anthropological Society of Nippon* 97(2):157–167.

Natori, M. (1989). A cladistic analysis of interspecific relationships of *Saguinus*. *Primates* 29(2):263–276.

Natori, M. (1990). Numerical analysis of the taxonomical status of *Callithrix kuhli* based on measurements of the postcanine dentition. *Primates* 31(4):555–562.

Natori, M., and Shigehara, N. (1992). Interspecific differences in lower dentition among eastern-Brazilian marmosets. *Journal of Mammalogy* 73(3):668–671.

Neitz, J., Neitz, M., and Jacobs, G. H. (1991). Spectral tuning of pigmens underlying red-green color vision. *Science* 252:971–974.

Neville, M. K. (1972a). The population structure of red howler monkeys (*Alouatta seniculus*). *Folia Primatologica* 17:56–86.

Neville, M. K. (1972b). Social relations within troops of red howler monkeys (*Alouatta seniculus*). *Folia Primatologica* 18:47–77.

Neville, M. K. (1974). "Carne de Monte" and its effect upon simian populations in Peru. Paper read at 73rd annual meeting, American Anthropological Association, Mexico City.

Neville, M. K. (1976a). The population and conservation of howler monkeys in Venezuela and Trinidad. In R. W. Thorington, Jr. and P. G. Heltne (eds.), *Neotropical Primates: Field Studies and Conservation* (pp. 101–108). Washington, DC: National Academy of Sciences.

Neville, M. K. (1976b). The red howler monkey troop as a social unit: Interactions among troops and with other stimuli. In E. Giles and J. Friedlaender (eds.), *Measures of Man* (pp. 72–108). Boston: Peabody Museum Press.

Neville, M. K., and Gunter, A. (1979). Howler monkey allogrooming. Paper presented at VIIth Congress, International Primatological Society, Bangalore, India.

Neville, M., Castro, N., Mármol, A., and Revilla, J. (1976). Censusing primate populations in the reserved area of the Pacaya and Samiria rivers, Department Loreto, Peru. *Primates* 17:151–181.

Neville, M. K., Glander, K. E., Braza, F., and Rylands, A. B. (1988). The howling monkeys, genus *Alouatta*. In R. A. Mittermeier, A. B. Rylands, A. F. Coimbra-Filho, and G. A. B. Fonseca (eds.), *Ecology and Behavior of Neotropical Primates*, Vol 2 (pp. 349–453). Washington DC: World Wildlife Fund.

Newcomer, M. W., and De Farcy, D. D. (1985). White-faced capuchin (*Cebus capucinus*) predation on a nestling coati (*Nasua narica*). *Journal of Mammalogy* 66(1):185–186.

Newman, J. D. (1985). Squirrel monkey communication. In L. A. Rosenblum and C. L. Coe (eds.), *Handbook of Squirrel Monkey Research* (pp. 99–126). New York: Plenum.

Newman, J. D., and Symmes, D. (1982). Inheritance and experience in the acquisition of primate acoustic behavior. In C. T. Snowdon, C. H. Brown, and M. R. Petersen (eds.), *Primate Communication* (pp. 259–278). New York: Cambridge University Press.

Newman, J. D., Lieblich, A. K., Talmage-Riggs, G., and Symmes, D. (1978). Syllable classification and sequencing in twitter calls of squirrel monkeys (*Saimiri sciureus*). *Zeitschrift für Tierpsychologie* 47:77–88.

Newman, J. D., Wamboldt, M. Z., Gelhard, R., and Minters, N. (1987). Characterization of separation-induced vocalizations in infant pygmy Marmosets. *American Journal of Primatology* 12(3):363.

Neyman, P. F. (1977). Aspects of the ecology and social organization of free-ranging cotton-top tamarins (*Saguinus oedipus*) and the conservation status of the species. In D. G. Kleiman (ed.), *The Biology and Conservation of the Callitrichidae* (pp. 39–71). Washington DC: Smithsonian Institution Press.

Neyman, P. F. (1980). Ecology and social organization of the cotton-top tamarin (*Saguinus oedipus*). Ph.D. dissertation, University of California, Berkeley.

Nishida, T., Hiraiwa-Hasegawa, M., Hasegawa, T., and Takahata, Y. (1985). Group extinction and female transfer in wild chimpanzees in the Mahale National Park, Tanzania. *Zeitschrift für Tierpsychologie* 67:284–301.

Nishimura, A. (1979). In search of woolly spider monkey. *Kyoto University Overseas Research Reports on New World Monkeys* 1:21–37.

Nishimura, A. (1987). Sociological characteristics of woolly monkeys (*Lagothrix lagotricha*) in the upper Caqueta, Colombia. *International Journal of Primatology* 8(5):521.

Nishimura, A. (1988a). Mating behavior of woolly monkeys (Lagothrix lagotricha) at La Macarena, Colombia. *Field Studies of New World Monkeys, La Macarena, Colombia* 1:19–27.

Nishimura, A. (1988b). Field studies of New World monkeys: past and present. *Monkey.* 223:4–9. [in Japanese].

Nishimura, A. (1990a). Mating behavior of woolly monkeys (*Lagothrix lagotricha*) at La Macarena, Colombia (II): Mating relationships. *Field Studies of New World Monkeys, La Macarena, Colombia* 3:7–12.

Nishimura, A. (1990b). A sociological and behavioral study of woolly monkeys, *Lagothrix lagotricha*, in the Upper Amazon. *Science and Engineering Reviews of Doshisha University* 31(2):87–121.

Nishimura, A., and Izawa, K. (1975). The group characteristics of woolly monkeys (*Lagothrix lagotricha*) in the upper Amazonian basin. In S. Kondo, M. Kawai, and A. Ehara (eds.), *Contemporary Primatology* (pp. 351–357). Basel: S. Karger.

Nishimura, A., Fonseca, G. A. B., Mittermeier, R. A., Young, A. L., Strier, K. B., and Valle, C. M. C. (1988). The muriqui, genus Brachyteles. In R. A. Mittermeier, A. B. Rylands, A. Coimbra-Filho, and G. A. B. Fonseca (eds.), *Ecology and Behavior of Neotropical Primates*, Vol 2 (pp. 577–610). Washington, DC: World Wildlife Fund.

Nishimura, A., Wilches, A., and Estrada, C. (1990). Reproductive behaviors of woolly monkeys, *Lagothrix lagotricha*, viewed from longterm studies. *Abstracts, XIIIth Congress of the International Primatological Society*, Nagoya, Japan, pp. 313–314.

Noback, C. R. (1975). The visual system of primates in phylogenetic studies. In P. Luckett and F. S. Szalay (eds.), *Phylogeny of the Primates* (pp. 199–218). New York: Plenum.

Noback, C. R., and Shriver, J. E. (1969). Encephalization and the lemniscal system during phylogeny. *Annals of the New York Academy of Science* 167:118–128.

Norconk, M. A. (1986). Interaction between primate species in a neotropical forest: Mixed species troops of *Saguinus mystax* and *S. fuscicollis* (Callitrichidae). Unpublished Ph.D. dissertation, University of California, Los Angeles.

Norconk, M. A., and Kinzey, W. G. (1990). Preliminary data on feeding ecology of *Pithecia pithecia* in Bolívar State, Venezuela. *American Journal of Primatology* 20(3):215.

Norconk, M. A., and Kinzey, W. G. (1992). Foraging patterns and troop fragmentation in bearded sakis and black spider monkeys. *Abstracts of the XIVth Congress of the International Primatological Society*, Strasbourg, France, p. 58.

Norconk, M. A., and Kinzey, W. G. (1993). Feast or famine? A comparison of bearded saki feeding ecology in terra firme and insular habitats. *American Journal of Physical Anthropology Supplement* 16:151–152.

Norconk, M. A., and Kinzey, W. G. (1994). Challenge of neotropical frugivory: Foraging patterns of spider monkeys and bearded sakis. *American Journal of Primatology* 34:171–183.

Norris, J. C. (1990a), The semantics of *Cebus olivaceus* alarm calls: Object designation and attribution. *American Journal of Physical Anthropology* 20(3):216.

Norris, J. C. (1990b). The semantics of *Cebus olivaceus* alarm calls: Object designation and attribution. Unpublished Ph.D. dissertation, University of Florida, Gainesville.

Nowak, R. M. (1991). *Walker's Mammals of the World*, 5th ed. Baltimore: Johns Hopkins University Press.

Nudo, R. J., and Masterton, R. B. (1990). Descending pathways to the spinal cord. III. Sites of origin of the corticospinal tract. *Journal of Comparative Neurology* 296:559–583.

Nunes, A. (1987). Notes on the feeding ecology of *Ateles b. belzebuth* in Maraca Island Ecological Station. *International Journal of Primatology* 8(5):480.

Oates, J. (1987). Food distribution and foraging behavior. In B. B. Smuts, D. L. Cheney, R. M. Seyfarth, R. W. Wrangham, and T. T. Struhsaker (eds.), *Primate Societies* (pp. 197–209). Chicago: University of Chicago Press.

O'Brien, T. G. (1988). Parasitic nursing behavior in the wedge-capped capuchin monkey (*Cebus olivaceus*). *American Journal of Primatology* 16(4):341–344.

O'Brien, T. G., and Robinson, J. G. (1987). The effects of group size and female rank on sex ratio at birth in capuchins *Cebus olivaceus*. *International Journal of Primatology* 8(5):499.

Odum, E. P. (1969). Strategy of ecosystem development. *Science* 164:262–270.

Odum, E. P. (1971). *Fundamentals of Ecology*. Philadelphia: W. B. Saunders.

Oftedal, O. T. (1991). The nutritional consequences of foraging in primates: The relationship of nutrient intakes to nutrient requirements. *Philosophical Transactions of the Royal Society of London Series B* 334:161–170.

Ogden, T. E. (1975). The receptor mosaic of *Aotus trivirgatus*: Distribution of rods and receptors. *Journal of Comparative Neurology* 163:193–202.

Ohnuki-Tierney, E. (1987). *The Monkey as Mirror: Symbolic Transformations in Japanese History and Ritual*. Princeton: Princeton University Press.

Oliveira, J. M. S., and Lima, M. G. (1981). Observaces sobre a ecologia, comportamento dos parauacus (*Pithecia pithecia*, Cebidae: Primates). Unpublished report to INPA, Manaus.

Oliveira, J. M. S., Lima, M. G., Bonvicino, C., Ayres, J. M., and Fleagle, J. G. (1985). Preliminary notes on the ecology and behavior of the Guianan saki (*Pithecia pithecia*, Linnaeus 1766; Cebidae, Primates). *Acta Amazônica* 15(1–2):249–263.

Olmos, F. (1994). Jaguar predation on muriqui *Brachyteles arachnoides*. *Neotropical Primates* 2(2):16.

Orlosky, F. J. (1973). Comparative dental morphology of extant and extinct Cebidae. Unpublished Ph.D. dissertation, University of Washington, Seattle.

Overdorff, D. (1988). Preliminary report on the activity cycle and diet of the red-bellied lemur (*Lemur rubriventer*) in Madagascar. *American Journal of Primatology* 16:143–153.

Overdorff, D. (1990). Flower predation and nectivory in *Lemur fulvus rufus* and *Lemur rubriventer*. *American Journal of Physical Anthropology* 81:276.

Overdorff, D. (1991). Ecological correlates to social structure in two prosimian primates in Madagascar: *Eulemur fulvus rufous* and *Eulemur rubriventer*. Ph.D. dissertation, Duke University.

Overdorff, D. (1991). Seasonal patterns of frugivory of *Lemur fulvus rufus* and *Lemur rubriventer* in Madagascar. *American Journal of Physical Anthropology* (Suppl. 12):139.

Overdorff, D. J. (1993). Ecological and reproductive correlates to range use in red-bellied lemurs (*Eulemur rubriventer*) and rufous lemurs (*Eulemur fulvus rufus*). In P. M. Kappeler and J. U. Ganzhorn (eds.), *Lemur Social Systems and Their Ecological Basis* (pp. 167–178). New York: Plenum.

Overdorff, D. J., and Rassmussen, M. A. (1995). Determinants of nighttime activity in "diurnal" lemurid primates. In L. Alterman, G. A. Doyle, and M. Kay Izard (eds.), *Creatures of the Dark* (pp. 61–74). New York: Plenum.

Oyama, T., Furusaka, T., and Kito, T. (1986). Color vision tests of Japanese and rhesus monkeys. In D. M. Taub and F. A. King (eds.), *Current Perspectives in Primate Biology* (pp. 253–269). New York: Van Nostrand Reinhold.

Paccagnella, S. G. (1986). Relatório sobre o censo da população de monos-carvoeiros do parque estadual de "Carlos Botelho." Unpublished report.

Packer, C. (1979). Male dominance and reproductive activity in *Papio anubis*. *Animal Behavior* 27:37–45.

Packer, O., Hendrickson, A. E., and Curcio, C. C. (1989). Photoreceptor topography of the retina in the adult pigtailed macaque (*Macaca nemestrina*). *Journal of Comparative Neurology* 288:165–183.

Padua, C. V. (1990). *1990 International Studbook for Black Lion Tamarin Leontopithecus chrysopygus*. São Paulo: Fundação Parque Zoologico de São Paulo.

Padua, C. V. (1992). Comparative study of four groups of black lion tamarins (*Leontopithecus chrysopygus*). *Abstracts of the XIVth Congress of the International Primatological Society*, Strasbourg, France, p. 94.

Pagel, M. D., and Harvey, P. H. (1989). Taxonomic differences in the scaling of brain on body weight among mammals. *Science* 244:1589–1593.

Pak, W. L., and O'Tousa, J. E. (1988). Molecular analysis of visual pigment genes. *Photochemistry and Photobiology* 47:877–882.

Parker, S. T., and Poti, P. (1990). The role of innate motor patterns in ontogenetic and experiential development of intelligent use of sticks in cebus monkeys. In S. T. Parker and K. R. Gibson (eds.), *"Language" and Intelligence in Monkeys and Apes. Comparative Developmental Perspectives* (pp. 219–243). New York: Cambridge University Press.

Patkay, S. (1992). Diseases of the Callitrichidae: A review. *Journal of Medical Primatology* 21(4):189–236.

Patton, J. L., Berlin, B., and Berlin, E. A. (1982). Aboriginal perspectives on a mammal community in Amazonian Peru: Knowledge and utilization patterns among the Aguaruna Jivaro. In M. A. Mares and H. H. Genoways (eds.), *Mammalian Biology in South America*. Pittsburgh: *University of Pittsburgh Pymatuning Laboratory of Ecology, Special Publications Series* 6:111–128.

Peetz, A. (1990). Soziale Kompensation, Ernährung und Habitatnutzung eines ohne Artgenossen freilebenden jungen Roten Brüllaffen (*Alouatta seniculus*). Unpublished Diplomarbeit thesis, Department of Biology, University of Bielefeld, Bielefeld, Germany.

Peetz, A. (in prep.). Habitatnutzung, Nahrungsökologie und soziale Organisation beim Rotrückensaki (*Chiropotes satanas chiropotes*) im Estado Bolívar, Venezuela." Unpublished Ph.D. thesis, Universität Bielefeld, Bielefeld, Germany.

Peetz, A., Norconk, M. A., and Kinzey, W. G. (1992). Predation by jaguar on howler monkeys (*Alouatta seniculus*) in Venezuela. *American Journal of Primatology* 28:223–228.

Pereira, M. E. (1993). Seasonality adjustment of growth rate and adult body weight in ringtailed lemurs. In P. M. Keppeler and J. U. Ganzhorn (eds.),

Lemur Social Systems and Their Ecological Basis (pp. 205–221). New York: Plenum.

Pereira, M. E., Klepper, A., and Simons, E. L. (1987). Tactics of care for young infants by forest living ruffed lemurs (*Varecia variegata variegata*): Ground nests, parking, and biparental guarding. *American Journal of Primatology* 13:129–144.

Peres, C. A. (1986a). Golden lion tamarin project. II. Ranging patterns and habitat selection in gold lion tamarins *Leontopithecus rosalia* (Linnaeus, 1766) (Callitrichidae, Primates). In M. Thiago de Mello (ed.), *A Primatologia no Brasil-2* (pp. 223–233). Brasilia: Instituto de Ciências Biológicas.

Peres, C. A. (1986b). Consequences of territorial defense in wild golden lion tamarins, *Leontopithecus rosalia*. *Primate Report* 14:234.

Peres, C. A. (1987). Effects of hunting on primate communities of western Brazilian Amazonia. *International Journal of Primatology* 8(5):492.

Peres, C. A. (1989a). Exudate-eating by wild golden lion tamarins, *Leontopithecus rosalia*. *Biotropica* 21(3):287–288.

Peres, C. A. (1989b). Costs and benefits of territorial defense in wild golden lion tamarins, *Leontopithecus rosalia*. *Behavioral Ecology and Sociobiology* 25(3):227–233.

Peres, C. A. (1990a). Effects of hunting on western Amazonian primate communities. *Biological Conservation* 54(1):47–59.

Peres, C. A. (1990b). A harpy eagle successfully captures an adult male red howler monkey. *Wilson Bulletin* 102:560–561.

Peres, C. A. (1991). Seed predation of *Cariniana micrantha* (Lecythidaceae) by brown capuchin monkeys in Central Amazonia. *Biotropica* 23(3):262–270.

Peres, C. A. (1993a). Notes on the ecology of buffy saki monkeys (*Pithecia albicans*, Gray 1860): A canopy seed-predator. *American Journal of Primatology* 31:129–140.

Peres, C. A. (1993b). Anti-predation benefits in a mixed-species group of Amazonian tamarins. *Folia Primatologica* 61(2):61–76.

Peres, C. A. (1993c). Structure and spatial organization of an Amazonian terra firme forest primate community. *Journal of Tropical Ecology* 9:259–276.

Peres, C. A. (1994). Diet and feeding ecology of gray woolly monkeys (*Lagothrix lagotricha cana*) in central Amazonia: Comparisons with other atelines. *International Journal of Primatology* 15(3):333–372.

Perloe, S. I. (1986). Conflict, affiliation, mating, and the effects of spatial confinement in a captive group of squirrel monkeys (*Saimiri sciureus*). In D. M. Taub and F. A. King (eds.), *Current Perspectives in Primate Social Dynamics* (pp. 89–98). New York: Van Nostrand Reinhold.

Perry, J., Bridgewater, D., and Horseman, D. (1972). Captive propagation: A progress report. *Zoologica* Fall 1972:109–117.

Perry, J. M., Izard, K. M., and Fail, P. A. (1995). An assessment of reproductive competence in captive mongoose lemurs (*Lemur mongoz*). *American Journal of Primatology* 9 (in press).

Perry, V. H., and Cowey, A. (1985). The ganglion cell and cone distributions in the monkey's retina: Implications for central magnification factors. *Vision Research* 25:1795–1810.

Persson, V. G., and Lorini, M. L. (1993). Notas sobre o micoleão-de-cara-preta,

Leontopithecus caissara Lorini and Persson, 1990, no sul do Brasil (Primates, Callithrichidae). In M. E. Yamamoto and M. B. C. de Souza (organizadoras), *A Primatologia no Brasil-4* (pp. 169–181). Salvador: Sociedade Brasileira de Primatologia.

Petito, L. A., and Marentette, P. F. (1991). Babbling in the manual mode: Evidence for the ontogeny of language. *Science* 251:1493–1496.

Petras, J. M. (1969). Some efferent connections of the motor and somatosensory cortex of simian primates and felid, canid and procyonid carnivores. *Annals of the New York Academy of Science* 167:469–505.

Petry von, H., Riehl, I., and Zucker, H. (1986). Energieumsatzmessungen an Weissbuschelaffchen (*Callithrix jacchus*). *Journal of Animal Physiology and Animal Nutrition* 55:214–224.

Phillips, K. A. (1994). Resource patch use and social organization in *Cebus capucinus*. *American Journal of Primatology* 33(3):233.

Pielou, E. C. (1991). *After the Ice Age: The Return of Life to Glaciated North America*. Chicago: University of Chicago Press.

Pilbeam, D. R. (1967). Man's earliest ancestors. *Science Journal* 3:14–53.

Pinder, L. (1986). Projeto mico-leão. III. Avaliação técnica de translocacão em *Leontopithecus rosalia* (Linnaeus, 1766) (Callitrichidae, Primates). In M. Thiago de Mello (ed.), *A Primatologia no Brasil-2* (pp. 235–241). Brasilia: Instituto de Ciências Biológicas.

Pinto, L. P. S., Costa, C. M. R., Strier, K. A., and Fonseca, G. A. B. (1991). Censo de primatas da reserva Biológica Augusto Ruschi, Santa Teresa, ES. *Resumos, XVIII Congresso Brasileiro de Zoologia* 394.

Pires, C. (1989). Exudate-eating by wild golden lion tamarins, *Leontopithecus rosalia*. *Biotropica* 21(3):287–288.

Pissinatti, A. (1992). Advances in veterinary and biomedical aspects of captive propagation. *Abstracts of the XIVth Congress of the International Primatological Society*, Strasbourg, France, p. 96.

Platt, M. L. (1994). Adaptive differences in spacial memory between lion tamarins and marmosets. *American Journal of Primatology* 33(3):234–235.

Plavcan, J. M., and Kay, R. F. (1988). Sexual dimorphism and dental variability in platyrrhine primates. *International Journal of Primatology* 9(3):169–178.

Podolsky, R. D. (1990). Effects of mixed-species association on resource use by *Saimiri sciureus* and *Cebus apella*. *American Journal of Primatology* 21(2):147–158.

Pokorny, J., Smith, V. C., Verriest, G., and Pinckers, A. J. L. G. (1979). *Congenital and Acquired Color Vision Defects*. New York: Grune & Stratton.

Pola, Y. V., and Snowdon, C. T. (1975). The vocalizations of pygmy marmosets (*Cebuella pygmaea*). *Animal Behavior* 23:826–842.

Pollock, J. I. (1975). Field Observations on *Indri indri*, a preliminary report. In I. Tattersall and R. Sussman (eds.), *Lemur Biology* (pp. 287–312). New York: Plenum.

Pollock, J. I. (1977). The ecology and sociology of feeding in *Indrii indrii*. In T. M. Clutton-Brock (ed.), *Primate Ecology* (pp. 37–68). London: Academic Press.

Pollock, J. I. (1979). Female dominance in *Indri indri*. *Folia Primatologica* 31:143–164.

Pollock, J. I. (1986). Primates and conservation priorities in Madagascar. *Oryx* 20:209–216.

Pollock, J. I. (1989). Intersexual relationships amongst prosimians. *Human Evolution* 4:133–143.

Polyak, S. L. (1957). *The Vertebrate Visual System*. Chicago: University of Chicago Press.

Pook, A. G. (1975). Breeding Goeldi's monkey (*Callimico goeldii*) at the Jersey Zoological Park. *Twelfth Annual Report of the Jersey Wildlife Preservation Trust* 17–20.

Pook, A. G. (1978a). A comparison between the reproduction and parental behaviour of the Goeldi's monkey (*Callimico goeldii*) and of the true marmosets (Callitrichidae). In H. Rothe, H. -J. Wolters, and J. P. Hearn (eds.), *Biology and Behaviour of Marmosets* (pp. 1–14). Gottingen: Eigenverlag Rothe.

Pook, A. G. (1978b). Some notes on the re-introduction into groups of six hand-reared marmosets of different species. In H. Rothe, H.-J. Wolters, and J. P. Hearn (eds.), *Biology and Behaviour of Marmosets* (pp. 155–159). Gottingen: Eigenverlag Rothe.

Pook, A. G., and Pook, G. (1979). A field study on the status and socioecology of the Goeldi's monkey (*Callimico goeldii*) and other primates in northern Bolivia. Unpublished Report to the New York Zoological Society.

Pook, A. G., and Pook, G. (1981). A field study of the socioecology of the Goeldi's monkey (*Callimico goeldii*) in northern Bolivia. *Folia Primatologica* 35:288–312.

Pook, A. G., and Pook, G. (1982). Polyspecific associations between *Saguinus fuscicollis*, *Saguinus labiatus*, *Callimico goeldii*, and other primates in North-Western Bolivia. *Folia Primatologica* 38:196–216.

Poole, T. B., and Box, H. O. (1986). A survey of non-human primate stocks in the U.K.: Implications for conservation and research. *Primate Eye* 29(suppl.):59–63.

Pope, B. L. (1966). The population characteristics of howler monkeys (*Alouatta caraya*) in northern Argentina. *American Journal of Physical Anthropology* 24:361–370.

Pope, B. L. (1968). Population characteristics. In M. R. Malinow (ed.), *Biology of the Howler Monkey (Alouatta caraya)* (pp. 13–30). Basel: S. Karger.

Pope, T. R. (1989). The influence of mating system and dispersal patterns on the genetic structure of red howler monkey populations. Unpublished Ph.D. thesis, University of Florida, Gainsville.

Pope, T. R. (1990). The reproductive consequences of male cooperation in the red howler monkey: Paternity exclusion in multi-male and single-male troops using genetic markers. *Behavioral Ecology and Sociobiology* 27(6):439–446.

Portman, O. W. (1970). Nutritional requirements of nonhuman primates. In R. Harris (ed.), *Feeding and Nutrition of Nonhuman Primates* (pp. 87–115). New York: Academic Press.

Posey, D. A. (1982). Keepers of the forest. *Garden* 6(1):18–24.

Power, M. L. (1991). Digestive function, energy intake and the response to dietary gum in captive callitrichids (marmosets, tamarins). Unpublished Ph.D. thesis, University of California, Berkeley.

Prance, G. T., and Lovejoy, T. E. (eds.), (1985), *Amazonia*. New York: Pergamon Press.

Prates, J. C., Gayer, S. M. P., Kunz, L. F., Jr., and Buss, G. (1987). Feeding habits of the brown howler monkey *Alouatta fusca clamitans* in Itapuã State Park (30°20'S; 50°55'W); RS; Brasil. *International Journal of Primatology* 8(5):534.

Prates, J. C., Gayer, S. M. P., Kunz, L. F., Jr., and Buss, G. (1990a). Feeding habits of the brown howler monkey *Alouatta fusca clamitans* (Cabrera, 1940)(Cebidae, Alouattinae) in Itapuã State Park: A preliminary report. *Acta Biologica Leopoldensia* 12(1):175–188.

Prates, J. C., Kunz, L. F., Jr., and Buss, G. (1990b). Comportamento postural e locomotor de *Alouatta fusca clamitans* (Cabrera, 1940) em floresta subtropical (Primates, Cebidae). *Acta Biologica Leopoldensia* 12(1):189–200.

Preston-Mafham, R. K. (1992). *Primates of the World*. New York: Facts on File.

Preuss, T. M., Beck, P. D., and Kaas, J. H. (1993). Areal, modular and connectional organization of visual cortex in a prosimian primate, the slow loris (*Nycticebus coucang*). *Brain Behavior and Evolution*. 42:321–335.

Price, E. C. (1990). Infant carrying as a courtship strategy of breeding male cotton-top tamarins. *Animal Behavior* 40(4):784–786.

Price, E. C. (1991). Competition to carry infants in captive families of cotton-top tamarins *Saguinus oedipus*). *Behaviour* 118(1–2):66–68.

Price, E. C. (1992a). Sex and helping: Reproductive strategies of breeding male and female cotton-top tamarins, *Saguinus oedipus*. *Animal Behavior* 43(5):717–728.

Price, E. C. (1992b). The costs of infant carrying in captive cotton-top tamarins. *American Journal of Primatology* 26(1):23–33.

Price, E. C., and McGrew, W. C. (1990). Cotton-top tamarins (*Saguinus (o.) oedipus*) in a semi-naturalistic breeding colony. *American Journal of Primatology* 20:1–12.

Price, E. C., and McGrew, W. C. (1991). Departures from monogamy in colonies of captive cotton-top tamarins. *Folia Primatologica* 57(1):16–27.

Pruetz, J. D., and Garber, P. A. (1991). Patterns of resource utilization, home range overlap, and intergroup encounters in moustached tamarin monkeys. *American Journal of Physical Anthropology Supplement* 12:146.

Pubols, B. H., and Pubols, L. M. (1971). Somatotopic organization of spider monkey somatic sensory cerebral cortex. *Journal of Comparative Neurology* 141:63–76.

Queiroz, H. L. (1992). A new species of capuchin monkey, genus *Cebus* Erxleben, 1777 (Cebidae Primates) from eastern Brazilian Amazonia. *Goeldiana, Zoologia* 15:1–13.

Rabinowitz, P. D., Coffin, M. F., and Falvey, D. (1983). The separation of Madagascar and Africa. *Science* 220:67–69.

Radinsky, L. (1982). Some cautionary notes on making inferences about relative brain size. In C. R. Noback and W. Montagna (eds.), *Primate Brain Evolution: Methods and Concepts* (pp. 29–38). New York: Plenum.

Rakic, P. (1976). Prenatal genesis of connections subserving ocular dominance in the rhesus monkey. *Nature (London)* 261:467–471.

Ramirez, M. (1980). Grouping patterns of the woolly monkey, *Lagothrix lagotricha*, at the Manu National Park, Peru. *American Journal of Physical Anthropology* 52:269.

Ramirez, M. (1988). The woolly monkey, genus *Lagothrix*. In R. A. Mittermeier, A. B. Rylands, A. Coimbra-Filho, and G. A. B. Fonseca (eds.), *Ecology and Behavior of Neotropical Primates*, Vol 2 (pp. 539–575). Washington, DC: World Wildlife Fund.

Ramirez, M. M. (1990). Feeding ecology and demography of the moustached tamarin (*Saguinus mystax*) in Northeastern Peru. Unpublished Ph.D. dissertation, City University of New York.

Ramirez, M., Freese, C., and Revilla, J. (1977). Feeding ecology of the pygmy marmoset, *Cebuella pygmaea*, in north-eastern Peru. In D. G. Kleiman (ed.), *The Biology and Conservation of the Callitrichidae* (pp. 91–104). Washington DC: Smithsonian Institution Press.

Ramirez-Cerquera, J. (1983). Reporte de una nueva especie de primates del género *Aotus* de Colombia. *Abstracts, Symposio sobre Primatologia en Latinoamerica.* Arequipa, Peru: IX Latin American Zoology Conference, October, 1983, p. 146.

Randolph, P. A., Randolph, J. C., Mattingly, K., and Foster, M. M. (1977). Energy costs of reproduction in the colton rat (*Sigmoden hispidos*). *Ecology* 58:31–45.

Rasmussen, D. R. (1981). Communities of baboon troops (*Papio cynocephalus*) in Mikumi National Park, Tanzania. *Folia Primatologica* 36:232–242.

Rasmussen, D. T. (1985). A comparative study of breeding seasonality and litter size in eleven taxa of captive lemurs (*Lemur* and *Varecia*). *International Journal of Primatology* 6:501–517.

Rasmussen, D. T. (1986). Anthropoid origins: A possible solution to the Adapidae-Omomyidae paradox. *Journal of Human Evolution* 15:1–12.

Rasmussen, D. T. (1990). The phylogenetic position of *Mahgarita stevensi*: Protoanthropoid or Lemuroid? *International Journal of Primatology* 11:437–467.

Rasmussen, D. T., and Simons, E. L. (1992). The paleobiology of the oligopithecines, the world's earliest known anthropoid primates. *International Journal of Primatology* 13:477–508.

Rasmussen, D. T., and Tan, C. L. (1992). The allometry of behavioral development: Fitting sigmoid curves to ontogenetic data for use in interspecific allometric analyses. *Journal of Human Evolution* 23:159–181.

Rasmussen, D. T., Bown, T. M., and Simons, E. L. (1992). The Eocene-Oligocene transition in continental Africa. In D. R. Prothero and W. A. Berggren (eds.), *Eocene-Oligocene Climatic and Biotic Evolution* (pp. 548–566) Princeton: Princeton University Press.

Rathbun, C. D. (1979). Description and analysis of the arch display in the golden lion tamarin, *Leontopithecus rosalia rosalia*. *Folia Primatologica* 32:125–148.

Rathbun, G. B., and Gache, M. (1980). Ecological survey of the night monkey, *Aotus trivirgatus*, in Formosa Province, Argentina. *Primates* 21:211–219.

Redford, K. H. (1989). The stakes in the game. *Orion: Nature Quarterly* 8(2):45–47.

Reichel-Dolmatoff, G. (1971). *Amazonian Cosmos: The Sexual and Religious Symbolism of the Tukano Indians.* Chicago: University of Chicago Press.

Remsen, J. V., and Parker, T. A. (1983). Contribution of river-created habitats to bird species richness in Amazonia. *Biotropica* 15(3):223–231.

Rettig, N. L. (1978). Breeding behavior of the harpy eagle (*Harpia harpyja*). *Auk* 95(4):629–643.

Richard, A. F. (1978). *Behavioral Variation: A Case Study of a Malagasy Lemur*. London: Bucknell University Press.

Richard, A. F. (1985a). Social boundaries in a Malagasy prosimian, the sifaka (*Propithecus verreauxi*). *International Journal of Primatology* 6:553–568.

Richard, A. F. (1985b). *Primates in Nature*. San Francisco: W.H. Freeman.

Richard, A. F. (in press). Sexual dimorphism and sexual selection: Evidence and implications from a behavioral study. *Journal of Human Evolution*.

Richard, A. F., and Dewar, R. E. (1991). Lemur ecology. *Annual Review of Ecological Systems* 22:145–175.

Richard, A. F., and Nicoll, M. (1987). Female social dominance and basal metabolism in a Malagasy primate, *Propithecus verreauxi*. *American Journal of Primatology* 12:309–314.

Richard, A. F., and Sussman, R. (1988). A framework for primate conservation in Madagascar. In R. W. Mittermeier and C. Marsh (eds.), *Primates in the Tropical Forest*. New York: Academic Press.

Richard, A. F., Rakotomanga, P., and Schwartz, M. (1991). Demography of *Propithecus verreauxi* at Beza Mahafaly: Sex ratio, survival and fertility 1984–1989. *American Journal of Physical Anthropology* 84:307–322.

Richman, D. P., Stewart, R. M., Hutchinson, J. W., and Caviness, V. S., Jr. (1975). Mechanical model of brain convolutional development. *Science* 189:18–21.

Ritchie, B. G., and Fragaszy, D. M. (1988). Capuchin monkey (*Cebus apella*) grooms her infant's wound with tools. *American Journal of Primatology* 16:345–348.

Rivero, M., and Arrendondo, O. (1991). *Paralouatta varonai*, a new Quaternary platyrrhine from Cuba. *Journal of Human Evolution* 21:1–11.

Roberts, M. (1994). Growth, development, and parental care in the Western Tarsier (*Tarsius bancanus*) in captivity: Evidence for a "slow" life history and nonmonogamous mating system. *International Journal of Primatology* 15(1):1–28.

Robinson, J. G. (1977). Vocal regulation of spacing in the titi monkey, *Callicebus moloch*. Unpublished Ph.D. dissertation, University of North Carolina.

Robinson, J. G. (1979a). Vocal regulation of use of space by groups of titi monkeys, *Callicebus moloch*. *Behavioral Ecology and Sociobiology* 5:1–15.

Robinson, J. G. (1979b). An analysis of the organization of vocal communication in the titi monkey, *Callicebus moloch*. *Zeitschrift für Tierpsychologie* 49:381–405.

Robinson, J. G. (1981a). Vocal regulation of inter- and intragroup spacing during boundary encounters in the titi monkey *Callicebus moloch*. *Primates* 22:161–172.

Robinson, J. G. (1981b). Spatial structure in foraging groups of wedge-capped capuchin monkeys *Cebus nigrivittatus*. *Animal Behaviour* 29:1036–1056.

Robinson, J. G. (1982). Vocal systems regulating within-group spacing. In C. T. Snowden, C. H. Brown, and M. R. Petersen, (eds.), *Primate Communication* (pp. 94–116). New York: Cambridge University Press.

Robinson, J. G. (1984a). Syntactic structures in the vocalizations of wedge-capped capuchin monkeys, *Cebus olivaceus*. *Behaviour* 90(1–3):46–79.

Robinson, J. G. (1984b). Diurnal variation in foraging and diet in the wedge-capped capuchin monkey, *Cebus olivaceus*. *Folia Primatologica* 43(4):216–228.

Robinson, J. G. (1986). Seasonal variation in use of time and space by the wedge-capped capuchin monkey *Cebus olivaceus*: Implications for foraging theory. *Smithsonian Contributions to Zoology*, No. 431 (pp. 1–60). Washington, DC: Smithsonian Institution Press.

Robinson, J. G. (1988a). Demography and group structure in wedge-capped capuchin monkeys, *Cebus olivaceus*. *Behaviour* 104(3/4):202–231.

Robinson, J. G. (1988b). Group size in wedge-capped capuchin monkeys *Cebus olivaceus* and the reproductive success of males and females. *Behavioral Ecology and Sociobiology* 23:187–197.

Robinson, J. G., and Janson, C. H. (1987). Capuchins, squirrel monkeys, and atelines: Socioecological convergence with Old World primates. In B. B. Smuts, D. L. Cheney, R. M. Seyfarth, R. W. Wrangham, and T. T. Struhsaker (eds.), *Primate Societies* (pp. 69–82). Chicago: University of Chicago Press.

Robinson, J. G., and O'Brien, T. G. (1991). Adjustments in birth sex ratio in wedge-capped capuchin monkeys. *American Naturalist* 138(5):1173–1186.

Robinson, J. G., Wright, P. C., and Kinzey, W. G. (1987). Monogamous cebids and their relatives: Intergroup calls and spacing. In B. Smuts, D. Cheney, R. Seyfarth, R. Wrangham, and T. Struhsaker (eds.), *Primate Societies* (pp. 44–53). Chicago: University of Chicago Press.

Rockwood, L. L., and Glander, K. E. (1979). Howling monkeys and leaf cutting ants: Comparative foraging in a tropical deciduous forest. *Biotropica* 11:1–10.

Roda, S. A., and Roda, S. (1987). Infanticide in a natural group of *Callithrix jacchus* (Callitrichidae, Primates). *International Journal of Primatology* 8(5):497.

Rodriguez-Luna, E., and Cortés-Ortiz, L. (1994). Translocacion y seguimiento de un grupo de monos *Alouatta palliata* liberado en una isla (1988–1994). *Neotropical Primates* 2(2):1–3.

Rodriguez Luna, E., Fa, J. E., Garcia Orduña, F., Silva Lopez, G., and Canales Espinoza, D. (1987). Primate conservation in Mexico. *Primate Conservation* 8:114–118.

Roeder, J. J., and Anderson, J. R. (1991). Urine washing in brown capuchin monkeys (*Cebus apella*): Testing social and nonsocial hypotheses. *American Journal of Primatology* 24(1):55–60.

Rohrs, M. (1959). Neue Ergebnisse und probleme der Allometrieforschung. *Zeitschrift für Wissenschaftliche Zoologie* 162:1–95.

Rondinelli, R., and Klein, L. L. (1976). An analysis of adult social spacing tendencies and related social interactions in a colony of spider monkeys, *Ateles geoffroyi*, at the San Francisco Zoo. *Folia Primatologica* 25:122–142.

Roosevelt, A. (1989). Lost civilization of the Lower Amazon. *Natural History* 89(2):74–82.

Rosa, M. G. P., Gattass, R., and Fiorani, M., Jr. (1988). Complete pattern of ocular dominance stripes in V1 of a New World monkey, *Cebus apella*. *Experimental Brain Research* 72:645–648.

Rosa, M. G. P., Gattass, R., and Soares, J. G. M. (1991). A quantitative analysis of cytochrome oxidase-rich patches in the primary visual cortex of *Cebus* monkeys: Topographic distribution and effects of late monocular enucleation. *Experimental Brain Research* 84:195–209.

Rose, K. D., and Fleagle, J. G. (1981). The fossil history of nonhuman primates in the Americas. In A. F. Coimbra-Filho and R. A. Mittermeier (eds.), *Ecology and Behavior of Neotropical Primates* (pp. 111–167). Rio de Janeiro: Academia Brasiliera de Ciencias.

Rose, L. M. (1994). Sex differences in diet and foraging behavior in white-faced capuchins (*Cebus capucinus*). *International Journal of Primatology* 15(1):95–114.

Rosenberger, A. L. (1977). *Xenothrix* and ceboid phylogeny. *Journal of Human Evolution* 6:461–481.

Rosenberger, A. L. (1978). Loss of incisor enamel in marmosets. *Journal of Mammalogy* 59:207–208.

Rosenberger, A. L. (1979). Phylogeny, evolution and classification of New World Monkeys (Platyrrhini, Primates). Ph.D. thesis. City University of New York.

Rosenberger, A. L. (1981). Systematics: the higher taxa. In A. F. Coimbra-Filho and R. A. Mittermeier (eds.), *Ecology and Behavior of Neotropical Primates*, Vol. 1 (pp. 9–27). Rio de Janeiro: Academia Brasileira de Ciencias.

Rosenberger, A. L. (1982). Supposed squirrel monkey affinities of the late Oligocene *Dolichocebus gaimanensis*. *Nature (London)* 298:202.

Rosenberger, A. L. (1983). Tale of tails: Parallelism and prehensibility. *American Journal of Physical Anthropology* 60:103–107.

Rosenberger, A. L. (1984). Fossil New World monkeys dispute the molecular clock. *Journal of Human Evolution* 13:737–742.

Rosenberger, A. L. (1992). The evolution of feeding niches in New World monkeys. *American Journal of Physical Anthropology* 88(4):525–562.

Rosenberger, A. L., and Coimbra-Filho, A. F. (1984). Morphology, taxonomic status and affinities of the lion tamarins, *Leontopithecus* (Callitrichinae, Cebidae). *Folia Primatologica* 42:149–179.

Rosenberger, A. L., and Kinzey, W. G. (1976). Functional patterns of molar occlusion in platyrrhine primates. *American Journal of Physical Anthropology* 45(2):281–298.

Rosenberger, A. L., and Stafford, B. J. (1994). Locomotion in captive *Leontopithecus* and *Callimico*: A multimedia study. *American Journal of Physical Anthropology* 94(3):379–394.

Rosenberger, A. L., and Strier, K. B. (1989). Adaptive radiation of the ateline primates. *Journal of Human Evolution* 18:717–750.

Rosenberger, A. L., Setoguchi, T., and Shigehara, N. (1990). The fossil record of callitrichine primates. *Journal of Human Evolution* 19(1–2):209–236.

Rosenberger, A. L., Hartwig, W. C., Takai, M., Setoguchi, T., and Shigehara, N. (1991a). Dental variability in *Saimiri* and the taxonomic status of *Neosaimiri fieldsi*, an early squirrel monkey from La Venta, Colombia. *International Journal of Primatology* 12(3):291–301.

Rosenberger, A. L., Setoguchi, T., and Hartwig, W. C. (1991b). *Laventiana annectens*, a new genus and species: Fossil evidence for the origins of callitrichine New World monkeys. *Proceedings of the National Academy of Science U.S.A.* 88:2137–2140.

Ross, C. (1988). The intrinsic rate of natural increase and reproductive effort in primates. *Journal of Zoology* 214(2):199–219.

Ross, C. (1994). The craniofacial evidence for anthropoid and tarsier relationships. In J. G. Fleagle and R. F. Kay (eds.), *Anthropoid Origins* (pp. 469–547). New York: Plenum.

Ross, E. B. (1976). Food taboos, diet, and hunting strategy: The adaptation to animals in Amazon cultural ecology. *Currrent Anthropology* 19(1):1–36.

Ross, R. A., and Giller, P. S. (1988). Observations on the activity patterns and social interactions of a captive group of blackcapped or brown capuchin monkeys (*Cebus apella*). *Primates* 29(3):307–317.

Rothe, H. (1990). Parental investment in the common marmoset (*Callithrix jacchus*). *XIIIth Congress of the International Primatological Society*, Nagoya and Kyoto, 18–24 July 1990, p. 110.

Rothe, H., and Darms, K. (1993). The social organization of marmosets: A critical evaluation of recent concepts. In A. B. Rylands (ed.), *Marmosets and Tamarins; Systematics, Behaviour, and Ecology* (pp. 176–199). Oxford: Oxford University Press.

Rothe, H., and Koenig, A. (1991). Variability of social organization in captive common marmosets (*Callithrix jacchus*). *Folia Primatologica* 57(1):28–33.

Rothwell, N. J., and Stock, M. J. (1985). Thermogenic capacity and brown adipose tissue activity in the common marmoset. *Comparative Biochemistry and Physiology* 81:683–686.

Rowe, M.H., Benevento, L. A., and Rezak, M. (1978). Some observations on the patterns of segregated geniculate inputs to the visual cortex in New World primates: An autoradiographic study. *Brain Research* 159:371–378.

Rowell, T. E., and Mitchell, B. J. (1991). Comparison of seed dispersal by guenons in Kenya and capuchins in Panama. *Journal of Tropical Ecology* 7:269–274.

Rudran, R. (1979). The demography and social mobility of a red howler (*Alouatta seniculus*) population in Venezuela. In J. Eisenberg (ed.), *Vertebrate Ecology in the Northern Neotropics* (pp. 107–126). Washington, DC: Smithsonian Institution Press.

Rumiz, D. I. (1990). *Alouatta caraya*: Population density and demography in northern Argentina. *American Journal of Primatology* 21(4):279–294.

Rumiz, D. I., Zunino, G. E., Obregozo, M. L., and Ruiz, J. C. (1986). *Alouatta caraya*: Habitat and resource utilization in northern Argentina. In D. M. Taub (ed.), *Current Perspectives in Primate Social Dynamics* (pp. 175–193). New York: Van Nostrand.

Runestad, J. A., and Teaford, M. F. (1990). Dental microwear and diet in Venezuelan primates. *American Journal of Physical Anthropology* 81(2):288–289.

Rusconi, C. (1933). Nuevos restos de monos fósiles del terciario antiguo de la Patagonia. *Anales de la Sociedad Rural Argentina* 116:286–289.

Russell, R. J. (1975). Body temperatures and behavior of captive cheirogaleids. In I. Tattersall and R. Sussman (eds.), *Lemur Biology* (pp. 193–208). New York: Plenum.

Rylands, A. B. (1979). Observações preliminares sobre o sagüi, *Callithrix humer-*

alifer intermedius (Hershkovitz, 1977) em Dardanelos, Rio Aripuanã, Mato Grosso. *Acta Amazonica* 9:589–602.

Rylands, A. B. (1980). The behavioural ecology of the golden-headed lion tamarin, *Leontopithecus rosalia chrysomelas*. Interim Report (September 1980) to the World Wildlife Fund, Washington, DC.

Rylands, A. B. (1981). Preliminary field observations on the marmoset *Callithrix humeralifer* intermedius (Hershkovitz, 1977) at Dardanelos, Rio Aripuanã, Mato Grosso. *Primates* 22:46–59.

Rylands, A. B. (1982). The behavior and ecology of three species of marmosets and tamarins (Callitrichidae, Primates) in Brazil. Unpublished Ph.D. dissertation, University of Cambridge.

Rylands, A. B. (1983). The behavioral ecology of the golden-headed lion tamarin, *L. chrysomelas*, and the marmoset *C. kuhli* (Callitrichidae, Primates). Unpublished Report to the World Wildlife Fund, Washington, DC.

Rylands, A. B. (1984a). Ecologia do mico leão, *Leontopithecus chrysomelas*, e o sagüi, *Callithrix kullii*, na Bahia. *Resumos, XI Congreso Brasileiro de Zoologia*, Belém, pp. 292–293.

Rylands, A. B. (1984b). Exudate-eating and tree-gouging by marmosets (Callitrichidae, Primates). In Chadwick and Suttin (eds.), *Tropical Rain-Forest* (pp. 155–168). Leeds: The Leeds Symposium, Leeds Philosophical and Literary Society.

Rylands, A. B. (1986a). Infant-carrying in a wild marmoset group, *Callithrix humeralifer*: Evidence for a polyandrous mating system. In M.T. de Mello (ed.), *A Primatologia no Brasil-2* (pp. 131–144). Campinas: Sociedade Brasileira de Primatologia.

Rylands, A. B. (1986b). Ranging behaviour and habitat preference of a wild marmoset group, *Callithrix humeralifer* (Callitrichidae, Primates). *Journal of the Zoological Society of London* (A) 210:489–514.

Rylands, A. B. (1987). Primate communities in Amazonian forests: Their habitats and food resources. *Experientia* 43:265–279.

Rylands, A. B. (1989). Sympatric Brazilian callitrichids: The black tufted-ear marmoset, *Callithrix kuhli*, and the golden-headed lion tamarin, *Leontopithecus chrysomelas*. *Journal of Human Evolution* 18:679–695.

Rylands, A. B. (1990). Scent marking behaviour of wild marmosets, *Callithrix humeralifer* (Callithrichidae, Primates). In D. W. Macdonald, D. Müller-Schwarze, and S. E. Natynczuk (eds.), *Chemical Signals in Vertebrates 5* (pp. 415–429). Oxford: Oxford University Press.

Rylands, A. B. (1993). The ecology of the lion tamarins, Leontopithecus: Some intrageneric differences and comparisons with other callitrichids. In A. B. Rylands (ed.), *Marmosets and Tamarins: Systematics, Behaviour, and Ecology* (pp. 296–313). Oxford: Oxford University Press.

Rylands, A. B., and de Faria, D. S. (1993). Habitats, feeding ecology, and home range size in the genus *Callithrix*. In A. B. Rylands (ed.), *Marmosets and Tamarins; Systematics, Behaviour, and Ecology* (pp. 262–272). Oxford: Oxford University Press.

Rylands, A. B., and Keuroghlian, A. (1988). Primate populations in continuous forest and forest fragments in central Amazonia. *Acta Amazônica* 18(3–4):291–307.

Rylands, A. B., Coimbra-Filho, A. F., and Mittermeier, R. A. (1993). Systematics, geographic distribution, and some notes on the conservation status of the Callitrichidae. In A. B. Rylands (ed.), *Marmosets and Tamarins; Systematics, Behaviour, and Ecology* (pp. 11–77). Oxford: Oxford University Press.

S. e Silva, J. de, Jr. (1991). Distribuição geográfica do cuxiú-preto (*Chiropotes satanas satanas* Hoffmansegg, 1807) na Amazônia Maranhense (Cebidae, Primates). In A. B. Rylands and A. T. Bernardes (eds.), *A Primatologia no Brasil-3* (pp. 275–284). Belo Horizonte: Fundação Bidiversitas.

Saffirio, G., and Hames, R. B. (1983). The forest and the highway. In impact of contact: Two Yanomama case studies. *Cambridge: Cultural Survival Occasional Papers* 11:1–52.

Saffirio, G., and Scaglion, R. (1982). Hunting efficiency in acculturated Yanomama villages. *Journal of Anthropological Research* 38(3):315–328.

Salo, J. (1986). River dynamics and diversity of Amazon lowland forest. *Nature (London)* 322:254–258.

Saltzman, W., Mason, W. A., and Mendoza, S. P. (1988). Synchronization in squirrel monkeys: Behavioral and physiological responses to isosexual pair formation. *American Journal of Primatology* 14(4):442–443.

Sanderson, I. T. (1957). *The Monkey Kingdom*. New York: Hanover House.

Sanford, R. L., Saldarriaga, J., Clark, K. E., Uhl, C., and Herrera, R. (1985). Amazon rain forest fires. *Science* 227:53–55.

Santos, C. V., French, J. A., and Otta, E. (1992). A comparative study of infant carrying behavior in callitrichid primates: *Callithrix* and *Leontopithecus*. *American Journal of Primatology* 27(1):56.

Santos, I. B., Mittermeier, R. A., Rylands, A. B., and Valle, C. M. C. (1987). The distribution and conservation status of primates in Southern Bahia, Brazil. *Primate Conservation* 8:126–142.

Sanz, V., and Nárquez, L. (1994). Conservacion del mono capuchino de Margarita (*Cebus apella* margaritae) en la Isla de Margarita, Venezuela. *Neotropical Primates* 2(2):5–8.

Sarich, V. M., and Cronin, J. E. (1980). South American mammal moelcular systematics, evolutionary clocks, and continental drift. In R. Ciochon and A. B. Chiarelli (eds.), *Evolutionary Biology of New World Monkeys and Continental Drift* (pp. 399–422). New York: Plenum.

Sassenrath, E. N., Mason, W. A., Fitzgerald, R. C., and Kenney, M. D. (1980). Comparative endocrine correlates of reproductive states in *Callicebus* (titi) and *Saimiri* (squirrel) monkeys. *Antropologia Contemporanea* 3:265.

Sauther, M. L. (1989). Antipredator behavior in troops of free-ranging *Lemur catta* at Beza Mahafaly Special Reserve, Madagascar. *International Journal of Primatology* 10:595–606.

Sauther, M. L. (1991). Reproductive behavior of free-ranging *Lemur catta* at Beza Mahafaly Special Reserve, Madagascar. *American Journal of Physical Anthropology* 84:463–477.

Savage, A., Shideler, S. E., Moorman, E. A., Ortuno, A., Whittier, C. A., Casey, K. K., and McKinney, J. (1992a). The reproductive biology of the white-faced saki (*Pithecia pithecia*) in captivity. *Abstracts of the XIVth Congress of the International Primatological Society*, Strasbourg, France, pp. 59–60.

Savage, A., Lasley, B. L., and Shideler, S. E. (1992b). Studying reproductive functioning in zoo animals: A case study of the white-faced saki program at the Roger Williams Park Zoo. *Abstracts of the XIVth Congress of the International Primatological Society*, Strasbourg, France, pp. 203–204.

Savage D. E. and Russell D. E. (1983) *Mammalian Paleofaunas of the World*. Reading, MA: Addison-Wesley.

Savage, J. L., Dronzek, L. A., and Snowden, C. T. (1987). Color discrimination by the cotton-top tamarin (*Saguinus oedipus oedipus*) and its relation to fruit coloration. *Folia Primatologica* 49:57–69.

Savage-Rumbaugh, S. (1988). A new look at ape language: Comprehension of vocal speech and syntax. In D. W. Leger (ed.), *Comparative Perspectives in Modern Psychology: Nebraska Symposium on Motivation* (pp. 201–255). Lincoln: University Nebraska Press.

Scanlon, C. E., and Chalmers, N. R. (1987). Infant development in wild common marmosets (*Callithrix jacchus jacchus*) at the ecological station of Tapacura. *International Journal of Primatology* 8(5):506.

Scanlon, C. E., Chalmers, N. R., and Monteiro da Cruz, M. A. O. (1988). Changes in the size, composition, and reproductive condition of wild marmoset groups (*Callithrix jacchus jacchus*) in North East Brazil. *Primates* 29(3):295–305.

Scanlon, C. E., Chalmers, N. R., and Monteiro da Cruz, M. A. O. (1989). Home range use and the exploitation of gum in the marmoset *Callithrix jacchus jacchus*. *International Journal of Primatology* 19:123–136.

Scanlon, C. E., Monteiro da Cruz, M. A. O., and Rylands, A. B. (1991). Exploração de exsudatos vegetais pelo sagui-comum, *Callithrix jacchus*. In A. B. Rylands and A. T. Bernardes (eds.), *A Primatologia no Brasil-3* (pp. 197–205). Belo Horizonte: Fundação Bidiversitas.

Schapiro, S., and Mitchell, G. (1986). Primate behavior in captive or confined free ranging settings. In G. Mitchell and J. Erwin (eds.), *Comparative Primate Biology Volume 2A: Behavior, Conservation, and Ecology* (pp. 93–139). New York: Alan R. Liss.

Schatz, G. E., and Malcomber, S. T. (in press). Botanical research at Ranomafana National Park: Baseline data for long-term ecological monitoring. *Ranomafana Symposium*.

Schlichte, H.-J. (1978). A preliminary report on the habitat utilization of a group of howler monkeys (*Alouatta villosa pigra*) in the National Park of Tikal, Guatemala. In G. G. Montgomery (ed.), *The Ecology of Arboreal Folivores* (pp. 551–559). Washington, DC: Smithsonian Institution Press.

Schlosser, M. (1911). Beitrage zur Kenntnis der Oligozanen Lansaugetiere aus dem Fayum, Aegypten. *Beitrage zur Palaeontologie Oesterreich-Ungarns Orients* 6:1–227.

Schmidt, L. H. (1973). Infections with *Plasmodium falciparium* and *Plasmodium vivax* in the owl monkey. Model systems for basic biological and chemotherapeutic studies. *Transactions of the Royal Society of Tropical Medicine and Hygiene* 67:446–474.

Schneider, H., Sampaio, M. I. C., Schneider, M. P. C., Ayres, J. M., Barroso, C. M. L., Hamel, A. R., Silva, B. T. F., and Salzano, F. M. (1991). Coat color and biochemical variation in Amazonian wild populations of *Alouatta belzebul*. *American Journal of Physical Anthropology* 85(1):85–93.

Schneider, H., Schneider, M. P. C., Harada, M. L., Stanhope, M., Czelusniak, J., and Goodman, M. (1993). Molecular phylogeny of the New World monkeys (Platyrrhini, Primates). *Molecular Phylogenetics and Evolution* 2:225–242.

Schneider, M. P. C., Sampaio, M. I. da C., Schneider, H., and Salzano, F. M. (1989). Genetic variability in natural populations of the Brazilian night monkey (*Aotus infulatus*). *International Journal of Primatology* 10(4):363–374.

Scholander, P. F., Hock R., Walters, V., and Irving, L. (1950). Adaptations to cold in arctic and tropical mammals and birds in relation to body temperature insulation and basal metabolic rate. *Biological Bulletin* 99:259–271.

Schön, Ybarra M. A. (1986). Loud calls of adult male red howling monkeys (*Alouatta seniculus*). *Folia Primatologica* 47:204–216.

Schwabe, G. H. (1968). Toward an ecological characterization of the South American continent. In E. J. Fittkau et al. (eds.), *Biogeography and Ecology in South America* (pp. 113–136). The Hague: Junk.

Schwartz, J. H., and Tattersall, I. (1985). Evolutionary relationships of the living lemurs and lorises (Mammalia, Primates) and their potential affinities with European Eocene Adapidae. *Anthropological Papers of the American Museum of Natural History* 60(1):1–100.

Seabra, H., Imana, J., Felfili, J. (1987). Structural analysis of gallery forest caretinga, habitat of *Callithrix penicillata*. *International Journal of Primatology* 8(5):522.

Seal, U. S., Ballou, J. D., and Padua, C. V. (1990). *Leontopithecus*: Population viability analysis, workshop report. Belo Horizonte: Captive Breeding Specialist Group (IUCN/SSC/CBSG), Species Survival Commission/IUCN.

Seidenstricker, J. (1983). Predation by *Panthera* cats and measures of human influence in habitats of South Asian monkeys. *International Journal of Primatology* 4(3):323–326.

Sekulic, R. (1982a). Behavior and ranging patterns of a solitary female red howler (*Alouatta seniculus*). *Folia Primatologica* 38:217–232.

Sekulic, R. (1982b). Daily and seasonal patterns of roaring and spacing in four red howler (*Alouatta seniculus*) troops. *Folia Primatologica* 39:22–48.

Sekulic, R. (1982c). The function of howling in red howler monkeys (*Alouatta seniculus*). *Behaviour* 81:38–54.

Sekulic, R. (1982d). Birth in free-ranging howler monkeys, (*Alouatta seniculus*). *Primates* 23:580–582.

Sekulic, R. (1983a). Male relationships and infant deaths in red howler monkeys (*Alouatta seniculus*). *Zeitschrift für Tierpsychologie* 61:185–202.

Sekulic, R. (1983b). The effect of female call on male howling in red howler monkeys (*Alouatta seniculus*). *International Journal of Primatology* 4:291–305.

Sekulic, R. (1983c). Spatial relationships between recent mothers and other troop members in red howler monkeys (*Alouatta seniculus*). *Primates* 24:475–485.

Sekulic, R., and Chivers, D. J. (1986). The significance of call duration in howler monkeys. *International Journal of Primatology* 7:183–190.

Sekulic, R., and Eisenberg, J. F. (1983). Throat-rubbing in red howler monkeys. In D. Muller-Schwartz and R. M. Silverstein (eds.), *Chemical Signals in Vertebrates*, Vol. 3 (pp. 347–350. New York: Plenum.

Selzer, B., and Pandya, D. N. (1978). Afferent cortical connections and architectonics of the superior temporal sulcus and surrounding cortex in the rhesus monkey. *Brain Research* 139:1–24.

Semendeferi, K. (1994). Evolution of the hominoid prefrontal cortex: A quantitative and image analysis of areas 13 and 10. Unpublished dissertation, University of Iowa.

Setoguchi, T., and Rosenberger, A. L. (1985). Miocene marmosets: First fossil evidence. *International Journal of Primatology* 6:615–625.

Setoguchi, T., and Rosenberger, A. L. (1987). A fossil owl monkey from La Venta, Colombia. *Nature (London)* 326(6114):692–694. [9][12]

Setoguchi, T., Watanabe, T., and Mouri, T. (1981). The upper dentition of *Stirtonia* (Ceboidea, Primates) from the Miocene of Colombia, South America and the origin of the postero-internal cusp of upper molars of howler monkeys (*Alouatta*). *Kyoto University Overseas Research Reports of N.W. Monkeys* 2:51–60.

Setoguchi, T., Takai, M., and Shigehara, N. (1991). Preliminary report of newly discovered upper dentition of *Neosaimiri* and its related form from the Miocene of Colombia, South America. In A. Ehara, T. Kimura, O. Takenaka, and M. Iwamoto (eds.), *Primatology Today* (pp. 527–530). Amsterdam: Elsevier Scientific Publishers.

Setz, E. Z. F. (1985). Estudo sobre a ecologia alimentar do parauacu (*Pithecia pithecia*, Cebidae, Primates) em um fragmento de floresta. Unpublished report to the World Wildlife Fund, Washington, DC.

Setz, E. Z. F. (1987). Comportamentos de alimentaco de *Pithecia pithecia* (Cebidae, Primates) em un fragmento florestal. *Resumos do XIV Congresso Brasileiro de Zoologia*, Juiz de Fora (MG), p. 165.

Setz, E. Z. F. (1988a). Forrageio de *Pithecia pithecia* (Cebidae, Primates) em um fragmento de floresta. *Anais do Simposio Internacional sobre Ecologia Evolutiva de Herbivoros Tropicais*, Campinas (SP), p. 60.

Setz, E. Z. F. (1988b). Feeding ecology of *Pithecia pithecia* (Pithecinae, Cebidae) in a forest fragment. *International Journal of Primatology* 8(5):543.

Setz, E. F. Z. (1991). Comportamentos de alimentação de *Pithecia pithecia* (Cebidae, Primates) em um fragmento florestal. In A. B. Rylands and A. T. Berardes (eds.), *A Primatologia no Brasil-3* pp. 327–330. Belo Horizonte: Fundação Bidiversitas.

Seuánez, H. N., Forman, L., Matayoshi, T., and Fanning, T. G. (1989). The *Callimico goeldii* (Primates, Platyrrhini) genome: Karyology and middle repetitive (Line-1) DNA sequences. *Chromosoma* 98(6):389–395.

Seuánez, H. N., Alves, G., Lima, M. M. C., de Souza Barros, R., Barroso, C. M. L., and Muniz, J. A. P. C. (1992). Chromosome studies in *Chiropotes satanas utahicki* Hershkovitz, 1985 (Cebidae, Platyrrhini): A comparison with *Chiropotes satanas chiropotes*. *American Journal of Primatology* 28(3):213–222.

Seyfarth, R. M. (1977). A model of social grooming among adult female monkeys. *Journal of Theoretical Biology* 65:671–698.

Sherman, P. T. (1991). Harpy eagle predation on a red howler monkey. *Folia Primatologica* 56:53–56.

Shideler, S. E., Savage, A., Ortoño, A. M., Moorman, E. A., and Lasley, B. L. (1994). Monitoring female reproductive function by measurement of fecal estrogen and progesterone metabolites in the white-faced saki (*Pithecia pithecia*). *American Journal of Primatology* 32(2):95–108.

Shipman, P., Bosler, W., and Davis, K. L. (1981). Butchering of giant geladas at an Acheulian site. *Current Anthropology* 22(3):257–268.

Shively, C., and Mitchell, G. (1986). Perinatal behavior of anthropoid primates. In G. Mitchell and J. Erwin (eds.), *Comparative Primate Biology, Vol. 2, Part A: Behavior, Conservation, and Ecology* (pp. 245–294). New York: Alan R. Liss.

Shoemaker, I. (1979). Reproduction and development of the black howler monkey *Alouatta caraya* at Colombia zoo. *International Zoo Yearbook* 19:15–155.

Sibley, C., and Ahlquist, J. E. (1984). The phylogeny of the hominoid primates as indicated by DNA-DNA hybridization. *Journal of Molecular Evolution* 20:2–15.

Silva, B. T. F., Sampaio, M. I. C., Schneider, H., Schneider, M. P. C., Montoya, E., Encarnacion, F., Callegari-Jacques, S. M., and Salzano, F. M. (1993). Protein electrophoretic variability in *Saimiri* and the question of its species status. *American Journal of Primatology* 29(3):183–194.

Silva, R. (1984). Elaboração e distribuição de dietas para calitriquideos em captiveiro. In M. T. de Mello (ed.), *A Primatologia no Brasil-1* (pp. 137–142). Brasilia: Sociedade Brasileira de Primatologia.

Silva-Lopes, G., Jimenez-Huerta, J., and Benitez-Rodriguez, J. (1987). Monkey populations in disturbed areas: A study on *Ateles* and *Alouatta* at Sierra de Santa Martha, Veracruz, Mexico. *American Journal of Primatology* 12(3):355–356.

Simons, D., and Holtkötter, M. (1986). Cognitive processes in cebus monkeys (*Cebus apella*) when solving problem-box tasks. *Folia Primatologica* 46:149–163.

Simons, E. L. (1965). New fossil apes from Eygpt and the initial differentiation of Hominoidea. *Nature (London)* 205:135–139.

Simons, E. L. (1967a). The earliest apes. *Scientific American* 217:28–35.

Simons, E. L. (1967b). New fossil apes from Egypt and the initial differentiation of Hominoidea. *Nature (London)* 205:135–139.

Simons, E. L. (1972). *Primate Evolution: An Introduction to Man's Place in Nature.* New York: Macmillan.

Simons, E. L. (1988). A new species of *Propithecus* (Primates) from Northeast Madagascar. *Folia Primatologica* 50:143–151.

Simons, E. L. (1989). Description of two genera and species of Late Eocene Anthropoidea from Egypt. *Proceedings of the National Academy of Science U.S.A.* 86:9956–9960.

Simons, E. L. (1990). Discovery of the oldest known anthropoidean skull from the Paleogene of Egypt. *Science* 247:507–509.

Simons, E. L. (1992). Diversity in the early Tertiary anthropoidean radiation in Africa. *Proceedings of the National Academy of Science U.S.A.* 9:10743–10747.

Simons, E. L. (1995). Skulls and anterior teeth of *Catopithecus* (Primates: Anthropoidea) from the Eocene and anthropoid origins. *Science* 268:1885–1888.

Simons, E. L., and Kay R. F. (1988). New material of *Quatrania* from Egypt with comments on the phylogenetic position of the Parapithecidae (Primates, Anthropoidea). *American Journal of Primatology* 15: 337–347.

Simons, E. L., and Rasmussen, D. T. (1991). The generic classification of Fayum Anthropoidea. *International Journal of Primatology* 12:163–178.

Simons, E. L., and Rasmussen, D. T. (1995). A Whole New World of Ancestors: Eocene Anthropoideans from Africa. *Evolutionary Anthropology* 3(4):128–138.

Simons, E. L., Godfrey, L. R., Buillaume-Randriamanantena, M. D., Chatrath, P. S., and Gagnon, L. M. (1990). Discovery of new giant subfossil lemurs of the Ankarana mountains of northern Madagascar. *Journal of Human Evolution* 19(3):311–319.

Simons, E. L., Godfrey, L. R., Jungers, W. L., Chatrath, P. S., and Rakotosamimanana, B. (1992). A new giant subfossil lemur, *Babakotia*, and the evolution of the sloth lemurs. *Folia Primatologica* 58(4):197–203.

Simons, E. L., Rasmussen, D. T., Bown, T. M., and Chatrath, P. S. (1994). The Eocene origin of anthropoid primates. In J. G. Fleagle and R. F. Kay (eds.), *Anthropoid Origins* (pp. 179–202). New York: Plenum.

Simons, E. L., Godfrey, L. R., Jungers, W. L., Chatrath, P. S., and Ravaoarisoa, J. (1995). A new species of Mesopithecus (Primates, *Paleaopropitheecidae*) from Northern Madagascar. *International Journal of Primatology* 16(4):653–682.

Simpson, B. B., and Haffer, J. (1978). Speciation patterns in the Amazon forest biota. *Annual Review of Ecology and Systematics* 9:497–518.

Simpson, G. G. (1945). The principles of classification and a classification of mammals. *Bulletin of the American Museum of Natural History* 85:1–35.

Simpson, G. G. (1980). *Splendid Isolation: The Curious History of South American Mammals*. New Haven: Yale University Press.

Simpson, M. J. A. (1973). Social grooming of male chimpanzees. In J. H. Crook and R. P. Michael (eds.), *Comparative Ecology and Behavior of Primates* (pp. 411–505). New York: Academic Press.

Sivertsen, D. W., Wright, P. C., and Trocco, T. (1982). Calls of free-ranging night monkeys in Peru. *American Journal of Physical Anthropology* 57(2):228.

Smith, E. O., and Whitten, P. L. (1988). Triadic interactions in savannah-dwelling baboons. *International Journal of Primatology* 9:409–424.

Smith, H. S., Newman, J. D., and Symmes, D. (1982a). Vocal concomitants of affiliative behavior in squirrel monkeys. In C. T. Snowdon, C. H. Brown, and M. R. Petersen (eds.), *Primate Communication* (pp. 30–49). New York: Cambridge University Press.

Smith, H. S., Newman, J. D., Hoffman, H. J., and Fetterly, K. (1982b). Statistical discrimination among vocalizations of individual squirrel monkeys. *Folia Primatologica* 37:267–279.

Smith, J. D. (1970). The systematic status of the black howler monkey, *Alouatta pigra* Lawrence. *Journal of Mammalogy* 51(2):358–369.

Smith, K. S. (1986). Dominance and mating strategies of chacma baboons, *Papio ursinus*, in the Okavango Delta, Botswana. Unpublished Ph.D. dissertation, University of California, Davis.

Smith, W. J., Pawlukiewicz, J., and Smith, S. L. (1978). Kinds of activities associated with singing patterns of the yellow-throated vireo. *Animal Behavior* 26:862–884.

Smuts, B. B. (1985). *Sex and Friendship in Baboons*. New York: Aldine de Gruyter.

Smuts, B. B., Cheney, D. L., Seyfarth, R. M., Wrangham, R. W., and Struhsaker, T. T. (1987). *Primate Societies*. Chicago: University of Chicago Press.

Smuts, B. B., and Watanbe, J. M. (1990). Social relationships and ritualized greetings in adult male baboons (*Papio cynocephalus anubis*). *International Journal of Primatology* 11:147–172.

Snodderly, D. M. (1978). Color discriminations during food foraging by a New World monkey. In D. J. Chivers and J. Herbert (eds.), *Recent Advances in Primatology, Vol. 1, Behaviour* (pp. 369–371. London: Academic Press.

Snodderly, D. M. (1979). Visual discrimination encountered in food foraging by a neotropical primate: Implications for the evolution of color vision. In E. H. Burtt (ed.), *The Behavioral Significance of Color* (pp. 237–279). New York: Garland.

Snowdon, C. T. (1983). Breeding endangered callitrichids in captivity. In M. T. de Mello (ed.), *A Primatologia no Brasil-1* (pp. 97–106). Brasilia: Sociedade Brasileira de Primatologia.

Snowdon, C. T. (1987). A naturalistic view of categorical perception. In S. Harnad (ed.), *Categorical Perception* (pp. 332–354. New York: Cambridge University Press.

Snowdon, C. T. (1988). A comparative approach to vocal communication. In D. W. Leger (ed.), *Comparative Perspectives in Modern Psychology, Nebraska Symposium on Motivation* (pp. 145–199). Lincoln: University of Nebraska Press.

Snowdon, C. T. (1989). Vocal communication in New World monkeys. *Journal of Human Evolution* 18:611–633.

Snowdon, C. T. (1990). Mechanisms maintaining monogamy in monkeys. In D. A. Dewsbury (ed.), *Contemporary Issues in Comparative Psychology* (pp. 225–251). Sunderland: Sinauer Associates.

Snowdon, C. T. (1993). A vocal taxonomy of the callitrichids. In A. B. Rylands (ed.), *Marmosets and Tamarins; Systematics, Behaviour, and Ecology* (pp. 78–94). Oxford: Oxford University Press.

Snowdon, C. T., and Cleveland, J. (1980). Individual recognition of contact calls in pygmy marmosets. *Animal Behavior* 28:717–727.

Snowdon, C. T., and Cleveland, J. (1984). "Conversations" among pygmy marmosets. *American Journal of Primatology* 7:15–20.

Snowdon, C. T., and Elowson, A. M. (1995). Ontogeny of primate vocal communication In T. Nishida, W. C. McGrew, P. Marler, M. Pickford, and F. B. M. de Waal (eds.), *Topics in Primatology: Vol. 1, Human Origins* (pp. 279–290). Tokyo: University of Tokyo Press.

Snowdon, C. T., and Hodun, A. (1981). Acoustic adaptations in pygmy marmoset contact calls: Locational cues vary with distance between conspecifics. *Behavioral Ecology and Sociobiology* 9:295–300.

Snowdon, C. T., and Pola, Y. V. (1978). Interspecific and intraspecific responses to synthesized pygmy marmoset vocalizations. *Animal Behavior* 26:196–206.

Snowdon, C. T., and Soini, P. (1988). The tamarins, genus *Saguinus*. In R. A. Mittermeier, A. B. Rylands, A. F. Coimbra-Filho, and G. A. B. da Fonesca (eds.), *Ecology and Behavior of Neotropical Primates*, Vol. 2 (pp. 223–298). Washington, DC: World Wildlife Fund.

Snowdon, C. T., and Suomi, S. J. (1982). Paternal behavior in primates. In H. Fitzgerald, J. Mullins, and P. Gage (eds.), *Child Nurturance, Vol. 3: Studies of Development in Nonhuman Primates* (pp. 63–108). New York: Plenum.

Snowdon, C. T., Cleveland, J., and French, J. A. (1983). Responses to context- and individual-specific cues in cotton-top tamarin long calls. *Animal Behavior* 31:99–111.

Snowdon, C. T., Coe, C. L., and Hodon, A. (1985). Population recognition of infant isolation peeps in the squirrel monkey. *Animal Behavior* 33:xx.

Snowdon, C. T., Hodun, A., Rosenberger, A. L., and Coimbra-Filho, A. F. (1986). Long-call structure and its relation to taxonomy in lion tamarins. *American Journal of Primatology* 11(3):253–261.

Snyder, P. A. (1974). Behavior of *Leontopithecus rosalia* (golden lion marmoset) and related species: A review. *Journal of Human Evolution* 3:109–122.

Soini, P. (1972). The capture and commerce of live monkeys in the Amazonian region of Peru. *International Zoo Yearbook* 12:26–36.

Soini, P. (1982). Ecology and population dynamics of the pygmy marmoset, *Cebuella pygmaea*. *Folia Primatologica* 39:1–21.

Soini, P. (1986). A synecological study of a primate community in the Pacaya-Samiria National Reserve, Peru. *Primate Conservation* 7:63–71.

Soini, P. (1987a). Ecology of *Cebuella*. *International Journal of Primatology* 8(5):437.

Soini, P. (1987b). Sociosexual behavior of a free-ranging *Cebuella pygmaea* (Callitrichidae, Platyrrhini) troop during postpartum estrus of its reproductive female. *American Journal of Primatology* 13:223–230.

Soini, P. (1987c). Ecology of *Lagothrix lagotricha* on the Rio Pacaya, Northeastern Peru. *International Journal of Primatology* 8(5):420.

Soini, P. (1987d). Ecology of the saddle-back tamarin *Saguinus fuscicollis illigeri* on the Rio Pacaya, northeastern Peru. *Folia Primatologica* 49:11–32.

Soini, P. (1988). The pygmy marmoset, Genus *Cebuella*. In Mittermeier, R. A., Rylands, A. B., Coimbra-Filho, A. F., and da Fonseca, G. A. B. (eds.), *Ecology and Behavior of Neotropical Primates*, Vol. 2. Washington DC: World Wildlife Fund.

Soini, P. (1993). The ecology of the pygmy marmoset, *Cebuella pygmaea*: Some comparisons with two sympatric tamarins. In A. B. Rylands (ed.), *Marmosets and Tamarins; Systematics, Behaviour, and Ecology* (pp. 257–261). Oxford: Oxford University Press.

Soini, P., and Soini, M. (1982). Distribución geografica y ecologia poblacional de *Saguinus mystax* (Primates, Callitrichidae). *Informe de Pacaya No. 6* (pp. 1–41). Iquitos, Peru: Ordeloreto, DRA/DFF.

Sokoloff, L. (1981). Circulation and energy metabolism. In G. J. Siegel, R. W. Albers, B. W. Agranoff, and R. Katzman (eds.), *Basic Neurochemistry* (pp. 471–495). Boston: Little, Brown.

Soma, H., and Kada, H. (1989). Placentation of the Goeldi's monkey, *Callimico*. *Erkrankungen der Zootiere* 31:93–98.

Southwick, C. H., and Smith, R. B. (1986). The growth of primate field studies. In G. Mitchell and J. Erwin (eds.), *Comparative Primate Biology Volume 2A: Behavior, Conservation, and Ecology* (pp. 73–91). New York: Alan R. Liss.

Spath, C. D. (1981). Getting to the meat of the problem: Some comments on protein as a limiting factor in the Amazon. *American Anthropologist* 83(2):377–379.

Spironelo, W. R. (1987). Range size of a group of *Cebus apella* in Central Amazonia. *International Journal of Primatology* 8(5):522.

Sponsel, L. E. (1981). The hunter and the hunted in the Amazon: An integrated biological and cultural approach to the behavior and ecology of human predation. Unpublished Ph.D. dissertation, Cornell University, Ithaca, NY.

Sponsel, L. E. (1986a). Amazon ecology and adaptation. *Annual Review of Anthropology* 15:67–97.

Sponsel, L. E. (1986b). La caceria de los Yekuana bajo una perspectiva ecologica. Montalban 5–29.

Sponsel, L. E. (1987). Cultural ecology and environmental education. *Journal of Environmental Education* 19(1):31–42.

Sponsel, L. E. (1989). Farming and foraging: A necessary complementarity in Amazonia? In S. Kent (ed.), *Farmers as Hunters: The Implications of Sedentism* (pp. 37–45). New York: Cambridge University Press.

Sponsel, L. E. (1992). The environmental history of Amazonia: Natural and human disturbances, and the ecological transition. In H. K. Steen and R. P. Tucker (eds.), *Changing Tropical Forests: Historical Perspectives on Today's Challenges in Central and South America* (pp. 233–251). Durham: Forest History Society.

Sponsel, L. E. (ed.), (1995). *Indigenous Peoples and the Future of Amazonia: An Ecological Anthropology of an Endangered World*. Tucson: University of Arizona Press.

Sponsel, L. E., and Loya, P. (1993). 'Rivers of hunger'?: Indigenous resource management in the oligotrophic ecosystems of the Rio Negro, Amazonas, Venezuela. In C. M. Hladik et al. (eds.), *Tropical Forests, People and Food: Biocultural Interactions and Applications* (pp. 435–446). Paris: UNESCO/Parthenon Publishing Group (UNESCO/MAB Series, Volume 15).

Spradley, J. P. (1979). *The Ethnographic Interview*. New York: Holt, Rinehart & Winston.

Srikosamatara, S. (1987). Group size in wedge-capped capuchin monkeys (*Cebus olivaceus*): Vulnerability to predators, intragroup and intergroup feeding competition. Unpublished Ph.D. Dissertation, University of Florida, Gainesville.

Stack, C. B. (1974). Sex roles and survival strategies in an urban black community. In M. Z. Rosaldo and L. Lamphere (eds.), *Woman, Culture and Society* (pp. 113–128). Stanford: Stanford University.

Stafford, B., Rosenberger, A. L., and Broadfield, D. C. (1992). Locomotor behavior in captive *Leontopithecus* and *Callimico. American Journal of Physical Anthropology* Suppl. 14:154–155.

Stallings, J. R. (1985). Distribution and status of primates in Paraguay. *Primate Conservation* 6:51–58.

Stallings, J. R. (1988). Small mammal communities in an eastern Brazilian park. Unpublished Masters Thesis, University of Florida, Gainesville.

Stallings, J. R., and Mittermeier, R. A. (1983). The black-tailed marmoset (*Callithrix argentata melanura*) recorded from Paraguay. *American Journal of Physical Anthropology* 4:159–163.

Stallings, J. R., and Robinson, J. G. (1991). Disturbance, forest heterogeneity and primate communities in Brazilian Atlantic forest park. In A. B. Rylands and A. T. Bernardes (eds.), *A Primatologia no Brasil-3* (pp. 357–368). Bello Horizonte: Sociadade Brasileira de Primatologia and Fundação Biodiversitas.

Stallings, J. R., West, L., Hahn, W., and Gamarra, I. (1989). Primates and their relation to habitat in the Paraguayan chaco. In K. H. Redford and J. F. Eisenberg (eds.), *Advances in Neotropical Mammalogy* (pp. 425–442). Gainesville: Sandhill Crane Press.

Starin, E. D. (1978). Food transfer by wild titi monkeys (*Callicebus torquatus torquatus*). *Folia Primatologica* 30:145–151.

Stark, N., Kinzey, W. G., and Pawlowski, P. (1980). Soil fertility and animal distribution. *Proceedings of the Symposium of Tropical Ecology, Kuala Lumpur, 1978; Tropical and Development,* pp. 101–111.

Stearns, M. (1993). *North American Regional Woolly Monkey Studbook—1993.* Glen Rose, TX: Fossil Rim Wildlife Center.

Stearns, M., White, B. C., Schneider, E., and Bean, E. (1988). Bird predation by captive woolly monkeys (*Lagothrix lagotricha*). *Primates* 29(3):361–367.

Stehli, F. G., and Webb, S. D. (eds.), (1985). *The Great American Biotic Exchange. Topics in Geobiology,* Vol. 4. New York: Plenum.

Stephan, H., Bauchot, R., and Andy, O. J. (1972). Data on size of the brain and of various brain parts in insectivores and primates. In C. R. Noback and W. Montagna (eds.), *The Primate Brain* (pp. 289–297). New York: Appleton-Century-Crofts.

Stephan, H., Frahm, H., and Baron, G. (1981). New and revised data on volumes of brain structures in insectivores and primates. *Folia Primatologica* 35:1–29.

Stephan, H., Baron, G., and Frahm, H. D. (1988). Comparative size of brains and brain components. In H. D. Steklis and J. Erwin (eds.), *Comparative Primate Biology, Vol. 4, Neurosciences* (pp. 1–38). New York: Alan R. Liss.

Sterling, E. J. (1993). Patterns of range use and social organization in aye-ayes *Daubentonis madagascariensis*) on Nosy Mangebe. In P. M. Kappeler and J. U. Ganzhorn (eds.), *Lemur Social Systems and Their Ecological Basis* (pp. 1–10). New York: Plenum.

Stern, J. T. (1971). Functional myology of the hip and thigh and its implications for the evolution of the erect posture. *Biblioteca Primatologica,* Vol. 14. Basel: S. Karger.

Stevenson, L. (1987). *Seven Theories of Human Nature.* New York: Oxford University Press.

Stevenson, M. F. (1978). Ontogeny of playful behaviour in family groups of the common marmoset. In D. J. Chivers and J. Herbert (eds.), *Recent Advances in Primatology, Vol. 1, Behaviour* (pp. 139–143). London: Academic Press.

Stevenson, M. F. (1986). Captive breeding of callitrichids: A comparison of reproduction and propagation in different species. In J. G. Else and P. C. Lee (eds.), *Primate Ecology and Conservation* (pp. 301–313). New York: Cambridge University Press.

Stevenson, M. F., and Poole, T. B. (1976). An ethogram of the common marmoset, *Callithrix jacchus jacchus*: General behavioural repertoire. *Animal Behavior* 24:428–451.

Stevenson, M. F., and Rylands, A. B. (1988). The marmosets, genus *Callithrix*. In R. A. Mittermeier, A. B. Rylands, A. Coimbra-Filho, and G. A. B. da Fonseca (eds.), *Ecology and Behavior of Neotropical Primates*, Vol. 2 (pp. 131–222). Washington, DC: World Wildlife Fund.

Steward, J. H. (ed.) (1944). Handbook of South American Indians. *Bulletin of the American Bureau of Ethnology* 143:1–7.

Stirton, R. A. (1951). Ceboid monkeys from the Miocene of Colombia. *University of California Publications, Bulletin of the Department of Geological Sciences* 28:315–356.

Stoltz, L. P. (1977). The population dynamics of baboons, *Papio ursinus* Kerr, 1792 in the Transvaal. Unpublished Ph.D. dissertation, University of Pretoria.

Stone, R. D. (1986). *Dreams of Amazonia*. New York: Viking/Penguin.

Stoner, K. E. (1991). Recovery of non-human primate populations at La Selva Biological Station, Costa Rica: Implications of future research. *American Journal of Physical Anthropology* Suppl. 12:169–170.

Strahl, S. D., and Brown, J. L. (1987). Geographic variation in social structure and behavior of *Aphelocoma ultramarina*. *Condor* 89:422–424.

Strayer, F. F., and Noel, J. M. (1986). The allocation of cohesive activity and diadic dominance differentials in captive groups of *S. sciureus*. *Primate Report* 14:234.

Stribley, J. A., French, J. A., and Inglett, B. J. (1987). Mating patterns in the golden lion tamarin (*Leontopithecus rosalia*): Continuous receptivity and concealed estrus. *Folia Primatologica* 49:137–150.

Strier, K. B. (1986a). The behavior and ecology of the woolly spider monkey, or muriqui (*Brachyteles arachnoides* E. Geoffroy 1806). Ph.D. thesis, Harvard University.

Strier, K. B. (1986b). Reprodução de *Brachyteles arachnoides* (Cebidae Primates). In M. T. de Mello (ed.), *A Primatologia no Brasil-2* (pp. 163–175). Brasilia: Instituto de Ciências Biológicas.

Strier, K. B. (1987a). Demographic patterns of one group of free-ranging woolly spider monkeys. *Primate Conservation* 8:73–74.

Strier, K. B. (1987b). Activity budgets of woolly spider monkeys, or muriquis (*Brachyteles arachnoides*). *American Journal of Primatology* 13(4):385–396.

Strier, K. B. (1987c). Socio-ecology of woolly spider monkeys, or muriquis (*Brachyteles arachnoides*). *American Journal of Physical Anthropology* 72(2):259.

Strier, K. B. (1987d). Ranging behavior of woolly spider monkeys, or muriquis, *Brachyteles arachnoides*. *International Journal of Primatology* 8(6):575–591.

Strier, K. B. (1988). Behavioral correlates of food patch size in muriquis (*Brachyteles arachnoides*) at Fazenda Montes Claros, Minas Gerais, Brazil. *American Journal of Physical Anthropology* 75(2):276.

Strier, K. B. (1989). Effects of patch size on feeding associations in muriquis (*Brachyteles arachnoides*). *Folia Primatologica* 52(1–2):70–77.

Strier, K. B. (1990a). New World primates, new frontiers: Insights from the woolly spider monkey, or muriqui (*Brachyteles arachnoides*). *International Journal of Primatology* 11(1):7–19.

Strier, K. B. (1990b). Demography, ecology and conservation: An example from Southeastern Brazil. *American Journal of Physical Anthropology* 81(2):302–303.

Strier, K. B. (1990c). Demography and social structure of one group of muriquis (*Brachyteles arachnoides*). *Abstracts, XIIIth Congress of the International Primatological Society*, Nagoya, Japan, p. 154.

Strier, K. B. (1990d). Development of social relationships in juvenile muriquis (*Brachyteles arachnoides*). *Abstracts, XIIIth Congress of the International Primatological Society*, Nagoya, Japan, p. 179.

Strier, K. B. (1991a). Demography and conservation of an endangered primate, *Brachyteles arachnoides*. *Conservation Biology* 5(2):214–218.

Strier, K. B. (1991b). Diet in one group of woolly spider monkeys, or muriquis (*Brachyteles arachnoides*). *American Journal of Primatology* 23(2):113–126.

Strier, K. B. (1992a). Causes and consequences of nonaggression in the woolly spider monkey, or muriqui (*Brachyteles arachnoides*). In J. Silverberg and J. Patrick Gray (eds.), *Aggression and Peacefulness in Humans and Other Primates* (pp. 100–116). New York: Oxford University Press.

Strier, K. B. (1992b). *Faces in the Forest, the Endangered Muriqui Monkeys of Brazil.* New York: Oxford University Press.

Strier, K. B. (1992c). The biology and conservation of woolly spider monkeys, or muriquis (*Brachyteles arachnoides*). *Abstracts of the XIVth Congress of the International Primatological Society*, Strasbourg, France, p. 95.

Strier, K. B. (1992d). Atelinae adaptations: Behavioral strategies and ecological constraints. *American Journal of Physical Anthropology* 88(4):515–524.

Strier, K. B. (1993a). Menu for a monkey. *Natural History* 103(2):34–43.

Strier, K. B. (1993b). Growing up in a patrifocal society: Sex differences in the spatial relations of immature muriquis (*Brachyteles arachnoides*). In M. E. Pereira and L. A. Fairbanks (eds.), *Juveniles: Comparative Socioecology*. New York: Oxford University Press.

Strier, K. B. (1994a). Myth of the typical primate. *Yearbook of Physical Anthropology* 37:233–271.

Strier, K. B. (1994b). Brotherhoods among atelins: Kinship, affiliation, and competition. *Behaviour* 130:151–167.

Strier, K. B., and Stuart, M. D. (1992). Intestinal parasites in the muriqui (*Brachyteles arachnoides*): Population variability, ecology, and conservation. *American Journal of Physical Anthropology* Suppl. 14:158–159.

Strier, K. B., and Ziegler, T. E. (1994). Insights into ovarian function in wild muriqui monkeys (*Brachyteles arachnoides*). *American Journal of Primatology* 32:31–40.

Strier, K. B., Mendes, F. D., Rimoli, J., and Rimoli, A. (1993). Demography and social structure in one group of muriquis (*Brachyteles arachnoides*). *International Journal of Primatology* 14:513–526.

Struhsaker, T. T., and Leakey, M. (1991). Prey selectivity by crowned hawk-eagles on monkeys in the Kibale Forest, Uganda. *Behavioral Ecology and Sociobiology* (in press).

Struhsaker, T. T., and Leland, L. (1987). Colobines: Infanticide by adult males. In B. Smuts, D. Cheney, R. Seyfarth, R. Wrangham, and T. Struhsaker (eds.), *Primate Societies* (pp. 83–97). Chicago: University of Chicago Press.

Strum, S. C. (1984). Why males use infants. In D. Taub (ed.), *Primate Paternalism* (pp. 146–185). New York: Van Nostrand Reinhold.

Strum, S. C. (1987). *Almost Human*. New York: Random House.

Stuart, M. D., Greenspan, L. L., Glander, K. E., and Clarke, M. R. (1990). A coprological survey of parasites of wild mantled howling monkeys, *Alouatta palliata palliata*. *Journal of Wildlife Diseases* 26(4):547–549.

Sussman, R. W. (1987). Species-specific dietary patterns in primates and human dietary adaptations. In W. Kinzey (ed.), *The Evolution of Human Behavior: Primate Models* (pp. 151–182). Albany: State University of New York.

Sussman, R. W. (1991). Demography and social organization of free-ranging *Lemur catta* in Beza Mahafaly Reserve, Madagascar. *American Journal of Physical Anthropology* 84:43–58.

Sussman, R. W., and Garber, P. A. (1987). A new interpretation of the social organization and mating system of the Callitrichidae. *International Journal of Primatology* 8(1):73–92.

Sussman, R. W., and Kinzey, W. G. (1984). The ecological role of the Callitrichidae: A review. *American Journal of Physical Anthropology* 64:419–449.

Sussman, R. W., and Raven, P. H. (1978). Pollination by lemurs and marsupials: An archaic coevolutionary system. *Science* 200:731–736.

Sussman, R. W., and Tattersall, I. (1976). Cycles of activity, group composition, and diet of *Lemur mongoz mongoz* Linnaeus, 1766 in Madagascar. *Folia Primatologica* 26:270–283.

Suzuki, K., and Nagai, H. (1986). Comparative morphology of digestive systems in primates: 4. Histochemistry of gastric mucosa. *Jinruigaku Zasshi/Journal of the Anthropological Society of Nippon* 94(2):236.

Swindler, D. R. (1976). *Dentition of Living Primates*. London: Academic Press.

Symington, M. M. (1987a). Sex ratio and maternal rank in wild spider monkeys: When daughters disperse. *Behavioral Ecology and Sociobiology* 20:421–425.

Symington, M. M. (1987b). Predation and party size in the black spider monkey, *Ateles paniscus chamek*. *International Journal of Primatology* 8(5):534.

Symington, M. M. (1987c). Long-distance vocal communication in *Ateles*: Functional hypotheses and preliminary evidence. *International Journal of Primatology* 8(5):475.

Symington, M. M. (1988a). Food competition and foraging party size in the black spider monkey (*Ateles paniscus chamek*). *Behaviour* 105(1–2):117–134.

Symington, M. M. (1988b). Demography, ranging patterns, and activity budgets of black spider monkeys (*Ateles paniscus chamek*) in the Manu National Park, Peru. *American Journal of Primatology* 15:45–67.

Symington, M. M. (1990). Fission-fusion social organization in *Ateles* and *Pan*. *International Journal of Primatology* 11(1):47–61.

Symmes, D., and Biben, M. (1988). Conversational vocal exchanges in squirrel monkeys. In D. Todt, P. Goedeking, and D. Symmes (eds.), *Primate Vocal Communication* (pp. 123–132). Berlin: Springer-Verlag.

Symmes, D., and Goedeking, P. (1988). Nocturnal vocalizations by squirrel monkeys (*Saimiri sciureus*). *Folia Primatologica* 51(2):143–148.

Symmes, D., Newman, J. D., Talmage-Riggs, G., and Lieblich, A. K. (1979).

Individuality and stability of isolation peeps in squirrel monkeys. *Animal Behavior* 26:1142–1152.

Symons, D. (1979). *The Evolution of Human Sexuality*. Oxford: Oxford University Press.

Szalay, F. S., and Delson, E. (1979). *Evolutionary History of the Primates*. New York: Academic Press.

Takai, M. (1994). New specimens of *Neosaimiri fieldsi*, a middle Miocene ancestor of the squirrel monkeys from La Venta, Colombia. *Journal of Human Evolution* 27(4):329–360.

Talmage-Riggs, G., Winter, P., Ploog, D., and Mayer, W. (1972). Effect of deafening on the vocal behavior of the squirrel monkey (*Saimiri sciureus*). *Folia Primatologica* 17:404–420.

Tan, C. L., and Wright, P. C. (1994). Behavioral development and mother-infant relations in *Propithecus diadema edwardsi* (Milne-Edward's sifaka), southeaster rain forest of Madagascar. *American Journal of Physical Athropology* Suppl. 18:194.

Tardif, S. D., and Harrison, M. L. (1990). Estimates of the energetic cost of infant transport in tamarins. *American Journal of Physical Anthropology* 81(2):306.

Tardif, S. D., Richter, C. B., and Carson, R. L. (1984). Effects of sibling rearing experience on future reproductive success in two species of Callitrichidae. *American Journal of Primatology* 6:377–380.

Tardif, S. D., Carson, R. L., and Gangaware, B. L. (1986). Comparison of infant care in family groups of the common marmoset (*Callithrix jacchus*) and the cotton-top tamarin (*Saguinus oedipus*). *American Journal of Primatology* 11:103–110.

Tardif, S. D., Harrison, M. L., and Simek, M. A. (1993). Communal infant care in marmosets and tamarins: Relation to energetics, ecology, and social organization. In A. B. Rylands (ed.), *Marmosets and Tamarins; Systematics, Behaviour, and Ecology* (pp. 220–234). Oxford: Oxford University Press.

Tarling, D. H. (1980). The geologic evolution of South America with special reference to the last 200 million years. In R. Ciochon and A. B. Chiarelli (eds.), *Evolutionary Biology of New World Monkeys and Continental Drift* (pp. 1–42). New York: Plenum.

Tattersall, I. (1982). *The Primates of Madagascar*. New York: Columbia University Press.

Tattersall, I. (1987). Cathemeral activity in primates: A definition. *Folia Primatologica* 49:200–202).

Tauber, A. (1991). *Homunculus patagonicus* Ameghino, 1891 (Primates, Ceboidea), Mioceno Temprano, De La Costa Atlantica Austral, Provincia De Santa Cruz, Republica Argentina. *Academia Nacional De Ciencias* 82:3–32.

Taylor, K. I. (1974). *Sanuma fauna: Prohibitions and Classifications*. Monografia 18. Caracas: Fundacion La Salle de Ciencias Naturales, Instituto Caribe Anthropologica y Sociologica.

Taylor, L., and Sussman, R. (1985). A preliminary study of kinship and social organization in a semi-free-ranging group of *Lemur catta*. *International Journal of Primatology* 6(6):601–614.

Teaford, M. F., and Glander, K. E. (1991). Dental microwear in live, wild-trapped *Alouatta palliata* from Costa Rica. *American Journal of Physical Anthropology* 85(3):313–319.

Teaford, M. F., and Robinson, J. G. (1989). Seasonal or ecological differences in diet and molar microwear in *Cebus nigrivittatus*. *American Journal of Physical Anthropology* 80(3):391–401.

Terborgh, J. (1983). *Five New World Primates: A Study in Comparative Ecology.* Princeton: Princeton University Press.

Terborgh, J. (1985). *The Ecology of Amazon Primates.* In G. T. Prance and T. E. Lovejoy (eds.), *Amazonia* (pp. 284–304). New York: Pergamon Press.

Terborgh, J. (1986a). The social systems of New World primates: An adaptationist view. In J. Else and P. Lee (eds.), *Primate Ecology and Conservation* (pp. 199–211). New York: Cambridge University Press.

Terborgh, J. (1986b). Keystone plant resources in the tropical forest. In M. E. Soule (ed.), *Conservation Biology* (pp. 330–344). Sunderland, MA: Sinauer.

Terborgh, J. (1986c). *Community Aspects of Frugivory in Tropical Forests: Frugivores and Seed Dispersal.* Dordrecht: W. Junk Publishers.

Terborgh, J. (1990). Mixed flocks and polyspecific associations: Costs and benefits of mixed groups to birds and monkeys. *American Journal of Primatology* 21(2):87–100.

Terborgh, J. (1992). *Diversity and the Tropical Rain Forest.* New York: Scientific American Library.

Terborgh, J., and Goldizen, A. W. (1985). On the mating system of the cooperatively breeding saddle-backed tamarin (*Saguinus fuscicollis*). *Behavioral Ecology and Sociobiology* 16:293–299.

Terborgh, J., and Janson, C. H. (1986). The socioecology of primate groups. *Annual Review of Ecology and Systematics* 17:111–135.

Terborgh, J., and Stern, M. (1989). The surreptitious life of the saddle-backed tamarin. *American Scientist* 75:260–269.

Terborgh, J., and van Schaik, C. P. (1987). Convergence vs. nonconvergence in primate communities. In J. H. R. Gee and P. S. Giller (eds.), *Organization of Communities Past and Present* (pp. 205–226). Oxford: Blackwell.

Terrace, H. S., Petito, L. A., Saunders, R. J., and Bever, T. G. (1979). Can an ape create a sentence? *Science* 206:891–902.

Thomas, O. (1904). New *Callithrix, Midas, Felis, Rhipidomys* and *Proechimys* from Brazil and Ecuador. *Annual Magazine of Natural History* (Series 7)14:188–195.

Thompson, C., and Vinci, A. (1991). Recognition of visual representations of individual group members by *Cebus apella* monkeys. *American Journal of Primatology* 24(2):137.

Thorington, R. W., Jr. (1967). Feeding and activity of *Cebus* and *Saimiri* in a Colombian forest. In D. Starck, R. Schneider, and H.-J. Kuhn (eds.), *Progress in Primatology* (pp. 180–184). Stuttgart: Gustav Fischer Verlag.

Thorington, R. W., Jr. (1968a). Observations of squirrel monkeys in a Colombian forest. In L. A. Rosenblum and R. W. Cooper (eds.), *The Squirrel Monkey* (pp. 69–85). New York: Academic Press.

Thorington, R. W., Jr. (1968b). Observations of the tamarin, *Saguinus midas. Folia Primatologica* 9:85–98.

Thorington, R. W., Jr. (1985). The taxonomy and distribution of squirrel monkeys (*Saimiri*). In L. A. Rosenblum and C. L. Coe (eds.), *Handbook of Squirrel Monkey Research* (pp. 1–33). New York: Plenum Press.

Thorington, R. W., Jr., and Vorek, R. E. (1976). Observations of the geographic variation and skeletal development of *Aotus*. *Laboratory Animal Science* 26(6,II):1006–1021.

Thorington, R. W., Jr., Muckenhirn, N. A., Montgomery, G. G. (1976). Movements of a wild night monkey (*Aotus trivirgatus*). In R. W. Thorington, Jr. and P. G. Heltne (eds.), *Neotropical Primates, Field Studies and Conservation* (pp. 32–34). Washington, DC: National Academy of Science.

Thorington, R. W., Jr., Ruiz, J. C., and Eisenberg, J. F. (1984). A study of a black howling monkey (*Alouatta caraya*) population in northern Argentina. *American Journal of Primatology* 6:357–366.

Tilden, C. C., and Oftedal, O. T. (1995). The bioenergetics of reproduction in prosimian primates: Is it related to female dominance? In L. Alterman, G. A. Doyle, and M. K. Izard (eds.), *Creatures of the Dark*. New York: Plenum Press.

Todd, P. H. (1986). *Intrinsic Geometry of Biological Surface Growth. Lecture Notes in Biomathematics, 67*. New York: Springer-Verlag.

Tokuda, K. (1988). Some social traits of howling monkeys (*Alouatta seniculus*) in La Macarena, Colombia. *Field Studies of New World Monkeys, La Macarena, Colombia* 1:35–38.

Tooby, J., and DeVore, I. (1987). The reconstruction of hominid behavioral evolution through strategic modeling. In W. G. Kinzey (ed.), *The Evolution of Human Behavior: Primate Models* (pp. 183–237). Albany: SUNY Press.

Tootell, R. B., Hamilton, S. L., and Silverman, M. S. (1985). Topography of cytochrome oxidase activity in owl monkey cortex. *Journal of Neuroscience* 5:2786–2800.

Tootell, R. B. H., Hamilton, S. L., Silverman, M. S., and Switkes, E. (1988). Functional anatomy of macaque striate cortex. 1. Ocular dominance, binocular interactions and baseline conditions. *Journal of Neuroscience* 8:1500–1529.

Tootell, R. B., Born, R. T., and Ash-Bernal, R. (1993). Columnar organization in visual cortex in non-human primates and man. In B. Gulyas, D. Ottoson, and P. E. Roland (eds.), *Functional Organization of the Human Visual Cortex* (pp. 59–73). New York: Pergamon.

Torres de Assumpção, C. (1981). *Cebus apella* and *Brachyteles arachnoides* (Cebidae) as potential pollinators of *Mabea fistulifera* (Euphorbiaceae). *Journal of Mammalogy* 62:386–388.

Torres de Assumpção, C. (1983a). An ecological study of the primates of Southeastern Brazil, with a reappraisal of *Cebus apella* races. Unpublished Ph.D. dissertation, University of Edinburgh.

Torres de Assumpção, C. (1983b). Ecological and behavioral information on *Brachyteles arachnoides*. *Primates* 24:584–593.

Tovee, M. J., Mollon, J. D., and Bowmaker, J. K. (1992). The relationship between cone pigments and behavioural sensitivity in a New World monkey (*Callithrix jacchus jacchus*). *Vision Research* 32:867–878.

Travis, D. S., Bowmaker, J. K., and Mollon, J. D. (1988). Polymorphism of visual pigments in a callitrichid monkey. *Vision Research* 28:481–490.

Turtin, J. A., Ford, O. J., Bleby, J., Hall, B. M., and Whiting, R. (1978). Composition of the milk of the common marmoset (*Callittirixiacchus*) and milk substitutes used in hand-rearing programs, with special reference to fatty acids. *Folia Primatologica* 29:64–79.

Tutin, C. E. G. (1979). Mating patterns and reproductive strategies in a community of wild chimpanzees (*Pan troglodytes schweinfurthii*). *Behavioral Ecology and Sociobiology* 6:29–38.

Tyler, D. E. (1991). The evolutionary relationships of *Aotus*. *Folia Primatologica* 56:50–52.

Ueno, Y. (1991). Urine washing in tufted capuchin (*Cebus apella*): Discrimination between groups by urine-odor. In A. Ehara, T. Kimura, O. Takenaka, and M. Iwamoto (eds.), *Primatology Today* (pp. 297–300). Amsterdam: Elsevier Scientific Publishers.

Ungar, P. S. (1990). Incisor microwear and feeding behavior in *Alouatta seniculus* and *Cebus olivaceus*. *American Journal of Primatology* 20(1):43–50.

Urton, G., ed. (1985). *Animal Myths and Metaphors in South America*. Salt Lake City: University of Utah Press.

Valderrama, X., Srikosamatara, S., and Robinson, J. G. (1990). Infanticide in wedge-capped capuchin monkeys, *Cebus olivaceus*. *Folia Primatologica* 54:171–176.

Valle, C. M. C., Santos, I. B., Alves, M. C., Machado Pinto, C. A., and Mittermeier, R. A. (1984). Algumas observações preliminares sobre o comportamento do mono (*Brachyteles arachnoides*) em ambiente natural (Fazenda Montes Claros, Municipio de Caratinga, Minas Gerais, Brasil). In M. T. de Mello (ed.), *A Primatologia no Brasil-1* (pp. 271–283). Brasilia: Instituto de Ciências Biológicas.

Van Couvering, J. A., and Harris, J. A. (1991). Late Eocene age of Fayum mammal faunas. *Journal of Human Evolution* 21:241–260.

Van Essen, D. C. (1985). Functional organization of primate visual cortex. In A. Peters and E. G. Jones (eds.), *Cerebral Cortex* Vol. 3 (pp. 259–329). New York: Plenum.

Van Horn, R. N. (1975). Primate breeding season: Photoperiodic regulation in captive *Lemur catta*. *Folia Primatologica* 24:203–220.

Van Horn, R. N. (1980). Seasonal reproductive patterns in primates. *Progress in Reproductive Biology* 5:181–221.

Van Horn, R. N., and Eaton, G. G. (1979). Reproductive physiology and behavior in prosimians. In G. A. Doyle and R. D. Martin (eds.), *The Study of Prosimian Behavior* (pp. 79–122). New York: Academic Press.

van Roosmalen, M. G. M. (1985). Habitat preferences, diet, feeding strategy and social organization of the black spider monkey (*Ateles paniscus paniscus* Linnaeus 1758) in Surinam. *Acta Amazonica* 15(3/4 Suppl.):1–238.

van Roosmalen, M. G. M. (1987a). Diet, feeding behaviour and social organization of the Guianan black spider monkey (*Ateles paniscus paniscus*). *International Journal of Primatology* 8(5):421.

van Roosmalen, M. G. M. (1987b). Seed predation in Guianan primates. *International Journal of Primatology* 8(5):433.

van Roosmalen, M. G. M. (in press). *Pithecia pithecia. Illustrated Monographs of Living Primates*. The Netherlands: Istitut voor Ontwikkelingsopdrachten.

van Roosmalen, M. G. M., and Klein, L. L. (1988). The spider monkeys, genus *Ateles*. In R. A. Mittermeier, A. B. Rylands, A. F. Coimbra-Filho, and B. A. B. Fonseca (eds.), *Ecology and Behavior of Neotropical Primates*, Vol 2 (pp. 455–537). Washington DC: World Wildlife Fund.

van Roosmalen, M. G. M., Mittermeier, R. A., and Milton, K. (1981). The bearded sakis, genus *Chiropotes*. In A. F. Coimbra-Filho and R. A. Mittermeier (eds.), *Ecology and Behavior of Neotropical Primates* (pp. 419–441). Rio de Janeiro: Academia Brasileira de Ciencias.

van Roosmalen, M. G. M., Mittermeier, R. A., and Fleagle, J. G. (1988). Diet of the northern bearded saki (*Chiropotes satanas chiropotes*): A neotropical seed predator. *American Journal of Primatology* 14(1):11–35.

van Schaik, C. P., and van Noordwijk, M. A. (1985). Interannual variability in fruit abundance and the reproductive seasonality in Sumatran long-tailed macaques (*Macaca fascicularis*). *Journal of Zoology London* 206:533–549.

van Schaik, C. P., and van Noordwijk, M. A. (1989). The special role of male *Cebus* monkeys in predation avoidance and its effect on group composition. *Behavioral Ecology and Sociobiology* 24:265–276.

van Schaik, C. P., van Noordwijk, M. A., Waisono, B., and Sutriono, E. (1983). Party size and early detection of predators in Sumatran forest primates. *Primates* 24(2):211–221.

Vecchio, A., and Miller, A. (1993). *1993 North American Regional Studbook for the White-faced Saki (Pithecia pithecia)*. Providence, RI: Roger Williams Park Zoo.

Vercauteren Drubbel, R., and Gautier, J. P. (1992). Nocturnal loud calls differing from diurnals in wild red howlers of French Guyana. *Abstracts of the XIVth Congress of the International Primatological Society*, Strasbourg, France, p. 128.

Vessey, S. H., Mortenson, B. K., and Muckenhirn, N. A. (1978). Size and characteristics of primate groups in Guyana. In D. J. Chivers and J. Herbert (eds.), *Recent Advances in Primatology, Vol. 1, Behavior* (pp. 187–188). London: Academic Press.

Vickers, W. T. (1984). The faunal component of lowland South American hunting kills. *Interciencia* 9(6):366–376.

Vickers, W. T. (1988). Game depletion hypothesis of Amazonian adaptation: Data from a native community. *Science* 239:1521–1522.

Vieira, C. C. (1944). Os símios do estado de São Paulo. *Papeis Avulsos de Zoologia (São Paulo)* 4:1–31.

Vieira, C. C. (1955). Lista remissiva dos mamíferos do Brasil. *Arquivos de Zoologia (São Paulo)* 8:341–374.

Vilensky, J., and Moore, A. M. (1991). Locomotor kinematics in *Cebus apella* and *Cebus capucinus*. *American Journal of Physical Anthropology* Suppl. 12:178.

Visalberghi, E. (1990). Tool use in Cebus. *Folia Primatologica* 54:146–154.

Visalberghi, E., and Fragaszy, D. M. (1987). Food-washing behaviour in tufted capuchin monkeys, *Cebus apella*, and crabeating macaques, *Macaca fascicularis*. *Animal Behaviour* 40(5):829–836.

Vivo, M. de (1991). Taxonomia de *Callithrix* Erxleben, 1777 (Callithrichidae: Primates). Belo Horizonte: Fundação Biodiversitas.

Walker, A. (1967). Patterns of extinction among the subfossil Madagascar lemuroids. In P. S. Martin and H. E. Wright (eds.), *Pleistocene Extinctions: The Search for a Cause* (pp. 425–432). New Haven: Yale University Press.

Walker, A. (1972). Prosimian locomotor behavior. In G. A. Doyle and R. D. Martin (eds.), *The Study of Prosimian Behavior* (pp. 543–566). New York: Academic Press.

Walker, A., and Teaford, M. F. (1989). Inferences from quantitative analysis of dental microwear. *Folia Primatologica* 53(1–4):177–189.

Walker, S. E. (1992). Positional behavior and habitat use in *Pithecia pithecia* and *Chiropotes satanas*. *Abstracts of the XIVth Congress of the International Primatological Society*, Strasbourg, France, p. 66.

Walker, S. E. (1993a). Qualitative and quantitative differences in leaping behavior between *Pithecia pithecia* and *Chiropotes satanas*. *American Journal of Physical Anthropology* Suppl. 16:202–203.

Walker, S. E. (1993b). Positional adaptations and ecology of the Pitheciini. Unpublished Ph.D. Dissertation, City University of New York.

Walker, S. E. (1994). Habitat use by *Pithecia pithecia* and *Chiropotes satanas*. *American Journal of Physical Anthropology* Suppl. 18:203.

Walker, S. E. (1995). Positional behavior and habitat use in *Chiropotes satanas* and *Pithecia pithecia*. *Proceedings, XIVth Congress of the International Primatological Society*.

Walls, G. L. (1942). *The Vertebrate Eye and Its Adaptive Radiation*. Bloomfield Hills, MI: Cranbrook Institute of Science.

Warneke, M. (1988). 200th *Callimico* born at Brookfield Zoo. *American Association of Zoological Parks and Aquariums Newsletter* 29(4):14.

Warneke, M. (1992). *Callimico goeldii: 1992 International Studbook*. Chicago: Chicago Zoological Society.

Warren, K. S., and Adel, A. F. M. (1984). *Tropical and Geographic Medicine*. Baltimore: Blackwell Scientific Publications.

Waser, P. M. (1982). The evolution of male loud calls among mangabeys and baboons. In C. T. Snowdon, C. H. Brown, and M. R. Petersen (eds.), *Primate Communication* (pp. 117–143). New York: Cambridge University Press.

Washburn, S. L. (1951). The analysis of primate evolution with particular reference to the origin of man. *Cold Spring Harbor Symposium of Quantitative Biology* 15:67–77.

Washburn, S. L. (1973). The promise of primatology. *American Journal of Physical Anthropology* 38:177–182.

Washburn, S. L., and DeVore, I. B. (1961). Social behavior of baboons and early man. In S. Washburn (ed.), *Social Life of Early Man* (pp. 91–105). Chicago: Aldine-Atherton.

Watson, L. M., and Petto, A. J. (1988). Infant adoption and reintroduction in common marmosets (*Callithrix jacchus*). *Laboratory Primate Newsletter* 27(3):1–3.

Webster, M. J., Ungerleider, L. G., and Bachevalier, J. (1991). Connections of the inferior temporal areas TE and TEO with medial temporal-lobe structures in infant and adult monkeys. *Journal of Neuroscience* 11:1095–1116.

Welker, W. (1990). Why does cerebral cortex fissure and fold? A review of determinants of gyri and sulci. In E. G. Jones and A. Peters (eds.), *Cerebral Cortex*, Vol. 8B (pp. 3–136). New York: Plenum.

Welker, C., Höhmann, H. and Schäfer-Witt, (1990). Significance of kin relations and individual preferences in the social behaviour of *Cebus apella*. *Folia Primatologica* 54:166–170.

Weller, R. E., and Steele, G. E. (1992). Cortical connections of subdivisions of inferior temporal cortex in squirrel monkeys. *Journal of Comparative Neurology* 324:37–66.

Westergaard, G. C. (1991). Hand preference in the use and manufacture of tools by tufted capuchin (*Cebus apella*) and lion-tailed macaque (*Macaca silenus*) monkeys. *Journal of Comparative Psychology* 105(2):172–176.

Westergaard, G. C., and Fragaszy, D. (1987). Self-treatment of wounds by a capuchin monkey (*Cebus apella*). *Human Evolution* 1(6):557–562.

Westergaard, G. C., and Suomi, S. J. (1993). Hand preference in capuchin monkeys varies with age. *Primates* 34(3):295–299.

Wheeler, R. J. (1990). Behavioral characteristics of squirrel monkeys at Bartlett Estate, Fort Lauderdale. *Florida Scientist* 53(4):312–316.

White, B., Miles, J., Stearns, M., Schneider, E., and Taylor, S. (1988a). Social correlates of urinary cortisol in captive woolly monkeys. *American Journal of Primatology* 14(4):451.

White, B., Stearns, M., Schneider, E., and Taylor, S. (1988b). An index of social standing for captive woolly monkeys. *American Journal of Primatology* 14(4):451.

White, B., Dew, S., Miles, P., Steffen, J., and Thomas, D. (1989a). Creatinine and specific gravity as corrections for estimating urinary excretion of cortisol in woolly monkeys (*Lagothrix lagotricha*). *American Journal of Primatology* 18(2):170.

White, B., Dew, S., Prather, J., Schneider, E., Taylor, S., and Stearns, M. (1989b). Chest-rubbing ritual in captive woolly monkeys (*Lagothrix lagotricha*): Gender differences in environmental marking. *American Journal of Primatology* 18(2):169.

White, F. J., Burton, A., Buchholz, S., and Glander, K. (1989). Social organisation, social cohession and group size of wild and captive black and white ruffed lemurs. *American Journal of Primatology* 18(2):170.

White, F. J., Overdorff, D. J., Balko, E. A., and Wright, P. C. (1995). Distribution of ruffed lemurs (*Varecia variegata*) in Ranomafana National Park, Madagascar. *Folia Primatologica* 64:124–131.

Whitehead, J. M. (1986). Development of feeding selectivity in mantled howling monkeys, *Alouatta palliata*. In J. G. Else and P. C. Lee (eds.), *Primate Ontogeny, Cognition and Social Behaviour* (pp. 105–117). New York: Cambridge University Press.

Whitehead, J. M. (1987). Vocally-mediated reciprocity between neighbouring groups of mantled howling monkeys, *Alouatta palliata* palliata. *Animal Behavior* 35:1615–1627.

Whitehead, J. M. (1989). The effect of the location of a simulated intruder on responses to long-distance vocalizations of mantled howling monkeys, *Alouatta palliata* palliata. *Behaviour* 108:73–103.

Widowski, T. M., Ziegler, T. E., Elowson, A. M., and Snowdon, C. T. (1990). The role of males in the stimulation of reproductive function in female cotton-top tamarins, *Saguinus o. oedipus*. *Animal Behavior* 40(4):731–741.

Wiener, S. G., Atha, K., and Levine, S. (1986). Influence of early experience on the development of fear in the squirrel monkey. *Primate Report* 14:29.

Wikler, K. C., and Rakic, P. (1990). Distribution of photoreceptor subtypes in the retina of diurnal and nocturnal primates. *Journal of Neuroscience* 10:3390–3401.

Willard, M. J., and Young, R. (1988). Capuchin monkeys as aides for quadriplegics. *Journal of Rehabilitation Research and Development* 25(Suppl.):136.

Willard, M. J., Dana, K., Stark, L., Owen, J., Zazula, J., and Corcoran, P. (1982). Training a capuchin (*Cebus apella*) to perform as an aide for a quadriplegic. *Primates* 23(4):520–532.

Williams, L. (1974). *Monkeys and the Social Instinct: An interliving study from The Woolly Monkey Sanctuary*. Looe (Cornwall), UK: Monkey Sanctuary Publications.

Williams, L. E., and Abee, C. R. (1988). Aggression with mixed age-sex groups of Bolivian squirrel monkeys following single animal introductions and new group formations. *Zoo Biology* 7(2):139–145.

Williams, R. W., and Herrup, K. (1988). The control of neuron number. *Annual Review of Neuroscience* 11:423–453.

Wilson, C. G. (1976). Food sharing behavior in primates: Another species added. *Arkansas Academy of Science Proceedings* 30:95–96.

Wilson, C. G. (1977). Gestation and reproduction in golden lion tamarins. In D. G. Kleiman (ed.), *The Biology and Conservation of the Callitrichidae* (pp. 191–192). Washington, DC: Smithsonian Institution Press.

Wilson, D. E. (1990). Mammals of La Selva Costa Rica. In A. H. Gentry (ed.), *Four Neotropical Rainforests* (pp. 273–286). New Haven: Yale University Press.

Wilson, D. E. (1991). Mammals of La Selva, Costa Rica. In A. H. Gentry (ed.), *Four Neotropical Rainforests* (pp. 273–286). New Haven: Yale University Press.

Wilson, E. O. (1975). *Sociobiology: The New Synthesis*. Cambridge: Harvard University Press.

Wilson, E. O. (1988). *Biodiversity*. Washington, DC: National Academy of Sciences.

Wilson, E. O. (1989). Threats to biodiversity. *Scientific American* 261(3):108–117.

Wilson, P. J. (1983). *Man, The Promising Primate: The Conditions of Human Evolution*. New Haven: Yale University Press.

Winge, H. (1895). Jordfundne og nulevende aber (primates) fra Lagôa Santa, Minas Gerais, Brasilien. *E. Mus. Lundii (Copenhagen)* 2:1–45.

Winter, P. (1969). Dialects in squirrel monkeys: Vocalizations of the roman arch type. *Folia Primatologica* 10:216–229.

Winter, P. (1972). Observations on the vocal behaviour of free-ranging squirrel monkeys. *Zeitschrift für Tierpsychologie* 31:1–7.

Winter, P., Handley, P., Ploog, D., and Schott, D. (1973). Ontogeny of squirrel monkey calls under normal conditions and under acoustic isolation. *Behaviour* 47:230–239.

Wislocki, G. B. (1939). Observations on twinning in marmosets. *American Journal of Anatomy* 64:445–448.

Wojcik, J. F., and Heltne, P. G. (1978). Tail marking in *Callimico goeldii*. In D. J.

Chivers and J. Herbert (eds.), *Recent Advances in Primatology, Vol. 1, Behavior* (pp. 507–509). London: Academic Press.

Wolf, R. H., Harrison, R. M., and Martin, T. W. (1975). A review of reproductive patterns in New World monkeys. *Laboratory Animal Science* 25:814–821.

Wolfe, L. D. (1987). *Field Primatology, A Guide to Research*. New York: Garland.

Wolfheim, J. H. (1983). *Primates of the World: Distribution, Abundance, and Conservation*. Seattle: University of Washington Press.

Wolters, H. J. (1978). Some aspects of role taking behaviour in captive family groups of the cotton-top tamarin *Saguinus oedipus oedipus*. In H. Rothe, H. J. Wolters, and J. P. Hearn (eds.), *Biology and Behaviour of Marmosets* (pp. 259–278). Gottingen: Eigenverlag Hartmut Rothe.

Wrangham, R. W. (1979a). On the evolution of ape social systems. *Social Science Information* 18:335–386.

Wrangham, R. W. (1979b). Sex differences in chimpanzee dispersion. In D. A. Hamburg and E. R. McCown (eds.), *The Great Apes* (pp. 481–489). Reading, MA: Benjamin-Cummings.

Wrangham, R. W. (1980). An ecological model of female-bonded primate groups. *Behaviour* 75:262–299.

Wrangham, R. W. (1987). Evolution of social structure. In B. Smuts, D. Cheney, R. Seyfarth, R. Wrangham, and T. Struhsaker (eds.), *Primate Societies* (pp. 282–296). Chicago: University of Chicago Press.

Wright, E. M., Jr., and Bush, D. E. (1977). The reproductive cycle of the capucin (*Cebus apella*). *Laboratory Animal Science* 27:651–654.

Wright, P. C. (1978). Home range, activity pattern, and agonistic encounters of a group of night monkeys (*Aotus trivirgatus*) in Peru. *Folia Primatologica* 29:43–55.

Wright, P. C. (1979). Patterns of grooming behavior in *Callicebus* and *Aotus*. *American Journal of Physical Anthropology* 50(3):494.

Wright, P. C. (1981). The night monkeys, genus *Aotus*. In A. F. Coimbra-Filho and R. A. Mittermeier (eds.), *Ecology and Behavior of Neotropical Primates*, Vol. 1 (pp. 211–240). Rio de Janeiro: Academia Brasileira de Ciências.

Wright, P. C. (1984). Biparental care in *Aotus trivirgatus* and *Callicebus moloch*. In M. Small (ed.), *Female Primates: Studies by Women Primatologists* (pp. 59–75). New York: Alan R. Liss.

Wright, P. C. (1985). The costs and benefits of nocturnality for the night monkey (*Aotus trivirgatus*). Unpublished Ph.D. dissertation, City University of New York, New York.

Wright, P. C. (1986). Ecological correlates of monogamy in *Aotus* and *Callicebus*. In J. Else and P. Lee (eds.), *Primate Ecology and Conservation* (pp. 159–167). Cambridge: Cambridge University Press.

Wright, P. C. (1987). Diet and ranging patterns of *Propithecus diadema edwardsi* in Madagascar. *American Journal of Physical Anthropology* 72(2):271.

Wright, P. C. (1988a). Lemurs lost and found. *Natural History* 97(7):56–66.

Wright, P. C. (1988b). Social behavior of *Propithecus diadema edwardsi* in Madagascar. *American Journal of Physical Anthropology* 75(2):289.

Wright, P. C. (1989). The nocturnal primate niche in the New World. *Journal of Human Evolution* 18(7):635–658.

Wright, P. C. (1990). Patterns of paternal care in primates. *Internatiional Journal of Primatology* 11(2):89–102.

Wright, P. C. (1992a). The monogamous community: Intergroup interactions in *Aotus* and *Callicebus*. *Abstracts of the XIVth Congress of the International Primatological Society*, Strasbourg, France, pp. 57–58.

Wright, P. C. (1992b). How to build a national park: Primate ecology rainforest conservation and economic development in Madagascar. *Evolutionary Anthropology* 1:25–33.

Wright, P. C. (1992c). Primate ecology, rainforest conservation, and economic development: Building a national park in Madagascar. *Evolutionary Anthropology* 1:25–33.

Wright, P. C. (1993). Variations in male-female dominance and offspring care in non-human primates. In B. D. Miller (ed.), *Sex and Gender Hierarchies* (pp. 127–145). Cambridge: Cambridge University Press.

Wright, P. C. (1994a). The behavior and ecology of the owl monkey. In J. F. Baer, R. E. Weller, and Ibulairiu Kakoma (eds.), *Aotus: The Owl Monkey* (pp. 97–112). New York: Academic Press.

Wright, P. C. (1994b). Night watch on the Amazon. *Natural History* 103(5):44–51.

Wright, P. C. (1995). Demography and life history of free-ranging *Propithecus diadema edwardsi* in Ranomafana National Park, Madagascar. 16(5)835–853.

Wright, P. C. (1996). The future of biodiversity in Madagascar: A view from Ranomafana National Park. In B. D. Patterson and S. M. Goodman (eds.), *Environmental Change in Madagascar*. Washington, DC: Smithsonian Institute. In press.

Wright, P. C., and Martin, L. M. (1995). Predation, pollination and torpor in two nocturnal prosimians: *Cheirogaleus major* and *Microcebus rufus* in the rainforest of Madagascar. In L. Alterman, G. Doyle, and K. Izard (eds.), *Creatures of the Dark: Then Nocturnal Prosimians* (pp. 45–60). New York: Plenum Press.

Wright, P. C., and White, F. J. (1990). The rare and the specialized: Conservation needs of rain forest primates in Madagascar. *American Journal of Physical Anthropology* 81(2):320.

Wright, P. C., Tan, C. L., and Rakotonirina, G. (1995). Comparison of infant development and reproduction in new world and Malagasy primates. *XVth Congress of the International Primatological Society, Bali, Indonesia: Handbook and Abstracts*, p. 187.

Wyss, A. R., Flynn, J. J., Norell, M. A., Swisher, C. C., III, Charrier, R., Novacek, M. J., and McKenna, M. C. (1993). South America's earliest rodent and recognition of a new interval of mammalian evolution. *Nature (London)* 365:434–437.

Yamamoto, M. E. (1993). From dependence to sexual maturity: The behavioural ontogeny of Callitrichidae. In A. B. Rylands (ed.), *Marmosets and Tamarins; Systematics, Behaviour, and Ecology* (pp. 235–254). Oxford: Oxford University Press.

Yokoyama, S., and Yokoyama, R. (1989). Molecular evolution of human visual pigment genes. *Molecular Biology and Evolution* 6:186–197.

Yoneda, M. (1981). Ecological studies of *Saguinus fuscicollis* and *Saguinus labiatus* with reference to habitat segregation and height preference. *Kyoto University Overseas Research Reports, New World Monkeys*, pp. 43–50.

Yoneda, M. (1984a). Comparative studies on vertical separation, foraging behavior and traveling mode of saddle-back tamarins (*Saguinus fuscicollis*) and red chested moustached tamarins (*Saguinus labiatus*) in northern Bolivia. *Primates* 25:414–442.

Yoneda, M. (1984b). Ecological study of the saddle-backed tamarin (*Saguinus fuscicollis*) in Northern Bolivia. *Primates* 25:1–12.

Yoneda, M. (1988). Habitat utilization of six species of monkeys in Rio Duda, Colombia. *Field Studies of New World Monkeys, La Macarena, Colombia* 1:39–45.

Yoneda, M. (1990). The difference of tree size used by five cebid monkeys in Macarena Colombia. *Field Studies of New World Monkeys, La Macarena, Colombia* 3:13–18.

Yost, J. A., and Kelley, P. M. (1983). Shotguns, blowguns, and spears: The analysis of technological efficiency. In R. B. Hames and W. T. Vickers (eds.), *Adaptive Responses of Native Amazonians* (pp. 189–224). New York: Academic Press.

Young, A. L., Richard, A. F., and Aiello, L. C. (1990). Female dominance and maternal investment in strepsirhine primates. *The American Naturalist* 135(4):473–488.

Ziegler, T. E., Snowdon, C. T., Baker, A., and Warneke, M. (1987). Endocrine profiles during conceptive cycles and pregnancy in the Goeldi's monkey. *International Journal of Primatology* 8(5):545.

Ziegler, T. E., Snowdon, C. T., and Warneke, M. (1989). Postpartum ovulation and conception in Goeldi's monkey, *Callimico goeldii*. *Folia Primatologica* 52(3–4):206–210.

Zilles, K., Armstrong, E., Schlaug, G., and Schleicher, A. (1986). Quantitative cytoarchitectonics of the posterior cingulate cortex in primates. *Journal of Comparative Neurology* 253:514–524.

Zilles, K., Armstrong, E., Moser, K. H., Schleicher, A., and Stephan, H. (1989). Gyrification in the cerebral cortex of primates. *Brain Behavior and Evolution* 34:143–150.

Zingeser, M. R. (1973). Dentition of *Brachyteles arachnoides* with reference to alouattine and atelinine affinities. *Folia Primatologica* 20:351–390.

Zoloth, S. R., Petersen, M. R., Beecher, M. D., Green, S., Marler, P., Moody, D. B., and Stebbins, W. C. (1979). Species-specific perceptual processing of vocal sounds by monkeys. *Science* 204:870–872.

Zunino, G. E. (1989). Habitat, dieta y actividad del mono aullador negro (*Alouatta caraya*) en el noreste Argentina. *Boletin Primatologico Latinoamericano* 1(1):74–97.

Index

FOUNDATIONS OF HUMAN BEHAVIOR
An Aldine de Gruyter Series of Texts and Monographs

SERIES EDITOR

Monique Borgerhoff Mulder, *University of California, Davis*
Marc Hauser, *Harvard University*